Data Analysis
in Astronomy II

ETTORE MAJORANA INTERNATIONAL SCIENCE SERIES
Series Editor:
Antonino Zichichi
European Physical Society
Geneva, Switzerland

(PHYSICAL SCIENCES)

Recent volumes in the series:

Volume 17   PHYSICS AT LEAR WITH LOW-ENERGY
            COOLED ANTIPROTONS
            Edited by Ugo Gastaldi and Robert Klapisch

Volume 18   FREE ELECTRON LASERS
            Edited by S. Martellucci and Arthur N. Chester

Volume 19   PROBLEMS AND METHODS FOR LITHOSPHERIC
            EXPLORATIONS
            Edited by Roberto Cassinis

Volume 20   FLAVOR MIXING IN WEAK INTERACTIONS
            Edited by Ling-Lie Chau

Volume 21   ELECTROWEAK EFFECTS AT HIGH ENERGIES
            Edited by Harvey B. Newman

Volume 22   LASER PHOTOBIOLOGY AND PHOTOMEDICINE
            Edited by S. Martellucci and A. N. Chester

Volume 23   FUNDAMENTAL INTERACTIONS IN LOW-ENERGY
            SYSTEMS
            Edited by P. Dalpiaz, G. Fiorentini, and G. Torelli

Volume 24   DATA ANALYSIS IN ASTRONOMY
            Edited by V. Di Gesù, L. Scarsi, P. Crane,
            J. H. Friedman., and S. Levialdi

Volume 25   FRONTIERS IN NUCLEAR DYNAMICS
            Edited by R. A. Broglia and C. H. Dasso

Volume 26   TOKAMAK START-UP: Problems and Scenarios
            Related to the Transient Phases of a
            Thermonuclear Fusion Reactor
            Edited by Heinz Knoepfel

Volume 27   DATA ANALYSIS IN ASTRONOMY II
            Edited by V. Di Gesù, L. Scarsi, P. Crane,
            J. H. Friedman, and S. Levialdi

A Continuation Order Plan is available for this series. A continuation order will bring delivery of each new volume immediately upon publication. Volumes are billed only upon actual shipment. For further information please contact the publisher.

Data Analysis in Astronomy II

Edited by

V. Di Gesù
University of Palermo and
Institute of Cosmic Physics and Informatics/CNR
Palermo, Italy

L. Scarsi
University of Palermo and
Institute of Cosmic Physics and Informatics/CNR
Palermo, Italy

P. Crane
European Southern Observatory
Garching/Munich, Federal Republic of Germany

J. H. Friedman
Stanford University
Stanford, California

and

S. Levialdi
University of Rome
Rome, Italy

Plenum Press • New York and London

Library of Congress Cataloging in Publication Data

International Workshop on Data Analysis in Astronomy (2nd: 1986: Erice, Sicily)
 Data analysis in astronomy II.

 (Ettore Majorana international science series. Physical sciences; v. 27)
 "Proceedings of the Second International Workshop on Data Analysis in Astronomy, held April 17–30, 1986, in Erice, Sicily, Italy"—T.p. verso.
 1. Astronomy—Data processing—Congresses. I. Di Gesù, V. II. Title. III. Title: Data analysis in astronomy 2. IV. Title: Data analysis in astronomy two. V. Series.
 QB51.3.E43I58   1986   523'.0028'5   86-25456
 ISBN 0-306-42473-8

Proceedings of the Second International Workshop on Data Analysis in Astronomy,
held April 17–30, 1986, in Erice, Sicily, Italy

© 1986 Plenum Press, New York
A Division of Plenum Publishing Corporation
233 Spring Street, New York, N.Y. 10013

All rights reserved

No part of this book may be reproduced, stored in a retrieval system, or transmitted in any form or by any means, electronic, mechanical, photocopying, microfilming, recording, or otherwise, without written permission from the Publisher

Printed in the United States of America

PREFACE

The II international workshop on "Data Analysis in Astronomy" was intended to provide an overview on the state of the art and the trend in data analysis and image processing in the context of their applications in Astronomy. The need for the organization of a second workshop in this subject derived from the steady growing and development in the field and from the increasing cross-interaction between methods, technology and applications in Astronomy.

The book is organized in four main sections:

- Data Analysis Methodologies
- Data Handling and Systems dedicated to Large Experiments
- Parallel Processing
- New Developments

The topics which have been selected cover some of the main fields in data analysis in Astronomy. Methods that provide a major contribution to the physical interpretation of the data have been considered. Attention has been devoted to the description of the data analysis and handling organization in very large experiments. A review of the current major satellite and ground based experiments has been included. At the end of the book the following 'Panel Discussions' are included:

- Data Analysis Trend in Optical and Radio Astronomy
- Data Analysis Trend in X and Gamma Astronomy
- Problems and Solutions in the Design of Very Large Experiments
- Trend on Parallel Processing Algorithms

These contributions in a sense summarize the 'live' reaction of the audience to the various topics.

The success of the meeting has been the result of the coordinated effort of several people from the organizers to those who presented a contribution and/or took part in the discussion. We wish to thank the members of the workshop scientific committee Prof. M. Capaccioli, Prof. G.A. De Biase,

Prof. E.J. Schreier, Prof. G. Sedmak, Prof. A. Zichichi, and of the local organizing committee Dr. R. Buccheri and Dr. M.C. Maccarone together with Miss P. Savalli and Dr. A. Gabriele of the E.Majorana Center for their support and unvaluable help in arranging the workshop.

P.Crane
V.Di Gesù
J.H.Friedman
S.Levialdi
L.Scarsi

SESSION CHAIRMEN

Session:

Data Analysis Methodologies   M.Capaccioli

Data Handling and Systems dedicated
to Large Experiments   P.Crane

Parallel Processing   V.Cantoni

New Developments   S.Levialdi

CONTENTS

DATA ANALYSIS METHODOLOGIES

A Statistician's View of Data Analysis 3
   J.H. Friedman

Cluster Photometry: Present State of the Art and Future
Developments 17
   I.R. King

Clustering Techniques and their Applications 31
   F. Murtagh

Problems and Solutions in Surface Photometry 45
   J.L. Nieto

Classification of Low-Resolution Stellar Spectra via
Template Matching - A Simulation Study 61
   H.-M. Adorf

An Approach to the Astronomical Optical Data Compression 71
   L. Caponetti, G.A. De Biase, L. Distante

Coded Aperture Imaging 77
   E. Caroli, J.B. Stephen, A. Spizzichino, G. Di Cocco, L. Natalucci

Study of Pulsar Ligth Curves by Cluster Analysis 87
   V. Di Gesu, R. Buccheri, B. Sacco

Multivariate Cluster Analysis of Radio Pulsar Data 97
   M.C. Maccarone, R. Buccheri, V. Di Gesù

Automatic Processing of Very Low-Dispersion Spectra 109
   P. Schuecker, H. Horstmann, C.C. Volkmer

DATA HANDLING SYSTEMS DEDICATED TO LARGE EXPERIMENTS

VLA: Methodological and Computational Requirements 119
   P.K. Moore

Perspective on Data Analysis for the Space Telescope 127
   P.M.B. Shames

Present and Planned Large Groundbased Telescopes: an Overview of some
Computer and Data Analysis Applications Associated with their Use ... 141
   H.J. Smith

Data Analysis for the ROSAT Mission 155
   H.U. Zimmermann, R. Gruber, G. Hasinger, J. Paul, J. Schmitt,
   W. Voges

Preparing Analysis of Hubble Space Telescope Data in Europe 165
   H.-M. Adorf, D. Baade, K. Banse

COMPASS: the COMPTEL Processing and Analysis Software System 171
   R. Diehl, G. Simpson, T. Casilli, V. Schoenfelder, G. Lichti,
   H. Steinle, B. Swanenburg, H. Aarts, A. Deerenberg, W. Hermsen,
   K. Bennett, C. Winkler, M. Snelling, J. Lockwood, D. Morris, J. Ryan

The Strasbourg Astronomical Data Centre (CDS) and the Setting Up
of a European Astronomical Data Network 181
   A. Heck

An Indexing Algorithm for Large Multidimensional Arrays 187
   P. Moore

The European Scientific Data Archive for the Hubble Space Telescope .. 193
   G. Russo, A. Richmond, R. Albrecht

PARALLEL PROCESSING

Cellular Machines: Theory and Practice 203
   M.J.B. Duff

The PAPIA Project 211
   S. Levialdi

Languages for Parallel Processors 225
   A.P. Reeves

The Massively Parallel Processor: a Highly Parallel Scientific
Computer 239
   A.P. Reeves

Parallel Processing: from Low- to High-Level Vision 253
   S.L. Tanimoto

Low Level Languages for the PAPIA Machine 263
   O. Catalano, G. Di Gaetano, V. Di Gesù, G. Gerardi, A. Machì,
   D. Tegolo

NEW DEVELOPMENTS

Expert Systems for Data Analysis 273
   J.M. Chassery

An Unified View of Artificial Intelligence and Computer Vision 285
   D. Dutta Majumder

Data Storage and Retrieval in Astronomy 305
   F. Ochsenbein

Vector Computers in Astronomical Data Analysis 315
   D. Wells

PANEL DISCUSSIONS

Data Analysis Trend in Optical and Radio Astronomy 327
   I.R. King

Data Analysis Trends in X-Ray and Gamma-Ray Astronomy 343
   H.U. Zimmermann

Problems in Large Scale Experiments.................................. 359
   E.J. Schreier

Trends in Parallel Processing Applications........................... 377
   M.J.B. Duff

Index 397

DATA ANALYSIS METHODOLOGIES

Chairman : M. Capaccioli

A STATISTICIAN'S VIEW OF DATA ANALYSIS

Jerome H. Friedman

Department of Statistics
Stanford Linear Accelerator Center
Stanford University

ABSTRACT

A brief overview of statistical data analysis is provided with a view towards examining its role in the analysis of astronomy data.

INTRODUCTION

A question that is often asked in any experimental or observational science is whether statistical considerations are useful in the analysis of its data. This is a question that can only be answered by the scientists who understand the data as well as mechanisms and instruments that produce it. In order to help answer this question it is useful to know how data and data analysis is viewed by people who regard themselves as statisticians. It is the purpose of this report to give a necessarily brief overview of statistical data analysis as viewed and practiced by statisticians.

First, it is important to understand what statisticians do not regard as data analysis, but which is never-the-less an important aspect of data understanding. This is the process of data reduction. In this phase the raw data from the detectors (telescopes, counters) are reduced to move useful and understandable quantities (such as images). The software (and sometimes hardware) that perform this task are simply regarded as computing engines that transform the raw data to forms that are more convenient for further calculations. Although statistical considerations may be involved in the development of these systems, they are usually dominated by considerations specific to the scientific field and the particular instruments that produce the data.

It is the further calculations that interest statisticians. That is, how to discover from the (refined) data, the properties of the systems under study that produced the data (stars, galaxies, etc.), and deduce statistically meaningful statements about them, especially in the presence of uncertain measurements.

Statistics can be viewed as the science that studies randomness. Central to statistical data analysis is the notion of the random variable or measurement. This is a measured quantity for which repeated observations (measurements) produce different values that cannot be (exactly) predicted in advance. Instead of a single value, repeated measurements will produce a distribution of values. The origin of the randomness can be due to random measurement errors associated with the instruments, or it could be a consequence of the fact that the measured quantity under consideration depends upon other quantities that are not (or cannot be) controlled – ie., held constant. In either case, a random variable is one for which we cannot predict exact values, only relative probabilities among all possible values the variable can assume.

The distribution of relative probabilities is quantified by the <u>probability density</u> function $p(X)$. Here X represents a value from the set of values that the variable can take on, and the function $p(X)$ is the relative probability that a measurement will produce that value. By convention the probability density function is required to have the properties

$$p(X) \geq 0 \text{ and } \int p(X)\, dX = 1$$

as X ranges over all of its possible values. Under the assumption that X is a random variable, the most information that we can ever hope to know about its future values is contained in its probability density function. It is the purpose of observation or experimentation to use repeated measurements of the random variable X to get at the properties of $p(X)$. It is the purpose of theory to calculate $p(X)$ from various mathematical (physical) models to compare with observation.

It is seldom the case that only one measurement is made on each object under study. Usually several simultaneous measurements of different quantities are made on each object, each of these measurements being a random variable. In this case we can represent each observation as an n-vector of measurements

$$X_1, X_2, \ldots, X_n \tag{1}$$

where n is the number of simultaneous measurements performed on each object. We call the collection of measurements (1) a vector-valued random variable of dimension n.

Statistics as a discipline has several divisions. One such division depends upon whether one decides to study each of the random variables separately—ignoring their simultaneous measurement—or whether one uses the data (collection of simultaneous measurements) to try to access the relationships (associations) among the variables. The former approach is known as <u>univariate statistics</u> which reduces to studying each random variable X_i, and its corresponding probability density $P_i(X_i)$, separately and independently of the other variables.

The latter approach is known as <u>multivariate statistics</u>. Central to it is the motion of the <u>joint probability density</u> function

$$p(X_1, X_2, \ldots, X_n) \tag{2}$$

which is the relative probability that the simultaneous set of values $X_1, X_2, \ldots, X_n$ will be observed. In multivariate statistics one tries to get at the properties of the joint probability density function (2) based on repeated observation of simultaneous measurements.

Another division in the study of statistics is between parametric (model dependent) and nonparametric (model independent) analysis. We begin with a little notation. Let
$$\underline{X} = (X_1, X_2, \ldots, X_n)$$
be an n – dimensional vector representing the simultaneous values of the n measurements made on each object. In <u>parametric statistics</u> the (joint) probability density function is assumed to be a member of a parameterized family of functions,
$$p(\underline{X}) = f(\underline{X}; \underline{a}), \tag{3}$$
where $\underline{a} = (a_1, a_2, \ldots, a_p)$ is a set of parameters, the values of which determine the particular member of the family. In parametric statistics the problem of determining the (joint) probability density function reduces to the determination of an appropriate set of values for the parameters. The parameterized family chosen for the analysis can come from intuition, theory, physical models, or it may just be a convenient approximation.

<u>Nonparametric statistics</u>, on the other hand, does not specify a particular functional form for the probability density, $p(\underline{X})$. It's properties are inferred directly from the data. As we will see, the histogram can be considered an example of a (univariate) probability density estimate.

Generally speaking, parametric statistical methods are more powerful than nonparametric methods provided that the <u>true</u> underlying probability density function is actually a member of the chosen parameterized family of functions. If not, parametric methods lose their power rapidly as the truth deviates from the assumptions, and the more robust nonparametric methods become the most powerful.

The final division we will discuss is between exploratory and confirmatory data analysis. With <u>exploratory data analysis</u> one tries to investigate the properties of the probability density function with no preconceived notions or precise questions in mind. The emphasis here is on detective work and discovering the unexpected. Standard tools for exploratory data analysis include graphical methods and descriptive statistics. <u>Confirmatory</u> data analysis, on the other hand, tries to use the data to either confirm or reject a specific preconceived hypothesis concerning the system under study, or to make precise probabilistic statements concerning the values of various parameters of the system.

For the most part this paper will concentrate on confirmatory aspects of data analysis with a few exploratory techniques (associated with nonparametric analysis) coming at the end.

Mini-Introduction to Estimation Theory

In estimation, we assume that our data, consisting of N observations, is a random sample from an infinite population governed by the probability density function $p(\underline{X})$. Our goal is to make inferences about $p(\underline{X})$. In parametric estimation we would like to infer likely values for the parameters. In nonparametric estimation we want to infer $p(\underline{X})$ directly.

Consider a parametric estimation problem. Here we have a data set $\{\underline{X}_i\}_{i=1}^N$ considered as a random sample from some (joint) probability density function $p(\underline{X})$ which is assumed to be a member of a parameterized family of functions $f(\underline{X};a)$ characterized (for this example) by a single parameter a. Our problem is to infer a likely value for (ie. estimate) a.

Let

$$Y = \phi(\underline{X}_1, \underline{X}_2, \ldots, \underline{X}_N) \tag{4}$$

be a function of N vector valued random variables. Here ϕ represents (for now) an arbitrary function. Since any function of random variables is itself a random variable, Y will be a random variable with its own probability density function $p_N(Y;a)$. This probability density function will depend on the joint probability density of the $\underline{X}_i$, $f(X_i;a)$, and through this on the (true) value of the parameter a. It will also depend on the sample size N. Suppose it were possible to choose the function ϕ in (4) so that $p_N(Y;a)$ is large only when the value of Y is close to that of a, and small everywhere else (provided the $\underline{X}_i$ follow $p(\underline{X}) = f(\underline{X};a)$). If this were the case then we might hope that when we evaluate ϕ for our particular data set that the value for Y so obtained would be close to that of a. A function of N random variables is called a "<u>statistic</u>" and its value for a particular data set is called an "<u>estimate</u>" (for a).

As an example of how it is possible to construct statistics with the properties described above, consider the <u>method of moments</u>. Define

$$G(a) = \int g(\underline{X}) p(\underline{X}) d\underline{X} = \int g(\underline{X}) f(\underline{X};a) d\underline{X} \tag{5}$$

where $g(\underline{X})$ is an arbitrary function of a single (vector valued) random variable. The quantity $G(a)$ is just the average of the function of $g(\underline{X})$ with respect to the probability density $p(\underline{X})$. Its dependence on the value of a is a consequence of the fact that $p(\underline{X}) = f(\underline{X};a)$ depends upon the value of a. Now, the law of large numbers (central limit theorem) tell us that

$$Z = \Theta(\underline{X}_1, \underline{X}_2, \ldots \underline{X}_N) = \frac{1}{N} \sum_{i=1}^N g(\underline{X}_i) \tag{6a}$$

has a normal (Gaussian) distribution

$$p_N(Z;a) = \frac{1}{\sqrt{2\pi}\sigma_N} \exp\left\{-\tfrac{1}{2}[Z - G(a)]^2/\sigma_N^2\right\} \tag{6b}$$

centered at $G(a)$, with standard deviation

$$\sigma_N = \frac{1}{\sqrt{N}}\{\int [g(\underline{X}) - G(a)]^2 f(\underline{X};a)d\underline{X}\}^{\frac{1}{2}} \qquad (6c)$$

as the sample size becomes large. That is, the sample mean (of $g(\underline{X})$) has a Gaussian distribution centered at the true mean with a standard deviation that becomes smaller as N grows larger ($\sim \frac{1}{\sqrt{N}}$). Therefore, for large enough N, likely values of Z will always be close to $G(a)$, and Z is a good statistic for estimating $G(a)$. If $g(\underline{X})$ is chosen so that $G(a)$ is not too wild a function of a, it then follows that

$$Y = G^{-1}(Z) = G^{-1}[\frac{1}{N}\sum_{i=1}^{N} g(\underline{X}_i)]$$

will be a good statistic for estimating the value for the parameter a.

Note that in this development the moment function $g(\underline{X})$ is fairly arbitrary. Therefore, this method can be used to construct a great many statistics for estimating the (same) parameter a. Some of these estimators will be better than others. The field of statistics is concerned to a large degree with finding good estimators (statistics for estimation).

Statisticians rate the quality of estimators on the basis of four basic properties: consistency, efficiency, bias, and robustness. <u>Consistency</u> concerns the property of the estimator as the sample size N becomes arbitrarily large. In particular an estimator (4) is said to be <u>consistent</u> if

$$\lim_{N \to \infty} p_N(Y;a) = \delta(Y - a)$$

where δ is the Dirac delta function. For a consistent estimator, the estimate becomes more and more accurate as the sample size increases. Note that (6) implies that moment estimates are consistent provided that the bracketed quantity in (6c) is finite (second central movement of $g(\underline{X})$).

<u>Efficiency</u> is concerned with the properties of the estimator for finite N. The efficiency of an estimator is inversely related to its <u>expected-squared-error</u>

$$ESE_N(Y) = \int (Y - a)^2 f_N(Y;a)dY.$$

This is the average-squared distance of the estimate from the truth. Note that if the estimator is consistent, then $\lim_{N \to \infty} ESE_N(Y) = 0$. The <u>relative efficiency</u> of two estimators Y and Z is defined as the inverse ratio of their corresponding expected squared errors,

$$RE_N(Y,Z) = ESE_N(Z)/ESE_N(Y).$$

<u>Bias</u> is concerned with whether or not the average value of a statistic is equal to the true value of the parameter it is estimating. In particular, the <u>bias</u> of an

estimator is defined to be

$$B_N(Y) = \int Y f_N(Y;a) dY - a.$$

This is just the difference between the average value of the statistic and the truth. Note that if an estimator is consistent then $\lim_{N \to \infty} B_N(Y) = 0$. An estimator for which $B_n(Y) = 0$ for all N is said to be <u>unbiased</u>. Generally speaking unbiased estimators are preferred if all other things are equal. However, all other properties are seldom equal. In particular, the efficiency of the best unbiased estimators is generally lower than that for the best biased estimators in a given problem. **Unbiased estimators are almost never best in terms of expected-squared-error.**

<u>Robustness</u> concerns the sensitivity of an estimator to violations in the assumptions that went in to choosing it. In parametric statistics the assumptions center on the particular parameterized family (3) assumed to govern the probability density of the (random) variables comprising the data. For a given parametric family there is usually an optimal estimator for its parameters (in terms of efficiency). However, it is often the case that the efficiency of such an optimal estimator degrades badly if the true probability density deviates only slightly from the closest member of the assumed parameterized family. <u>Robust</u> estimators generally have a little less efficiency than the optimal estimator in any given situation (if the true density were known), but maintain their relatively high efficiency over a wide range of different parameterized forms for probability density functions. Robust estimators are generally preferred since it is often impossible to know for certain that the assumed parametric form for the probability density is absolutely correct.

As an example of robustness consider estimating the center of a symmetric distribution. If the probability density corresponding to the distribution were Gaussian, then the sample mean is the most efficient estimator. If, however, the distribution has higher density than the Gaussian for points far away from the center (fat tails), then the efficiency of the mean degrades badly. The sample median, on the other hand, is less efficient than the mean for Gaussian data (relative efficiency approximately 0.64) but has much higher efficiency for fat tailed distributions.

Although the method of moments described above can be (and often is) used to construct estimators, it is not the favorite way among statisticians. Far and away the most popular method is that of <u>maximum likelihood</u>. By definition, the relative probability of simultaneously observing the set of values $\underline{X} = (X_1, X_2, \ldots, X_N)$ is the value of the joint probability density function $p(\underline{X})$. Let $\underline{X}_i (i = 1, N)$ be one of the observations in our data set. The relative probability of observing this observation (before we actually observed it) was $p(\underline{X}_i)$. If we believe that all of our N observations were <u>independently</u> drawn from a population governed by $p(\underline{X})$, then the relative probability of seeing all N of our observations (again in advance of actually seeing them) is simply the product of the probabilities for seeing the individual observations. Thus the relative probability among all possible data sets that we would have seen, the set of data that we actually saw, is

$$L_N(a) = \Pi_{i=1}^N p(\underline{X}_i) = \Pi_{i=1}^N f(\underline{X}_i; a).$$

This expression is known as the likelihood function. It is a function of the parameter a through the dependence of the probability density function on this parameter. The principal of maximum likelihood estimation is to choose as our parameter estimate that value that maximizes the probability that we would have seen the data set that we actually saw, that is the value that makes the realized data set most likely. Let $\hat{a}$ be the maximum likelihood estimate of the parameter a. Then,

$$L_N(\hat{a}) = \underset{a}{Maximum}\, L_N(a).$$

In practice it is usually more convenient to maximize the logarithm of the likelihood function

$$\omega_N(a) = log L_N(a) = \sum_{i=1}^N log f(\underline{X}_i; a)$$

since it achieves its maximum at the same value.

As an example of maximum likelihood estimation, suppose

$$f(X;a) = \frac{1}{\sqrt{2\pi}\sigma} \exp\left\{-\tfrac{1}{2}(X-a)^2/\sigma^2\right\}$$

for a single random variable X and we wish to estimate the parameter a from a sample of size N. The logarithm of the likelihood function is

$$\omega_N(a) = \sum_{i=1}^N log f(X_i; a) = -\frac{1}{2\sigma^2}\sum_{i=1}^N (X_i - a)^2 - N log(\sqrt{2\pi}\sigma).$$

Taking the first derivative with respect to a and setting it equal to zero, yields the solution

$$Y_{ML}^{(a)} = \frac{1}{N}\sum_{i=1}^N X_i = \bar{X}$$

which is the sample mean. Thus, the sample mean is the maximum likelihood estimate for the center of a Gaussian distribution.

If we want the maximum likelihood estimate for σ, the standard deviation of the Gaussian, we set the first derivative of ω_N with respect to σ equal to zero gives the solution

$$Y_{ML}^{(\sigma)} = [\frac{1}{N}\sum_{i=1}^N (X_i - a)^2]^{1/2}$$

which depends on the value for a. However, we know that the likelihood solution for a, $\hat{a}$, is the sample mean $\bar{X}$ independent of σ, so making this substitution we have

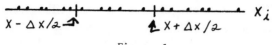
Figure 1

Histogram density estimate:

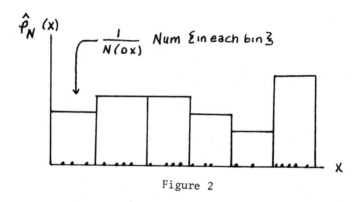
Figure 2

Rosenblatt (square kernal) estimate:

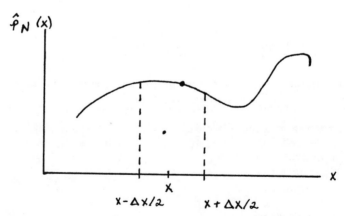
Figure 3. Overlapping Bins

$$Y_{ML}^{(\sigma)} = [\frac{1}{N}\sum_{i=1}^{N}(X_i - \bar{X})^2]^{\frac{1}{2}},$$

which is just the sample standard deviation.

In many (classic) cases it is possible to calculate in closed form the maximum likelihood estimate as was done for the simple case above. More often this is not possible and it is necessary to explicitly maximize the log-likelihood using numerical optimization techniques in order to obtain the maximum likelihood solution.

There is good reason for statisticians to like maximum likelihood estimation. First it always provides a prescription for parametric estimation. As long as one can compute the joint probability density given a set of parameter values, the likelihood function can be formed and maximized—either algebraically or numerically. The maximum likelihood estimate (MLE) can be shown to always be consistent. As the sample becomes large ($N \to \infty$), the MLE can be shown to have the highest possible efficiency. Also as the sample size becomes large, the distribution of the MLE estimate $\hat{a}$ can be shown to have a Gaussian distribution about the true value a

$$p_N(\hat{a}, a) = \frac{1}{\sqrt{2\pi}\sigma_{\hat{a}}} \exp\left\{-\tfrac{1}{2}(\hat{a}-a)^2/\sigma_{\hat{a}}^2\right\}$$

with

$$\sigma_{\hat{a}}^2 = 1/[\frac{\delta^2 \omega_N(a)}{\delta a^2}]_{a=\hat{a}}.$$

This information can be used to assign standard errors to maximum likelihood estimates.

There are a few drawbacks to maximum likelihood estimation. The estimates tend to be very non-robust. Also, if numerical optimization is used to obtain the MLE, it can be computationally expensive.

Nonparametric Probability Density Estimation.

In nonparametric estimation we assume that the data is a random sample from some (joint) probability density, but we do not assume a particular parameterized functional form. This is usually because—for the situation at hand—the correct functional form is simply unknown. The idea is to try to directly estimate the probability density of the population directly from the data in the absence of a specific parameterization. Such estimates are generally used for exploratory data analysis purposes.

Nonparametric density estimation is well developed only for the univariate case. Here we have a set of measurements $\{X_i\}_{i=1}^{N}$ presumed to be a random sample from some probability density function $p(X)$. Figure 1 illustrates a possible realized configuration of data on the real line. Consider an interval centered at a point X of width ΔX. The probability that one of our data points would have a value in this interval (before we actually observed it) is just the probability content of the interval

$$Prob\{X - \Delta X/2 \leq X_i \leq X + \Delta X/2\}$$

$$= \int_{X-\Delta X/2}^{X+\Delta X/2} p(\varsigma)d\varsigma$$

$$\simeq p(X)\Delta x.$$

The latter approximation assumes that the width of the interval is small. A reasonable estimate for the probability content of the interval, based on the data, is simply the fraction of data that lies in the interval. (This is in fact the MLE of this quantity), ie.

$$est[Prob\{\cdot\}] = \frac{1}{N} Num\{\cdot\}.$$

Combining these results yields an estimate for the probability density at X

$$\hat{p}_N(X) = \frac{1}{(\Delta X)N} Num\{X - \Delta/2 \leq X_i \leq X + \Delta/2\} \tag{7}$$

in terms of the number of counts in the interval of width ΔX centered at X. This result is central to two of the most popular methods of nonparametric density estimation—histograms and window estimates.

For the histogram density estimate the range of the data is divided into M bins or intervals (usually of equal width) and the density is estimated as a (different) constant within each bin using (7) (see Figure 2). The window or square kernel density estimate uses overlapping windows. At each point X for which a density estimate is required, a symmetric interval (window) centered at X of width ΔX is constructed and (7) is used to compute the density estimate (see Figure 3). The windows associated with close points will necessarily have a great deal of overlap.

For both these methods, there is an associated parameter that controls the degree of averaging that takes place. For the histogram estimate it is the number of bins, M. The larger this number, the less smooth the density estimate will become, but the better able it will be to capture narrow effects (sharp peaks) in the density. For the window estimate this trade-off is controlled by the width ΔX chosen for the window. The smaller the value of ΔX, the rougher the estimate will be, with the corresponding increase in sensitivity to narrow structure.

For multivariate $n > 1$ data, nonparametric density estimation becomes difficult. For two dimensions ($n = 2$) the straightforward generalizations of the histogram and window estimates involving rectangular bins or windows tend to have satisfactory performance. However, for higher dimensions ($n > 2$) performance degrades severely. This is due to the so-called "curse-of-dimensionality."

Consider a histogram density estimate in ten dimensions ($n = 10$). If we choose to have ten bins on each of the ten variables then there would be a total of 10^{10} bins. Clearly for any data set of reasonable size nearly all of these bins would be empty and the few that were not empty would generally contain only one count. Even with only two bins per variable (a very coarse binning) there would be over 1000 bins.

The window estimate suffers similarly. If for a uniform distribution in a ten dimensional unit cube, we wish our window (centered at each data point) to contain ten percent of the data points on the average, the edge length of the window would have to be approximately 0.8; that is, it would have to be 80% of the extent of the data on each variable. Clearly with such a window it would be impossible to detect all but the very coarsest structure of the probability density with such an estimate. Therefore, the most we can hope for is to be able to get a general idea of the joint probability density $p(X_1, X_2, \ldots, X_n)$ in high ($n > 2$) dimensional situations.

Cluster analysis is one approach for doing this. Here the goal is to try to determine if the joint density is very small nearly everywhere, except for a small number of isolated regions where it is large. This effect is known as clustering. Clustering algorithms attempt to determine when this condition exists and to identify the isolated regions.

Mapping the data to lower dimensional subspaces (usually one or two dimensional subspaces) and studying density estimates on the subspace is often a quite fruitful approach. Good nonparametric density estimation is possible in one and two dimensions. The trick is to perform the mapping in a way that preserves as much as possible the information contained in the full dimensional data set.

Let $\underline{X} = (X_1, X_2, \ldots, X_n)$ be a point in n-dimensions and $t = T(\underline{X})$ represent its mapping to one dimension. Here T is a single valued function of the n arguments $X_1, X_2, \ldots, X_n$. Since $\underline{X}$ is a (vector valued) random variable, t is also a random variable with a corresponding probability density function $p_T(t)$, that depends on the transformation function T. This (one-dimensional) probability density can be easily estimated and examined for different choices of transformations.

For a mapping onto two dimensions, one defines two transformation functions $t_1 = T_1(\underline{X}), t_2 = T_2(\underline{X})$ creating the random variables t_1, t_2 with joint distribution $p_{T_1, T_2}(t_1, t_2)$, depending on the choice of the transformation functions. Again, it is straightforward to estimate and examine the two-dimensional joint density of the mapped points t_1 and t_2. By performing judiciously chosen dimension reducing transformations and studying the corresponding density estimates, one can often gain considerable insight concerning the n-dimensional joint probability density $p(X_1, X_2, \ldots, X_n)$.

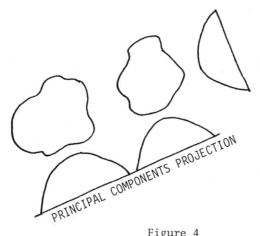

Figure 4

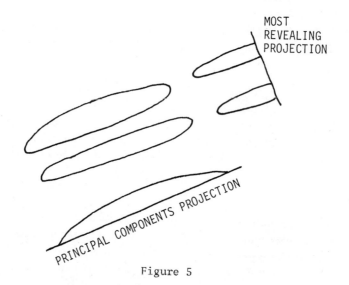

Figure 5

Generally the choice of mapping functions is guided by the intuition of the researcher using his knowledge of the data and the mechanisms that give use to it. There are also techniques that attempt to use the data itself to suggest revealing mappings to lower dimensions. The useful techniques so far developed involve only linear mapping functions

$$t = \sum_{j=1}^{N} a_j X_j = \underline{a}^T \underline{X} \qquad \text{(one − dimension)}$$

$$t_1 = \underline{a}_1^T \underline{X}, t_2 = \underline{a}_2^T \underline{X} \qquad \text{(two − dimensions)}$$

where the projection vectors $\underline{a}, \underline{a}_1, \underline{a}_2$ depend upon the data.

The most commonly used data driven mapping technique is based on principal components analysis. Here the basic notion is that projections (linear mappings) that most spread out the data are likely to be the most interesting. This concept is illustrated in Figure 4 for the case of mapping two-dimensional data to a one-dimensional subspace. Here there are two symmetrically shaped clusters separated in one direction. This direction is the one in which the (projected) data are most spread out, and is also the direction that reveals the existence of the clustering.

Principal components mapping can be fooled, however, as illustrated in Figure 5. Here the clusters are not symmetrically shaped, but are highly elliptical. The separation of the clusters is along the minor axes in the direction for which the pooled data is least spread out. Principal components in this case would choose the direction along the major axes (direction of most data spread) which in this case does not reveal the clustering.

This shortcoming of principal components mapping has lead to the development of projection pursuit mapping. Here, instead of finding mappings (projections) that maximize the spread of the data, one tries to find those mappings that maximize the information (negative entropy) defined as

$$I(\underline{a}) = -\int p_T(t) \, log \, p_T(t) dt$$

with $t = \underline{a}^T \underline{X}$, and $p_T(t)$ the probability density function of the projected data. This approach successfully overcomes the limitations of the projection pursuit approach but at the expense of additional computation.

Conclusion

The purpose of this report has been to give a broad (but necessarily quite shallow) overview of statistical data analysis. The intent was to introduce astronomers to the way statisticians view data so that they can judge whether increased familiarity with statistical concepts and methods will be helpful to them.

CLUSTER PHOTOMETRY: PRESENT STATE OF THE ART
AND FUTURE DEVELOPMENTS

Ivan R. King

Astronomy Department
University of California
Berkeley, CA 94720, U.S.A.

My assigned topic is a difficult one to cover. It is a broad topic; furthermore there are people in this room who know more than I do about the present state of the art, and I can't imagine how I or anyone else could be qualified to describe with any confidence what future developments are likely to take place.

Not only is the topic broad; it also has a logical structure that makes it difficult to arrange in the one-dimensional structure of a talk that proceeds in a straight line from beginning to end. Many of you deal with arrays that are perceptively called data cubes; here I have to deal with a logic cube. It is illustrated in Figure 1. The subject can be divided along each of three independent axes. One of them is the technique by which the photometry is carried out, the second is the type of photometry that is being done, and the third is the astronomical problem toward which the work is directed. I shall organize my talk pretty much according to these three axes, in this logical order of subordination. The main organization will be according to observational techniques, and within each of these I will try to take up the different kinds of photometry and their application to the diferent astronomical problems that may attract the *aficionados* of star clusters. In doing this I will deal chiefly with globular clusters, partly because I know them better, but mainly for the better-justified reason that they not only exemplify the problems that arise but in many cases they also require more strenuous effort or more delicate attention than do the problems of fields that are less crowded or less faint.

PHOTOGRAPHY

For many decades the workhorse of astronomical imaging was the photographic process. As a photometric technique, unfortunately, it was misused more often than it was used correctly, and for that reason it became widely distrusted. It is certainly true that the rule of *caveat emptor* applies, but there have been many honest goods in this marketplace, as a number of people in this room have shown with their own achievements.

The disadvantages of photography have now relegated much of it to historical interest, but it does nevertheless retain some advantages

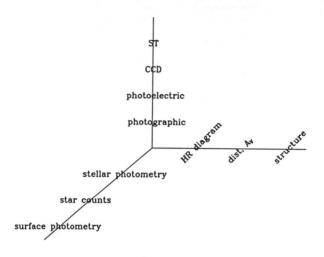

Fig.1. The "logic cube" of this talk.

that will keep it alive for a long time. The problems of photographs are well known. They have a low quantum efficiency, their response is non-linear, their absolute level of sensitivity is well-nigh impossible to control, and their low information capacity per unit area prevents them from reaching the faint magnitude limit of the electronic detectors that have replaced them. Yet photographs do have one advantage: they offer a format that has far more pixels than we are likely to see in any electronic detector for a long time to come. The 35-cm plates used in the large Schmidt cameras have, when one considers the pixel size that is appropriate to their resolving power, over 10^9 pixels, or more than 200 times the number offered even by the 2048×2048 CCDs that now hang tantalizingly on our horizon.

The big Schmidt cameras will remain our survey instruments, I believe, for the indefinite future. (Remember, a survey is not just taken once and for all; a couple of decades later a repeat of it will yield proper motions.) But I should get back to the star clusters. The reason why photography can still contribute to their study is that most of them are too big for the limited fields of the electronic detectors. Particularly for the study of the extended distribution of a cluster's stars, it would seem that only a photograph can cover a large enough area.

I say "it would seem" advisedly. In fact, faint star counts in globular clusters could be done more effectively with a set of overlapping CCD exposures than it can be done on a photograph. In some of the planning studies of instrumentation for the 10-meter Keck telescope, we calculated the relative abilities of a photographic plate and a CCD to map an area. The CCD won in two ways. With its repeated but shorter exposures it could cover an area faster than a photograph -- to the limiting magnitude that the photograph can reach. Furthermore, if the expenditure of time were not the question asked, the CCD could, by exposing longer, reach two or three magnitudes fainter than the photograph.

Why then do I say that photography is still needed for its wide-field survey ability? The answer is purely economic, and largely a question of manpower. We are reluctant to collect, reduce, and combine (or even worse, reproduce) the 100 CCD fields that would replace one survey photograph. So we will certainly go on using photographs for a while yet, and I should devote a few words to their proper use.

Assuming (as I hope one can) that the normally accepted processing procedures are followed, there is only one caution that I would emphasize. It is essential that every photograph that is taken have impressed on it a set of standard spots of known relative intensity. This is best done in the telescope itself, simultaneously with the celestial exposure. This is indeed the practice at many of the best-planned telescopes, including the Kitt Peak and Cerro Tololo 4-meter reflectors and the large Schmidts of the UK SERC, ESO, and Palomar Observatory.

The second prerequisite, which can be expensive and burdensome, is a good automatic microdensitometer. The PDS has been the workhorse of the world, but faster and more efficient machines are available, most notably the Cambridge Automatic Plate Machine.

In this connection, those who will be discussing the processing, storage, access to, and transmission of large quantities of data should consider the implications of files that each have 1.2×10^9 pixels, remembering that it takes 1600 fields to cover the sky and that we want several colors and perhaps a number of objective-prism surveys. Eventually we will get rid of photography, at the cost of the extra multi-image CCD effort that I have referred to, but the grand (?) total to which I have alluded remains (for, say, *UBVRI* plus two prism ranges) 1.3×10^{13}. It's a big universe out there.

The largest existing digitized photographic survey of which I know is the guide-star survey at the Space Telescope Science Institute. Although it is undersampled and only in one color, it will fill several hundred optical disks. Multiply by perhaps 30 for the full survey to which I referred above.

One final area to discuss in connection with photography is astrometry. What photography offers here is the tying together of a large sky area on a single rigid base. In the alternative that CCD frames offer, a large number of image overlaps have to be made, and there is a loss of accuracy at each overlap. The CCD images give a higher astrometric accuracy, because of their photometric superiority (King 1983), as already demonstrated by recent US Naval Observatory CCD parallaxes (Monet and Dahn 1983), but I have have never calculated the tradeoff between the increased accuracy in one image and the greater number of overlaps.

In one of their aspects photographs will always remain valuable in astrometry, however. Proper motions require a long baseline in time, and nearly all the old positional information that we have for star clusters is photographic. In this context CCD frames will eventually become important too, as time passes and some of them can properly be called old. A problem, however, is the lack of archiving, or in many cases even of cataloguing, of the CCD frames that are being produced in such large numbers at our major observatories. Their ability to produce proper motions in the future may very well be limited completely by the fact that no one can find them -- or even knows that they exist!

PHOTOELECTRIC PHOTOMETRY

Another traditional technique is largely outdated but still has a few things to offer: photoelectric photometry -- that is, the observation of a single small sky area with a photoelectric cell. Here I have to distinguish between the different photometric problems that face us. Where the question is relative magnitudes of individual stars, CCD photometry is just as accurate and infinitely faster, since it registers many stars simultaneously. The one place where I believe that photoelectric photometry still has something to offer us, however, is in setting magnitude zero points. CCD observers do indeed intersperse observations of standard stars, but it is my impression that this is usually not done with the same dedication that is shown by photoelectric observers who set out to make the best possible comparison between individual objects in different parts of the sky. Moreover, many CCD observations are made on non-photometric nights, when the CCD observations themselves are perfectly good but no meaningful comparison with standards is possible.

In this respect globular clusters are in a particularly fortunate position. The reason is that there is already a large body of carefully calibrated photoelectric observations of globular clusters, made through apertures of various sizes for the purpose of investigating the structures of the clusters. If one adds the light in a CCD frame out to the radius of any given photoelectric observation, the comparison immediately gives a magnitude zero point for the CCD frame. In this area Peterson (1986) has done a tremendous service by collecting and combining, from the scattered literature, all photoelectric observations of globular clusters. From these one can in principle derive a zero point for any CCD frame of any cluster that is in Peterson's catalog. The zero point will hold not only for surface photometry; through the point spread function it can be converted into a zero point for the star images too.

It may not, however, be necessary in the future to resort to this sort of calibration for CCD images. One of the difficulties of the existing standard stars is that, having been set up for photoelectric photometry, they tend to be relatively bright stars, separated from each other. A CCD can observe such standards only one at a time, and with exposures so short that they may be difficult to time accurately. Some standard fields are now being set up for CCD calibration (in clusters, of course), with a good number of fainter stars in each. When they are ready, the standardization of CCD photometry will be possible directly, the only restriction being that photometric sky will still be needed. (I am grateful to Charles Peterson for calling my attention to this development.)

There is a great drawback in both CCD and the older type of photometry, however, which leads me to bring up what I believe is the most serious outstanding problem of cluster photometry: color equation. Even with the old "comfortable" *UBV* standards, the uncertainties of color systems were serious; consider for example the excess-M-dwarf fiasco of the 1970's, which turned out to be due to an incorrect color equation in a set of photographic magnitudes (as shown by Faber *et al.* 1976). Today's situation is potentially much worse, because of the proliferation of detector-filter systems. We run the danger of having marvelously faint and accurate photometry but not knowing what it means in terms of the astrophysical quantities that matter. Let us not emulate the rich man who starved to death with his pockets full of gold, because he could not turn it into bread to eat. I shall return to this vital point later, when discussing the more extensive work that is now done with CCDs.

CCD PHOTOMETRY

The great majority of astronomical photometry being done today uses electronic detectors. I am no expert on the relative characteristics of detectors and will therefore concentrate on CCDs, which represent the bulk of today's imaging.

CCDs have such advantages, when compared with the previous photographic methods, that they seem like a dream. Their quantum efficiency is of the order of 1/2 -- which seems like unity to anyone brought up with the older detectors -- they are very nearly linear, and they have a very large capacity, in terms of maximum counts in a pixel, which determines both the dynamic range before saturation and, through the square-root law, the maximum accuracy that can be attained in a single observation. But perhaps most important from the point of view of this meeting is the fact that the output of all these electronic detectors is *digital*. It can be computer-processed immediately. Interestingly, many older observers have at first been disconcerted to return home with a box of magnetic tapes but no visible images -- until they consider the alternative of having a lot of pretty pictures that require tracing and a considerable amount of recondite witchcraft before they will yield anything quantitative.

But fear not -- CCDs have not made intelligent astronomers obsolete; CCDs do not produce quantitative results at the push of a button. Like any other scientific instrument, a CCD yields outputs that are a function of its inputs. As always, what separates the astronomers from the mere takers of pretty pictures is the ability to recover as accurately as possible the input intensities that went into the picture. Here one must get intimately involved with the crankiness of the CCD, which I have neither the ability nor the time to go into intensively here. Instead I will refer you to an excellent review article (Djorgovski 1984) and merely give a brief summary of the problems.

As a starting point, let us recognize that the ideal objective of astronomical photometry is to recover intensities whose accuracy is limited only by the square root of the number of detected photons. Here the CCD interposes its first crankiness: readout noise, which is introduced by the preamplifiers that extract its output. By convention this is always quoted as an rms equivalent -- which is to say that the actual noise is really the square of the number quoted. A CCD that has a readout noise of 15 electrons is a device that adds 225 noise counts to every number that is read out of it. Fortunately, however, in most cluster-photometry applications the readout noise is less than the noise of the sky signal in each pixel.

Another problem is threshold, sometimes confusingly called "fat zero" or "skinny zero"; the detector reacts in a less-than-linear way to the first few electrons received by each pixel. The cure for this trouble is to preflash the CCD -- thus adding to the combined sky and readout noise.

Still another set of problems of any CCD is of a cosmetic class. The detector is a chip of silicon whose pixels have been made as uniform and reliable as possible, but invariably some of them are bad. These have to be mapped -- hopefully once and for all for a given chip -- and in some way masked out or interpolated over.

In addition, each frame has its individual problems. One class of these is cosmic rays, for which a chip of silicon is an excellent detector. In ground-based frames these are individual blackened pixels, which are easily -- but laboriously -- removed. In Space Telescope images they can present a special problem, which I shall take up later.

Bright stars are another problem. They tend to "bleed" down the column that they are in, ruining a lot of other pixels; and if the star is very bright it can leave a residual image on the surface of the chip, which may take appreciable time to disappear. This sort of problem can be quite troublesome in globular clusters, where one would like to be able to work on stars of 23rd magnitude and fainter but it is almost impossible to avoid having stars in the field that are ten magnitudes brighter.

But by far the greatest practical problem in getting accurate results out of a CCD is the so-called "flat field." No chip can ever have a perfectly uniform sensitivity over its surface; so it is necessary to determine, in effect, the relative sensititivity of each individual pixel. Clearly this problem would be solved if we knew the response of the chip to a source of absolutely uniform illumination -- hence the name flat field, which refers to the array of multiplicative corrections that must be applied, pixel by pixel, to convert the observed counts into those that would have been made by a uniform detector. In practice, however, it is diabolically -- or at the very least, expensively -- difficult to find a perfect flat-field source. One possibility is to point at a uniformly illuminated surface inside the telescope dome, which is made even more flat by the fact that it is out of focus for the telescope. In practice, however, "dome flats" always seem to produce an illumination that is smooth but not totally uniform -- somewhat bowed across the whole extent of the field.

A second possibility is to use the night sky. Over an exposure of appreciable duration, it is quite uniform -- except of course for the stars and galaxies, which in this context are noise rather than signal. The usual technique for building up a "sky flat" is to combine a number of sky exposures, while deleting as well as possible all of the objects in them. For large objects this requires hand-deletion, but small objects can be eliminated by techniques of median smoothing. If many sky frames of sparse fields are available, a "vertical" median (i.e., a median of each individual pixel over all frames) will ignore the stars, cosmic rays, and other small objects. If not enough well-exposed sky frames are available, then median smoothing (or perhaps better, a determination of the mode) can be applied to a sub-area of a frame, where the median value will be characteristic of the sky and not be influenced by the small number of higher pixel values that result from the stars. The various sky frames, each one heavily smoothed, can then be averaged, for a more stable set of values.

An effective flatfielding technique is to combine a dome flat with a sky flat. (In principle a good sky flat would do the job, but in many cases the sky is not heavily enough exposed to have a high enough S/N for individual pixels.) The dome flat gives an excellent correction of small-scale pixel-to-pixel variations but is not flat on a large scale. The sky flat contains, after the median smoothing of large subareas, large-scale information only. This can be used to accomplish the long-scale flattening of the dome flat.

There is, in fact, one ready source of completely even illumination: twilight sky. It is unfortunately not directly suitable for flat-field exposures, because it has a very different color from the astronomical objects that we are trying to measure. Sadly, the flat field depends on color. It has to be determined separately for your CCD in each color band in which you are observing, and because of this very color dependence it has to be determined from objects of similar color to those that you want to measure.

Nevertheless I think that twilight sky may eventually provide us with our best flat fields. First, it may be possible to find a color-balancing filter that will make the very blue twilight sky look more like the color of a typical star (opposite to the effect of the standard filter that photographic studios use to make incandescent light look like daylight). Second, and more important, it may be possible to work out for flat fields the equivalent of the color equation that we use in allowing for differences in color sensitivity between different detectors. The color-dependence of the flat field is, in fact, just an abbreviated way of expressing the fact that each pixel has its own individual color equation. Indeed, I am sure that there will be profitable consequences in looking at a CCD as an array of simultaneous photoelectric photometers, and treating them accordingly -- but by mass data-processing techniques. This is an area that I believe deserves really serious consideration, because flat-fielding and color equation are the two most serious limitations in the accurate photometric use of CCDs.

The most insidious problem of flat fields is fringing; fortunately it occurs only under certain circumstances. The thin wafer of silicon that constitutes the light-sensitive part of a CCD is rather opaque to light of short wavelengths but becomes more transparent at longer wavelengths, where light can reflect off the back surface and create interference fringes. The worst fringes would obviously arise from monochromatic light; this is unfortunately the case for the infrared emission bands of the airglow, which furthermore change with time. In broad-band images the fringes coming from different emission wavelengths tend to be smoothed out, but in narrow-band infrared observations they can be a serious limitation. De-fringing of CCD images is an art that I cannot discuss here.

But this has been a long preamble; let us get to the clusters themselves. Our objective should be, as I have said, to achieve photometry whose accuracy is limited only by the Poisson noise in the number of counts. Let us assume that we have solved all the practical problems of the detector and are at this desirable level. There still remains the problem of finding the magnitudes of stars. In a noiseless environment this could be done by simply counting up the number of photons detected from the star. Noise cannot be avoided, however -- from the readout noise and especially from the sky. When noise is present, adding up all the light is not a good strategy, because the outer parts of a star image add much more to the noise than they do to the signal. The best strategy here is to fit, preferably by least squares, the known profile that applies to stars in the frame (the so-called point-spread function, or PSF). It can be determined with good accuracy from some of the brighter stars. Each of the fainter stars should have this same profile, with an unknown intensity factor that gives the star's relative magnitude. The least-squares solution actually has 4 unknowns, however.

The sky-background level must be fitted, as must the star's x and y coordinates -- so a free-of-charge by-product is astrometry! (For a discussion of the statistics of this problem, see King 1983.)

The approach that I have just described is that taken by DAOPHOT, a program package developed by Peter Stetson at Dominion Astrophysical Observatory. I believe that it is becoming a world standard for stellar photometry of digital images. Moreover, DAOPHOT has powerful crowded-field capacities that can be especially valuable in clusters. (As yet, there is no published description of this package, except for the excellent manual that is distributed with it; but Stetson is preparing a descriptive paper for *Publ. Astr. Soc. Pacific*. About 65 copies of DAOPHOT have been distributed to astronomical institutions around the world.)

There are a number of other steps to be taken before PSF-fitting can be applied. First, one must have a list of approximate positions of the stars to be measured. If this does not already exist, DAOPHOT can provide such a list, by means of a local-peak-finding procedure (which avoids galaxies and dirt-specks by first asking the user for the approximate full-width-half-maximum of proper star images). This process also produces a set of classification-quality magnitudes -- by Gaussian fitting, which although less accurate is quick.

In least-squares fitting it is important to use correct weights, and for this, one must treat the photon statistics correctly. Most CCD systems scale their output, for data-recording convenience, into an arbitrary "digital unit," and one must therefore know how many actual detected photons correspond to 1 DU.

Thus, in summary, in a new star field one first goes through the finding procedure. Then the remaining steps are the same as for all other frames of this field: (1) edit the finding list as necessary (to adjust the limiting magnitude, for example), (2) find the sky value in the neighborhood of each star, (3) determine the PSF for the frame, (4) determine magnitudes by PSF fitting.

One of the user-chosen parameters in PSF fitting is the size of the area around each image over which the fitting is to be done. In uncrowded fields this radius can have a generous size, although little is gained by setting it at more than a couple of times the half width at half maximum.

DAOPHOT has some special tricks for dealing with crowded fields. One can determine magnitudes for the brighter stars, and then ask DAOPHOT to subtract out the fitted-PSF representation of each of them. If the fitting was good, then these stars are cleanly removed and one can proceed to identify the fainter stars that were previously lost in the edges of the bright ones. Most important, DAOPHOT has a procedure for dealing simultaneously with the members of a group of neighboring stars, by making a simultaneous PSF-fit to each of the members of the group. In most regions, and especially in a globular cluster, this is the final stage of photometry.

In general, however, before measuring crowded star images one should be clear about one's objectives and aware of the dangers. The most obvious of the latter are photometric errors due to inclusion of unresolved companions and due to inability to determine a correct sky level next to a star in a crowded region. In general, crowded photometry is not going to be very accurate. One should therefore measure near the center of a cluster only when there is a strong reason for doing so, as for example in the search for the optical counterpart of the X-ray source in M15 (Aurière *et al.* 1984, Aurière *et al.* 1986). But if the object of a study is a color-magnitude array, it is far better to keep to the less-crowded regions in the outer parts of the cluster, even at the cost of the inclusion of more field stars.

As techniques improve, observers are reaching fainter and fainter magnitudes in globular clusters. Because CCDs have a high quantum efficiency, they register photons fast, and their limited capacity per pixel can be circumvented by adding up many frames, each of which has a relatively short exposure. Calculations of photon statistics suggest that a total exposure of 2 hours with a 4-meter-class telescope in good seeing should be capable of producing photometry with a signal-to-noise ratio of 5 at 26th magnitude. This goal has not yet been reached, although several observing groups are pushing toward it. The key to success is the ability to extract faint star images from the sky background, which must be determined to an accuracy of 1/2 % or better. What is needed is presumably better flat-fielding, avoidance of crowding, and careful local determinations of the inevitable background of faint stars.

Finally, let me mention again the problem of color equation, which limits our ability to interpret our photometry at any limiting magnitude. An important point to note in this connection is that there is a proliferation of color systems for CCDs, instead of the standardization that we so much need.

STAR COUNTS

In some studies related to the structure of clusters one does not need to go as far as detailed photometry; simple counts of the number of stars per unit area will suffice. Even star counts require some care, however. It is here that crowding is most dangerous; it has the tendency to cause some stars to be lost, thus distorting the star densities as a function of the very density that is supposedly being measured. Empirical studies have shown (King *et al.* 1968) that crowding effects set in when more than 1/50 of the area is occupied by star images. Such a region does not look at all crowded, but for purposes of star counting it is!

Another purpose to which star counts have been put is the determination of luminosity functions in clusters. One counts plates of different exposures and takes the number of stars as a function of the limiting magnitude of the count. This is actually a rather unreliable procedure, because it is nearly impossible to determine how faint the limiting magnitude of a count is. The only reliable way to determine a luminosity function is to make at least a rough photometric measure of every star, in order to be sure which of the magnitude classes to count it in. DAOPHOT, incidentally, offers an excellent means for doing this right, and it is less likely to be affected by a low level of crowding. (I am indebted to George Djorgovski for this remark.)

One more remark about luminosity functions: what is wanted in almost all cases is the luminosity function of the cluster as a whole, not just that of a limited region. Because of the dynamical segregation of stars by mass, the two functions are different. As a result luminosity functions are poorly known, especially for faint stars in globular

clusters (King 1971). To determine their total number, we would have
to extrapolate them into the unobservably crowded central regions; but
dynamical theory is too uncertain to allow this to be done reliably.
This is a problem that will have to be solved by direct observation, and
those observations will have to wait for the resolving power of Space
Telescope. (It pleases me particularly that the ST observations will
also go a long way toward resolving the dynamical uncertainties that I
referred to.)

SURFACE PHOTOMETRY

Since star counts cannot be made in the dense central parts of a
cluster, we resort to surface photometry for density information there.
This is a necessity, not a virtue; in determining densities, surface
photometry is inferior to star counts that go even moderately faint.
The reason is again a Poisson limitation. The density of stars in a
given area obviously has an uncertainty that depends on the square root
of the expected number of stars; but the light coming from that area
also has a corresponding statistical uncertainty, and it is worse. The
reason for this greater uncertainty is that the integrated light from
the stars of a cluster is strongly dominated by the contribution of
the red giants, whose effective number is quite small (King 1966). The
dominance of this contribution can be somewhat reduced by working in the
ultraviolet, where the red giants have relatively less influence, but
even there their statistical effect is damaging.

Yet until we achieve ST resolution we can study the densities of
the central regions of clusters only with surface photometry. As I
have already mentioned, a good deal of such work has already been done
with photoelectric photometers; but this is another area in which CCDs
can excel, both in observing efficiency and in the sometimes important
ability to remove the light of superposed foreground stars.

I shall describe CCD surface photometry of globular clusters as
it is done in a large program that is being carried out at Berkeley.
(For preliminary results, see Djorgovski and King [1986].) We begin
of course by cleaning the images: flattening, and removing defects and
cosmic rays.

Our next step is to determine the cluster center. This is an
important step, because a too-casual fixing of the center can lead
to serious distortions of the central profile. If, for instance, we
set on a local peak of brightness, this can create a false central
brightness peak. Conversely, if we miss the real center and choose a
fainter region nearby, we will create a false shoulder, at the annulus
that contains the real center. To avoid problems of this sort we
have developed an algorithm that we believe is likely to fix on the
true center of the cluster. It depends on cross-correlating the light
distribution with a mirror image of itself and finding the point that
gives the highest level of agreement. For this purpose we choose a
region large enough to have a good statistical significance and to
smooth over small irregularities -- about the size of the core radius.

We also remove, in an interactive procedure on a TV display, bright
stars that we consider to be foreground non-members. We draw a box
around each of these stars, and every pixel in the box is marked with a
flag that tells our reduction programs to ignore the pixel. (This step
cannot precede the determination of the center, however, because our

centering algorithm does not tolerate the omission of any pixels.)

Another step is the determination of the sky level. This is done in several boxes that we mark near the corners of the frame. For the sky level we take the modal pixel value, which is an excellent estimator of blank sky. This has the disadvantage, however, of ignoring the faint stars. Properly we should include the same number of background and foreground stars that would be found in an equivalent area of the cluster. The sky determination is also an inherently weak point in CCD surface photometry of clusters, because the field of a CCD is too small for even its corners to be outside a typical globular cluster; hence the "sky" areas actually include some cluster light. We intend to remedy this later in our study by using our preliminary photometric profile of each cluster to correct for this residual light. Hopefully these two shortcomings somewhat cancel each other at the present preliminary stage of our work. Djorgovski and I are still arguing about how to do the final job.

For the actual photometry we do just as a photoelectric photometer would have: we measure the total light in an area -- in this case, just by adding up the pixels -- and divide by the area, which in this case is just the number of valid pixels. Relative to the photoelectric photometer we have two advantages, however (in addition to speed): we can choose any spatial resolution that we wish, and (as already mentioned) we have eliminated obvious foreground stars.

One caution should be noted. Since it is necessary to add up the light of all the cluster stars, the exposure must be regulated so that none of the stars is saturated. (This restriction obviously does not apply to the foreground stars that are going to be ignored.) This turns out not to be a serious restriction; the dynamic range of a CCD is great enough that we can avoid saturating the centers of the brightest cluster stars and still register enough light to trace the diffuse brightness nearly to the edge of the frame.

Thus we choose annuli of scientifically desirable sizes, and find the average surface brightness in each. In this operation we have the additional opportunity of deriving an empirical value for the sampling error that I have referred to above. We simply divide each annulus into 8 sectors, and we then use the 8 residuals from the mean to give us a mean error for our mean surface brightness of the annulus.

In this way we derive a surface-brightness profile of the cluster, which then lacks only a magnitude zero point. This can be provided, as I have indicated, by comparing with photoelectric photometry from the literature, which we collect into a "growth curve" of magnitude versus radius. An equivalent curve can be derived from our radial CCD photometry, and comparison of the two gives us our zero point.

SPACE TELESCOPE

In the final part of this talk I should like to address the question of what the Hubble Space Telescope can do for the photometry of globular clusters. The answer lies almost completely in resolving power. It is not hard to predict how well ST will be able to resolve stars in a given crowded area. What I have done for this purpose, in general, is to sum the light of the stars in the luminosity function that is expected in the region to be observed. (I have done this for several different luminosity functions, appropriate for different populations.) It is then possible to go from the observed surface brightness of an area to the number of contributing stars at each absolute magnitude, and thus predict the star densities with which ST will have to cope. For globular clusters the densities turn out not to be excessive, except at the centers of those clusters that have dynamically collapsed cores. There, of course, one of the outcomes of the ST observations will be to see how small the remaining cores are and how high the central densities actually get. This can be done by reverting to surface photometry at those centers.

In the more normal clusters ST will resolve the stars well enough to measure them all, down to the limit to which it can do significant photometry in an exposure that lasts one orbital night. This corresponds to $S/N \sim 5$ at 26th magnitude.

It is not mere coincidence that this is the same specification to which I referred in discussing the limits of ground-based photometry. A ground-based CCD on a 4-meter telescope will reach about the same limit in two hours of good seeing as ST will in one orbit. Thus we have chosen in our work to use ST only in the central parts of the clusters; the outermost field in each cluster can be done from the ground, releasing ST time for other tasks that only ST can do.

Although I hate to sound like a broken record stuck in the same groove, it is time to mention color equation again. The most serious problem in interpreting ST observations of faint stars in globular clusters is going to be determining the transformation from ST magnitudes to the $M_{\rm bol}, \log T_e$ plane in which the theoreticians do their work.

Photometry with ST is going to present some interesting new problems. As in ground-based work, the best photometry with the Faint Object Camera (which is the high-resolution instrument with which the centers of globular clusters will be observed) will be done by PSF fitting. But now a new difficulty appears. The PSF of the FOC is determined almost completely by the diffraction pattern of the entrance pupil of ST, and the size of this pattern depends on wavelength. In the broad bands in which faint photometry is done, the effective PSF will be determined by a convolution of this variable PSF over the wavelength sensitivity function of the band -- but worse, the PSF will depend on the color of the star, since a redder star will get more contribution from the wider PSF that applies to the longer-wavelength side of the band. It appears to me that photometry will be a two-stage operation. First we will have to measure all the stars with a single PSF; then, with a knowledge of their colors we will have to remeasure each star with the PSF that is appropriate to its color.

The Wide Field/Planetary Camera also has its troubles. Each of the cameras has a field that is covered by four separate CCD chips, each with its own characteristics. Where this will make the most difference is in the broad infrared band in which a large fraction of the observations will be made. The filter used for this band, F785LP, defines only the short-wavelength cutoff; the long edge of the band is determined by the dropoff of the sensitivity curve of the CCD, which differs for every chip. Thus the different parts of the WF/PC will have 8 different I-band color equations.

Another potential problem turns out not to be serious. In the WFC mode, the pixel size is large relative to the PSF, so that a large fraction of the light of a star will be concentrated in one part of one pixel. Fortunately, tests made by the team have shown that sensitivity is uniform within a pixel. This will not necessarily be true for other CCDs, however; when a CCD is used in a badly undersampled mode, it may be that *intra*-pixel sensitivity differences will be a photometric problem.

Another problem of ST observations is going to be cosmic rays. Particularly when a star image can be concentrated largely in a single pixel, cosmic rays will be hard to distinguish *a priori*. Most WF/PC exposures will therefore have to be split into two parts, so that any "object" that does not appear on both parts can be rejected as a cosmic-ray hit. This has the unfortunate effect of doubling the number of readout-noise counts and lowering the signal-to-noise ratio correspondingly. For the FOC, on the other hand, cosmic rays are more of a nuisance than a detriment, since the FOC has a picture-data store that can be read out non-destructively and without noise, in the middle of an exposure.

Finally, let me mention one aspect of ST that makes it particularly valuable for cluster observations: it can observe at the same time with both the FOC and the WF/PC. The two cameras are 6.5 arcmin apart and can expose simultaneously. In a globular cluster this allows using the FOC in the most crowded regions of the center while pointing the WFC at a larger area farther out. And in other galaxies it allows us to get serendipitous pictures of globular clusters around the edge of the galaxy. Simultaneous exposures of this type will actually be made on all suitable occasions.

CONCLUSION

The data-reduction problems of cluster photometry are driven by the methods with which we are able to observe the clusters, and these keep changing. A review such as this one would have said quite different things at the previous Data-Analysis Workshop, and no doubt the subject will sound quite different at the next one.

REFERENCES

Aurière, M., Le Fèvre, O., and Terzan, A., 1984, *Astron. Astrophys.*, 138: 415.

Aurière, M., Maucherat, A., Cordoni, J.-P., Fort, B., and Picat, J.-P., 1986, (preprint).

Djorgovski, S., 1984, CCDs: Their Cause and Cure, in: "Proceedings of the Workshop on Improvements to Photometry," NASA Conference Publication 2350, Washington, D. C.

Djorgovski, S., and King, I. R., 1986, A Preliminary Survey of Collapsed Cores in Globular Clusters, *Astrophys. J. Letters*, 305: (in press).

Faber, S. M., Burstein, D., Tinsley, B. M., and King, I. R., Rediscussion of the Local Density of M Dwarf Stars, *Astron. J.*, 81: 45.

King, I. R., 1966, The Structure of Star Clusters, IV. Photoelectric Surface Photometry in Nine Globular Clusters, *Astron. J.*, 71: 276.

King, I. R., 1971, The Dynamics of Star Clusters, *Sky and Telescope*, 41: 139.

King, I. R., 1983, Accuracy of Measurements of Star Images on a Pixel Array, *Publ. Astron. Soc. Pacific*, 94: 163.

King, I. R., Hedemann, E. J., Hodge, S. M., and White, R. E., 1968, The Structure of Star Clusters, V. Star Counts in 54 Globular Clusters, *Astron. J.*, 73: 456.

Monet, D. G., and Dahn, C. C., 1983, CCD Astrometry. I. Preliminary Results from the KPNO 4-m/CCD Parallax Program, *Astron. J.*, 88: 1489.

Peterson, C. J., 1986, Globular Cluster Photometry Catalogue, *Astronomical Data Center Bulletin*, 1: (in press).

CLUSTERING TECHNIQUES AND THEIR APPLICATIONS

Fionn Murtagh (*)

Space Telescope - European Coordinating Facility
Karl-Schwarzschild-Str. 2
D-8046 Garching bei München, F.R.G.

INTRODUCTION

Solid progress has been made in clustering, and in adjacent fields such as nearest neighbour searching, in recent years. This article will overview recent research work which appears to be very promising, and will also pinpoint difficult issues arising. We hope that we can indicate both the exciting advances which have taken place in this area, and also convey the latent potential for further gains.

Automatic classification algorithms are used in widely different fields in order to provide a description or a reduction of data. A clustering algorithm is used to determine the inherent or natural groupings in the data, or to provide a convenient summarization of the data into groups. Although the term "classification" is often applied both to this area and to Discriminant Analysis, this paper will be solely concerned with unsupervised clustering, with no prior knowledge on the part of the analyst regarding group memberships. A recent, and very readable, textbook on Discriminant Analysis is James (1985), and a comprehensive survey of methods in this area is provided by Hand (1981).

Without precluding possible data recoding and other transformations, the objects to be classified have numerical measurements on a finite set of variables or attributes. Hence the analysis is carried out on the rows of an array or matrix. The objects, or rows of the matrix, can be viewed as vectors in multidimensional space (the dimensionality of this space being the number of variables or columns).

In this paper, we will survey widely-used clustering algorithms which have been employed for the interpretation and analysis of astronomical data. In sections 2 and 3 we will review an important method (the minimal spanning tree) and a representative of a widely used class of methods (iterative, optimization algorithms for non-hierarchical clustering). In section 4, we turn attention to hierarchical methods. Recent algorithmic advances in hierarchical algorithms have drawn this area very close to the area of nearest neighbour finding (or best match searching). These developments are surveyed in section 5. Section 6 discusses

---

(*) Affiliated to Astrophysics Division, Space Science Department, European Space Agency

applications to astronomical problems. Finally section 7 is looking to the future, foresees continued dynamism and expansion in this area.

MINIMAL SPANNING TREE

The minimal spanning tree (MST) is described in most texts on graph theory, and in many other areas besides (see Tucker, 1980; Graham and Hell, 1985). The following algorithm constructs a MST by a "greedy" or nearest neighbour approach.

Input: A set of points, with some dissimilarity (e.g. the Euclidean distance) defined on them.

Step 1: Select an arbitrary point and connect it to the least dissimilar neighbour. These two points constitute a subgraph of the MST.

Step 2: Connect the current subgraph to the least dissimilar neighbour of any of the members of the subgraph.

Step 3: Loop on Step 2, until all points are in the one subgraph: this, then, is the MST.

Step 2 agglomerates subsets of objects using the criterion of connectivity. The close relationship between the MST and the single linkage hierarchical clustering method is illustrated below in Figure 1.

Obtaining components from a MST is the problem addressed by Zahn (1971). An edge is said to be <u>inconsistent</u> if it is of length much greater than the lengths of other nearby edges. (Here, we use graph theoretic terminology, where <u>vertices</u> are points, and <u>edges</u> are interconnecting lines). Zahn applied these approaches to point pattern recognition, - obtaining what he termed "Gestalt patterns" among sets of planar points. The MST provides a useful starting point for undertaking such pattern recognition problems (see di Gesù and Sacco, 1983, for an application in gamma-ray astronomy; Barrow et al., 1985, for an application involving galaxy clustering; and Pirenne et al., 1985, for the analysis of interferograms).

The MST is also often suitable for outlier detection. Unusually large edge lengths will indicate the abnormal points (or data items) sought (Rohlf, 1975).

ITERATIVE OPTIMIZATION ALGORITHMS

Non-hierarchical clustering methods appear to be most used in astronomical applications (see Kruszewski, 1985; Jarvis and Tyson, 1981; Egret and Heck, 1983a,b; Egret et al., 1984; Heck et al., 1984; Heck et al., 1985; Tholen, 1984; the first two of these are concerned with object classification, following the scanning of digitized images; the last is concerned with asteroids; and the remainder are concerned with IUE spectral classification).

As a non-hierarchic strategy, the variance criterion has always been popular (some of the reasons for this are discussed below when dealing with the hierarchical method based on the variance criterion). We may, for instance, minimize the within-class variance

$$V_{opt} = \min_Q \sum_{p \in Q} \sum_{i \in p} \|\mathbf{i}-\mathbf{p}\|^2$$

where the partition Q consists of classes p of centre **p** (in block type), and we desire the minimum of this criterion over all possible partitions, Q. To avoid a nontrivial outcome (e.g. each class being singleton, giving zero totalled within-class variance), the number of classes (k) must be set.

A solution, - not necessarily optimal - uses iterative refinement, as follows.

Step 1: Arbitrarily define a set of k cluster centres.

Step 2: Assign each object to the cluster to which it is closest (using the Euclidean distance, $d^2(i,p) = \|i-p\|^2$).

Step 3: Redefine cluster centres on the basis of the current cluster memberships.

Step 4: If the total of within-class variances is better than at the previous iteration, then return to Step 2.

We have omitted in Step 4 a test for convergence (the number of iterations should not exceed, e.g., 12). Cycling is also possible between solution states. This algorithm could be employed on the results at level n-k of a hierarchic clustering (see below) in order to improve the partition found. It is however a suboptimal algorithm, - the minimal distance strategy used by this algorithm is clearly a <u>sufficient</u> but not a <u>necessary</u> condition for an optimal partition.

A difficulty with iterative algorithms, in general, is the requirement for parameters to be set in advance (Anderberg, 1973, describes a version of the ISODATA iterative clustering method which requires 7 preset parameters). As a broad generalization, it may thus be asserted that iterative algorithms ought to be considered when the problem is clearly defined in terms of numbers and other characteristics of clusters; but that hierarchical routines often offer a more general-purpose and user-friendly option.

TRADITIONAL HIERARCHICAL CLUSTERING ALGORITHMS

Most published work on clustering has employed hierarchical algorithms (three quarters, according to Blashfield and Aldenderfer, 1978). Among astronomical applications, we have Heck et al., 1977; Egret and Heck, 1983b; Davies et al., 1982; Tholen, 1984; Cowley and Henry, 1979; Cowley, 1983; Mennessier, 1983, 1985; and Materne, 1978; these are concerned with IUE spectra, asteroids, chemical aspects of stars, light curves, and galaxies.

Much work in the past has used the single linkage hierarchical method, in part because of its greater computational efficiency. Most practitioners agree that undesirable "chaining" makes this method less than suitable in many applications. Apart from computational advantages, the single link method is conceptually simple, and is closely related to the MST. On the other hand, the minimum variance method (often referred to as Ward's method) appears to be best for the objective of summarizing data. The following attributes of this method may be noted:

1) Each cluster when using the minimum variance criterion may be represented by the cluster centre. Thus a convenient summarization and interpretation of a cluster is at hand (unlike the single link method, where a cluster may be graphically portrayed as a subgraph,

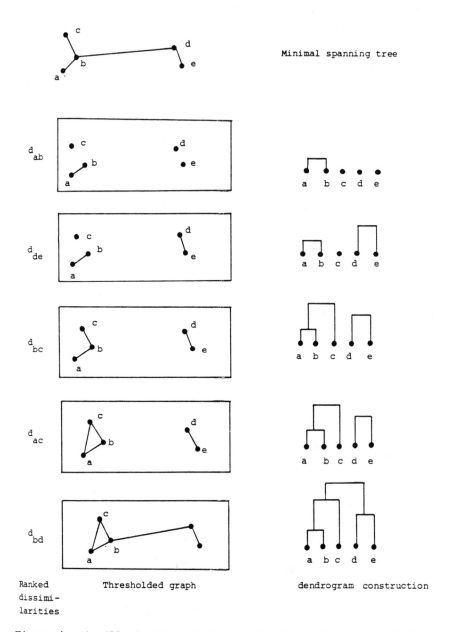

Figure 1. An illustration of the construction of a single linkage dendrogram and its associated MST.

but where it is more difficult to obtain a satisfactory representative member).

2) The name "minimum variance" is explained as follows. We seek to agglomerate two clusters at each stage of the agglomeration, such that these clusters are as close as possible; i.e. their combined variance is as small as possible. But the variance of the original set of data-vectors cannot change, and it is known that the variance between classes plus the sum of variances within classes equals the total variance (Huyghen's theorem in classical mechanics). Thus minimizing the variance within clusters amounts to maximizing variance between clusters. That is to say, we simultaneously optimize class compactness while maximizing relative class isolation.

The minimum variance method shares with all other agglomerative methods the same basic algorithm (cf. Fig. 1):

Step 1: determine the two objects which are closest; merge them into a cluster;

Step 2: determine the next two closest objects and/or clusters; merge them;

Step 3: repeat the last step until only one cluster (containing all initial objects) remains.

Agglomerative methods differ in how the merger, or agglomeration, is defined. The Lance-Williams formula is one way of describing varied hierarchical methods. If i and j are two objects or clusters, which are merged into a new cluster k, and if l is any "outsider" object or cluster, then the dissimilarity between k and l in the case of the minimum variance method is defined to be:

[1] $d^2(k,l) = a(i) \, d^2(i,l) + a(j) \, d^2(j,l) - b \, d^2(i,j)$
$a(i) = (n(i)+n(l))/(n(k)+n(l))$, $b = n(l)/(n(k)+n(l))$,
$n(k) = n(i)+n(j)$

where n(.) is the number of objects associated with the cluster (initially 1). The corresponding formula for the single link method is

[2] $d^2(k,l) = \frac{1}{2} d^2(i,l) + \frac{1}{2} d^2(j,l) - \frac{1}{2} |d^2(i,l) - d^2(j,l)|$

which (it may be verified by looking at a few simple geometric examples) is identical to

$d^2(k,l) = \min \{ d^2(i,l) , d^2(j,l) \}$

The above update formulas apply for dissimilarities (a weaker notion than distances: the triangular inequality in particular does not hold). If the Euclidean distance is the initially-chosen dissimilarity, an alternative formulation of the minimum variance criterion is possible. Instead of updating a set of dissimilarities following each agglomeration, the two agglomerated objects or clusters are replaced by their centre of gravity; and a new dissimilarity is used between such a new cluster and the preexisting objects /clusters. The formulas here are respectively:

[3] Cluster centre: $(n(i) \, \mathbf{i} + n(j) \, \mathbf{j})/(n(i) + n(j))$
where **i** (in block type) is an object or cluster vector;

[4] Dissimilarity: n(i) n(j) d²(i,j) / (n(i) + n(j))

When d^2, above, is the squared Euclidean distance then the hierarchies produced using expression [1] on the one hand, and expressions [3] and [4] on the other, give identical results.

FAST CLUSTERING THROUGH FASTER NN-FINDING

A recent theme in hierarchic clustering research has been the use of nearest neighbour chain (NN-chain) and reciprocal NN (RNN) approaches for implementing the clustering algorithms (see Murtagh, 1983a; 1985). Using cluster centres, each stage of hierarchical clustering is identical: we have a set of points in multidimensional space, and we seek the closest pair. Initially we are given n such points, and at the end of the sequence of agglomerations which are representable by a dendrogram or hierarchy, we have one point only. Each such point nominates a unique other point as its NN (we can usually ignore the situation of tied closest neighbours and make an arbitrary choice if necessary). Note that the NN of the NN is not necessarily the originally-considered point. When such is the case, we have a pair of reciprocal NNs (RNNs: see Fig. 2). Implementations of the cluster centre algorithm can now be informally specified as follows.

a   b   c   d   b=NN(a), c=NN(b), d=NN(c), c=NN(d);
                                      c and d are RNNs.
Figure 2. An example of a NN-chain

Parallel clustering

Step 1: Determine the NNs of all points, and thereby the RNNs.

Step 2: Agglomerate all RNN pairs, replacing them with the cluster centres.

Step 3: Redetermine the NNs of all points, followed by the RNNs. Return to step 2 until only one point remains.

NN-chain based clustering

Step 1: From an arbitrary point, determine its NN, then the NN of this point, and continue until the sequence of points halts with a RNN pair.

Step 2: Agglomerate the RNN pair, replacing them with the cluster centre.

Step 3: Now continue the NN-chain from the point prior to the agglomerated points. Reenter step 1 while at least 2 points remain.

The NN-chain based approach is illustrated in Fig. 2: note that a NN-chain is well-defined - the links must decrease, it must terminate after a finite number of links, and must end in a RNN pair. As mentioned before, it is a simple matter to impose arbitrary resolutions on tied situations. It is also straightforward to rephrase Step 3 of the NN-chain based algorithm in order to allow for the first two points considered constituting RNNs.

It must be stressed that the above algorithms yield identical hierarchies to the more traditional algorithms based on dissimilarities or on the use of cluster centres. A general condition for this to be the case is that if i and j are agglomerated into k, and if l is an outsider object/cluster, then k cannot be closer to l than was either i or j:

[5] $d(i,j) \leq \inf \{d(i,l), d(j,l)\} \Rightarrow \inf \{d(i,l), d(j,l)\} \leq d(k,l)$

Note that d is a dissimilarity, which for the minimum variance method is defined by expression [1] (section 4). If k were closer to l than i and j then it is conceivable that an agglomeration could subsequently take place between k and l with $d(k,l) < d(i,j)$, thereby causing an "inversion" in the hierarchy and causing the parallel clustering and NN-chain based algorithms, above, to become ill-defined. It can be verified that the dissimilarity defined by expression [1] for the minimum variance method satisfies expression [5], thereby ensuring that the above algorithms are well-defined (see Murtagh, 1985 for proof).

Although little work has been undertaken in implementing clustering algorithms on parallel machine architectures the parallel algorithm above would appear to lend itself to possible implementations in this area (e.g. a machine such as the ICL Distributed Array Processor where RNNs could be obtained by different processors and agglomerations carried out independently). The NN-chain based algorithm has been of greater interest to date however: it has worst case computational complexity of $O(n^2)$, which together with $O(n)$ storage requirements, make it equal to the single link method in worst case performance.

Furthermore, such approaches allow the incorporation of fast NN-searching routines (Murtagh, 1984). Such fast NN-finding algorithms include the following techniques: local search through mapping all points onto a regular grid structure (Rohlf, 1978; Murtagh, 1983b), multidimensional binary search trees (Bentley and Friedman, 1978) and bounding (Fukunaga and Narendra, 1975). We will briefly look at each of these in turn.

The grid-mapping approach involves taking the integer part of the coordinates of the point, following division by a grid cell dependent constant.

A MDBST is simply a crude clustering (see Fig. 3). It allows approximate $O(\log n)$ search time but is not feasible for dimensions greater than about 8 (say; see Bentley and Friedman, 1978).

The triangular inequality is satisfied by distances: let any arbitrary group of points have as centre g and distance to the most outlying point, $r(g)$. Then if δ is the current NN distance of point x, the group of all points associated with g may be rejected from consideration if we have $d(x,g) - r(g) \geq \delta$. Another similar rejection rule for point y is $|d(x,g) - d(y,g)| \geq \delta$. Note that $d(x,y)$ does not have to be calculated in order to verify this inequality, and instead $O(n)$ distance calculations only (i.e. $d(y,g)$ for all points y) are required. These rules were incorporated by Fukunaga and Narendra (1975) into a crude hierarchic clustering of the data set (subdividing the data into 3 subsets, and repeating this for 3 levels), whose only desirable property was to be efficient.

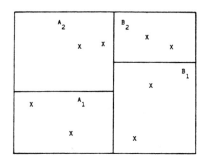 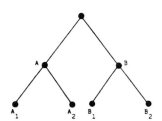

Figure 3. A multidimensional binary search tree (MDBST) in the plane.

ANNOTATED BIBLIOGRAPHY OF ASTRONOMICAL WORK

In the following short descriptions of published articles, we will not take into account the use of Principal Components Analysis or Discriminant Analysis for arriving at a classification.

The first two of the following references present surveys of clustering methods; references 3 to 5 deal with star-galaxy separation, using digitized image data; references 6 to 11 deal with spectral classification, - the prediction of spectral type from photometry; references 12 to 17 deal with taxonomy construction (references 12 to 14 with asteroids, references 15 and 16 with stars, and reference 17 with stellar light curves); references 18 and 19 deal with the use of methods related to the single linkage hierarchical method or the minimal spanning tree; references 20 and 21 deal with studies in the area of gamma and X-ray astronomy; reference 22 deals with galaxy classification; reference 23 concerns lunar geology; reference 24 concerns cosmic sources; reference 25 concerns galaxy clustering; and reference 26 concerns interferogram analysis.

1. Bijaoui (1979).

2. Zandonella (1979). The foregoing present surveys of clustering methods.

3. Kurtz (1983). Kurtz lists a large number of parameters - and functions of these parameters - which have been used to differentiate stars from galaxies.

4. Kruszewski (1985). The Inventory routine in MIDAS has a non-hierarchical iterative optimization algorithm. It can immediately work on up to 20 parameters, determined for each object in a scanned image.

5. Jarvis and Tyson (1981). An iterative minimal distance partitioning method is employed in the FOCAS system to arrive at star/galaxy/ other classes.

6. Heck et al. (1977). A photometric catalogue of uvbyβ measurements is used. The single and complete linkage hierarchical methods - among others - are used, with a number of metrics. Using 2849 stars, about 8% inconsistency with the MK spectral type classification is pointed out.

7. Egret and Heck (1983a). The "Nuées dynamiques" method is used, - this is a variance minimizing partitioning technique, which looks for the clusters which remain stable through successive optimization passes of the algorithm. 4645 stars are used here, with improved parameters.

8. Egret and Heck (1983b). Low dispersion IUE spectra of 267 stars are used, where each are characterised on 16 normalized fluxes: these correspond to significant features in the spectra, derived from a morphological analysis. The single and complete linkage methods, together with a variance minimizing partitioning method, are used.

9. Egret et al. (1984). The variables are given by the median fluxes in binned spectra, and maximum deviations from the median flux in emission and in absorption represent the "lines". A partitioning algorithm is used, and stars close to the group centres are examined.

10. Heck et al. (1984). A novel weighting procedure is discussed, which adjusts for the symmetry/asymmetry of the spectrum. The spectral type and luminosity classes are checked against the groups found, where the ultimate aim is a procedure for automatically classifying IUE spectral data.

11. Heck et al. (1985). Among other results, it is found that UV standard stars are located in the neighbourhood of the centres of gravity of groups found, thereby helping to verify the algorithm implemented.

12. Davies et al. (1982). Physical properties of 82 asteroids are used. The dendrogram obtained is compared with other classification schemes based on spectral characteristics or colour-colour diagrams. The clustering approach is justified also in being able to pinpoint objects of particular interest for further observations; and in allowing new forms of data - e.g. from broadband infrared photometry - to be quickly incorporated into the overall approach of classification-construction.

13. Tholen (1984). Between 400 and 600 asteroids using good-quality multi-colour photometric data are analysed.

14. Carusi and Messaro (1978). An unusual clustering method, termed the G-mode central method, is used; this method is also used in references 23 and 24. Significant classes of asteroids are confirmed, but doubt is cast on the significance of small sized families reported from other sources.

15. Cowley and Henry (1979). Forty stars are used, characterised on the strength with which particular atomic spectra - the second spectra of yttrium, the lanthanides, and the iron group - are represented in the spectrum. Stars with very similar spectra end up correctly grouped; and anomolous objects are detected. Clustering using lanthanides, compared to clustering using iron group data, gives different results for A_p stars. This is not the case for A_m stars, which thus appear to be less heterogeneous. The need for physical explanations are thus suggested.

16. Cowley (1983). About twice the number of stars, as used in the previous reference, are used here. A greater role is seen for chemical explanations of stellar abundances and/or spectroscopic patterns over nuclear hypotheses.

17. Mennessier (1983, 1985). Light curves - the variation of luminosity with time in a wavelength range - are analysed. Standardization is applied, and then three hierarchical methods. "Stable clusters" are sought from among all of these. The object is to define groups of variable stars with similar physical properties, and the use of near-infrared appears better than the visual for this.

18. Huchra and Geller (1982). The single linkage hierarchical method, or the minimal spanning tree, have been rediscovered many times - see, for instance, Graham and Hell (1985). In this study, a close variant is used for detecting groups of galaxies using three variables, - two positional variables and redshift.

19. Feitzinger and Braunsfurth (1984). In an extended abstract, the use of linkages between objects is described.

20. di Gesù and Sacco (1983). In this and the following works, the Minimal Spanning Tree or fuzzy set theory - which, is clear from the article titles - are applied to point pattern distinguishing problems involving gamma and X-ray data. For a rejoinder to the foregoing reference, see Dubes and Hoffman (1986).

21. di Gesù, Sacco and Tobia (1980), di Gesù and Maccarone (1983, 1984, 1986), di Gesù (1984) and de Biase, di Geù and Sacco (1986).

22. Materne (1978). Ward's minimum variance hierarchical method is used.

23. Coradini et al. (1986); Bianchi et al. (1980a,b). The G-mode method (see references 14 and 24 also) is used.

24. Giovanelli et al. (1981). X-ray sources and dwarf novae are studied.

25. Barrow et al. (1985). The MST is used for galaxy clustering studies.

26. Pirenne et al. (1985). An MST-based approach, implemented in ESO's MIDAS image processing system, is used to automate the analysis of interferograms.

THE FUTURE

In any review of clustering, we could not hope to be thoroughly comprehensive. In particular, in this article, we have not dealt with the application of fuzzy clustering in X-ray and gamma astronomy (see di Gesù and Maccarone, 1984; 1986). However, we have described methods which are implemented in major software packages, and which are also now available in the European Southern Observatory's MIDAS image processing system.

We have briefly dealt with the exciting advances in clustering algorithms, achieved by means of fast nearest neighbour searching. This increases interest in clustering in the areas of information retrieval and database access, when fast best match algorithms are implemented (cf. Perry and Willett, 1983). And, of course, the vast predicted amount of data to be provided by the Space Telescope makes research in this area both promising and probably necessary (see Heck et al., 1984).

Open problems remain: algorithmic efficiency (with the ultimate goal of applying clustering techniques to the ever-growing quantity of data in astronomical databases); and the wider application of clustering in particular, and multivariate statistical and pattern recognition techniques in general, to astronomical research problems. It has in the past been said that clustering techniques are descriptive rather than inferential. Astronomy, as a descriptive science, would appear to be ideally suited to such methods.

REFERENCES

Anderberg, M.R., 1973, "Cluster Analysis for Applications", Academic Press, New York.
Barrow, J.D., Bhavsar, S.P. and Sonoda, D.H., 1985, Minimal spanning trees, filaments and galaxy clustering, Monthly Notices of the Royal Astronomical Society, 216, 17:35.
Bentley, J.L., and Friedman, J.H., 1978, Fast algorithms for constructing minimal spanning trees in coordinate spaces, IEEE Transactions on Computers, C-27, 97:105.
Bianchi, R., Coradini, A., and Fulchignoni, M., 1980a, The statistical approach to the study of planetary surfaces, The Moon and the Planets, 22, 293:304.
Bianchi, R., Butler, J.C., Coradini, A., and Gavrishin, A.I., 1980b, A classification of lunar rock and glass samples using the G-mode central method, The Moon and the Planets, 22, 305:322.
de Biase, G.A., di Gesù, V. and Sacco, B., 1986, Detection of diffuse clusters in noise background, preprint.
Bijaoui, A., 1979, in Ballereau, D. (ed.), Classification Stellaire, Compte Rendu de l'Ecole de Goutelas, Observatoire de Meudon, Meudon.
Blashfield, R.K., and Aldenderfer, M.S., 1978, The literature on cluster analysis, Multivariate Behavioral Research, 13, 271:295.
Carusi, A., and Massaro, E., 1978, Statistics and mapping of asteroid concentrations in the proper elements' space, Astronomy and Astrophysics Supplement Series, 34, 81:90.
Coradini, A., Fulchignoni, M., and Gavrishin, A.I., 1976, Classification of lunar rocks and glasses by a new statistical technique, The Moon, 16, 175:190.
Cowley, C.R., and Henry, R., 1979, Numerical taxonomy of Ap and Am stars, The Astrophysical Journal, 233, 633:643.
Cowley, C.R., 1983, Cluster analysis of rare earths in stellar spectra, in Statistical Methods in Astronomy, European Space Agency Special Publication SP-201, pp. 153:156.
Davies, J.K., Eaton, N., Green, S.F., McCheyne, R.S., and Meadows, A.J., 1982, The classification of asteroids, Vistas in Astronomy, 26, 243:251.
Dubes, R.C. and Hoffman, R.L., 1986, Remarks on some statistical properties of the minimum spanning forest, Pattern Recognition, 19, 49:53.
Egret, D., and Heck, A., 1983a, Prediction of spectral classification from photometric data. Application to the MK spectral classification and the Geneva photometric system. Progress report, in Statistical Methods in Astronomy, European Space Agency Special Publication SP-201, pp. 149:152.
Egret, D., and Heck, A., 1983b, Preliminary results of a statistical classification of ultraviolet stellar spectra, in Statistical Methods in Astronomy, European Space Agency Special Publication SP-201, pp. 59:63.
Egret, D., Heck, A., Nobelis, Ph., and Turlot, H.C., 1984, Statistical classification of ultraviolet stellar spectra (IUE satellite): progress report, Bulletin d'Information, Centre de Données de Strasbourg, No. 26.

Feitzinger, J.V. and Braunsfurth, E., 1984, The spatial distribution of young objects in the Large Magellanic Cloud - a problem of pattern recognition, in van den Berg, S. and de Boer, K.S. (eds.), <u>Structure and Evolution of the Magellanic Clouds</u>, IAU, 93:94.

Fukunaga, K., and Narendra, P.M., 1975, A branch and bound algorithm for computing k-nearest neighbours, <u>IEEE Transactions on Computers</u>, C-**24**, 750:753.

di Gesù, V., Sacco, B. and Tobia, G., 1980, A clustering method applied to the analysis of sky maps in gamma-ray astronoma, <u>Memorie della Società Astronomica Italiana</u>, 517:528.

di Gesù, V. and Sacco, B., 1983, Some statistical properties of the minimum spanning forest, <u>Pattern Recognition</u>, **16**, 525:531.

di Gesù, V. and Maccarone, M.C., 1983, A method to classify celestial shapes based on the possibility theory, id Sedmak, G. (ed.), ASTRONET 1983 (Convegno Nazionale Astronet, Brescia, Published under the auspices of the Italian Astronomical Society), 355:363.

di Gesù, V., 1984, On some properties of the KNN and KMST, Stanford Linear Accelerator Report, SLAC-PUB-3452.

di Gesù, V. and Maccarone, M.C., 1984, Method to classify spread shapes based on possibility theory, Proceedings of the 7th International Conference on Pattern Recognition, Vol. 2, IEEE Computer Society, New York, 869:871.

di Gesù, V. and Maccarone, M.C., 1986, Features selection and possibility theory, <u>Pattern Recognition</u>, **19**, 63:72.

Giovanelli, F., Coradini, A., Lasota, J.P., and Polimene, M.L., 1981, Classification of cosmic sources: a statistical approach, <u>Astronomy and Astrophysics</u>, **95**, 138:142.

Graham, R.L., and Hell, P., 1985, On the history of the minimum spanning tree problem, <u>Annals of the History of Computing</u>, **7**, 43:57.

Hand, D.J., 1981, "<u>Discrimination and Classification</u>", Wiley, New York.

Heck, A., Albert, A., Defays, D., and Mersch, G., 1977, Detection of errors in spectral classification by cluster analysis, <u>Astronomy and Astrophysics</u>, **61**, 563:566.

Heck, A., Egret, D., Nobelis, Ph., and Turlot, J.C., 1984, Statistical classification of IUE stellar spectra by the Variable Procrustean Bed approach, in <u>4th European IUE Conference</u>, European Space Agency Special Publication SP-218.

Heck, A., Egret, D., Nobelis, Ph., and Turlot, J.C., 1985, Statistical confirmation of the UV spectral classification system based on IUE low-dispersion stellar spectra, preprint.

Heck, A., Murtagh, F., and Ponz, D., 1985, The increasing importance of statistical methods in astronomy, <u>The Messenger</u>, No. 41, 22:25.

Huchra, J.P. and Geller, M.J., 1982, Groups of galaxies. I. Nearby groups, <u>The Astrophysical Journal</u>, **257**, 423:437.

James, M., 1985, "Classification Algorithms", Collins, London.

Jarvis, J.F., and Tyson, J.A., 1981, FOCAS: faint object classification and analysis system, <u>The Astronomical Journal</u>, **86**, 476:495.

Kruszewski, A., 1985, Object searching and analyzing commands, in <u>MIDAS - Munich Image Data Analysis System</u>, European Southern Observatory Operating Manual No. 1, Chapter 11.

Kurtz, M.J., 1983, Classification methods: an introductory survey, in <u>Statistical Methods in Astronomy</u>, European Space Agency Special Publication SP-201.

Materne, J., 1978, The structure of nearby clusters of galaxies. Hierarchical clustering and an application to the Leo region, <u>Astronomy and Astrophysics</u>, **63**, 401:409.

Mennessier, M.O., 1983, A cluster analysis of visual and near-infrared light curves of long period variable stars, in <u>Statistical Methods in Astronomy</u>, European Space Agency Special Publications SP-201, pp. 81:84.

Mennessier, M.O., 1985, A classification of miras from their visual and near-infrared light curves: an attempt to correlate them with their evolution, Astronomy and Astrophysics, **144**, 463:470.

Murtagh, F., 1983a, A survey of recent advances in hierarchical clustering algorithms, The Computer Journal, **26**, 354:359.

Murtagh, F., 1983b, Expected time complexity results for hierarchic clustering algorithms which use cluster centres, Information Processing Letters, **16**, 237:241.

Murtagh, F., 1984, A review of fast techniques for nearest neighbour searching, COMPSTAT 1984, Physica-Verlag, Wien, pp. 143:147.

Murtagh, F., 1985, "Multidimensional Clustering Algorithms", COMPSTAT Lectures Volume 4, Physica-Verlag, Wien.

Perry, S.A., and Willett, P., 1983, A review of the use of inverted files for best match searching in information retrieval systems, Journal of Information Science, **6**, 59:66.

Pirenne, B., Ponz, D. and Dekker, H., 1985, Automatic analysis of interferograms, The Messenger, No. 42, 2:3.

Rohlf, F.J., 1975, Generalization of the gap test for the detection of multivariate outliers, Biometrics, **31**, 93:101.

Rohlf, F.J., 1978, A probabilistic minimum spanning tree algorithm, Information Processing Letters, **7**, 44:48.

Tholen, D.J., 1984, Asteroid taxonomy from cluster analysis of photometry, PhD Thesis, University of Arizona.

Tucker, A., 1980, "Applied Combinatorics", Wiley, New York.

Zahn, C.T., 1971, Graph-theoretical methods for detecting and describing Gestalt clusters, IEEE Transactions on Computers, **C-20**, 68:86.

Zandonella, A., 1979, Object classification: some methods of interest in astronomical image analysis, in Sedmak, G. et al. (eds.), Image Processing in Astronomy, Osservatorio Astronomico di Trieste, Trieste, 304:318.

PROBLEMS AND SOLUTIONS IN SURFACE PHOTOMETRY

Jean-Luc Nieto

Observatoire du Pic du Midi et de Toulouse
14 Avenue Edouard Belin, 31400 Toulouse

Summary

We concentrate on two extreme problems in surface photometry that require drastically different approaches:

i) Detection (and measurement) of faint surface-brightness levels, such as the faint extensions of galaxies: The problem is difficult because of the contribution of the sky background and the inhomogeneity of the receptor.

ii) Study of crowded fields or high-luminosity gradient regions, such as the nuclei of galaxies, the cores of globular clusters, etc...: The problem, requiring the highest possible resolution, is difficult because of the degradation of the images by several factors, notably the atmosphere.

We discuss several new techniques allowing us to solve these problems:

i) A high accuracy for faint-surface brightness levels may be obtained by systems such as (or similar to) CCDs in the Scanning Mode,

ii) The image quality can be improved by using active optics systems or instruments able to record "instant" images (e.g. Photon Counting Detectors in the Time Resolved Imaging Mode or CCDs in the Cinematographic Mode).

INTRODUCTION

At the previous Erice workshop on Data Analysis in Astronomy, M. Capaccioli (1985) presented several aspects related to two-dimensional photometry. Concentrating on galaxies in his contribution, he divided his paper essentially into three parts: i) nuclear regions, ii) intermediate regions, iii) faint outer regions of galaxies, for which he discussed technical questions and proposed some solutions to these questions.

It would be very difficult to supersede his contribution by adopting the same approach. It is probably more useful to expand it with some particular aspects treated there. Even if the field of surface photometry has been subjected to deep structural changes with the advance of a permanently improving technology (e.g. Nieto, 1985), essentially the same problems remain. This allows us to concentrate this paper on some specific techniques in surface photometry (SF) that are now possible with new instruments, proposing so new solutions to these old problems.

The two most difficult problems in SF are essentially:
i) The detection and measurement of faint levels above the sky background in extended objects.
Several important astrophysical questions are attached to this problem. They concern for instance the outer regions of galaxies, the diffuse light background in clusters of galaxies, as well as e.g. dwarfs in globular clusters, quasar extensions, etc...
ii) The study of regions that are quite affected by the degradation of the images (Point-Spread- Function).
Except for the fact that a high resolution is always useful, it is crucial to reduce as much as possible the degradation of the images for some astrophysical questions, such as those related to crowded fields (e.g. central regions of globular clusters, or stellar studies in nearby galaxies), high-luminosity gradient objects (e.g. central regions of galaxies, compact galaxies,...) or quite remote objects (in cosmological questions for instance), etc...

Theses two problems will be discussed in Sections 1 and 2 respectively. Section 3 will be devoted to a specific experiment aiming at improving the image quality with Photon-Counting-Detectors.

1. MEASURING FAINT LEVELS: PROBLEMS AND SOLUTIONS

1a) The characteristics of the problems

First, the light coming from any celestial object, whether extended or not, is superposed to (and, in our case, much fainter than) the sky brightness, on each element of resolution (e.g. pixel) of the receptor.

Second, each pixel of the receptor has its own sensitivity. This sensitivity can vary slowly from one side of the field to another: this is the case for emulsions of photographic plates or for photocathode-mounted receptors (electronographic cameras or photon counting systems). This is quite different for CCDs where each pixel has a value independent of that of its neighbors. Consequently the measurement of a signal may be quite erroneous if the most possible perfect correction for the inhomogeneous sensitivity of the receptor is not achieved. Afterwards the extraction of a faint signal would require the most precise determination of the sky contribution.

1b) Solutions: the Flat-Field procedure

It is apparently an easy trick to obtain the whole sensitivity map for most receptors whose sensitivity is reproducible, like CCDs or Photon Counting Systems: It is sufficient to observe an area of the sky void of extended or bright objects, generally a clear sky at sunrise or sunset, since the sky intensity is quite constant in the field of view covered by the receptor. Another technique, but controversial among experts, is an exposure of the illuminated dome. This is the flat-field procedure.

b1) Photographic Plates

For photographic plates however, this is not possible (emulsions, like matches, can be used just once) so that several tricks are applied relying on the always possible (but difficult) interpolation of the sky densities in the region of interest. One of them that has become classical is the Numerical Mapping Technique (Jones et al., 1967) where the sky background is determined in the "innerfield" after an iterative procedure in the "outerfield" rejecting first the discrete sources (e.g. stars and defects) and fitting after with a high degree polynomial the smoothly evolving component. This requires the outerfield to be large enough (at least ten times the innerfield), and free of undesirable objects so that the background densities can be measured. Therefore the study of the outer region of a very large galaxy (relative to the plate size) or of a galaxy located in a dense group or in a cluster is almost impossible.

b2) CCDs

In the case of CCDs, flat-fielding is not such a straightforward matter. A signal Sij on pixel (i,j) of the frame is affected by parasiting effects such as an offset αij, a dark current per unit time βij, a readout noise δij and the unavoidable shot noise φij (e.g. Fort, 1985) so that:

$$Sij = \eta ij \; \phi ij \; \tau + \alpha ij + \beta ij \; \tau + \delta ij + \varphi ij.$$

The flux ϕij is then determined by:

$$\phi ij = (Sij - \alpha ij) / \eta ij \; \tau$$

The accuracy on ϕij would depend only on dark current, shot noise and readout noise, if αij and ηij can be accurately determined, and if we neglect the residual fringe pattern due to the emission lines of the sky. αij, the offset value, can be determined by very short exposures in the dark. Each exposure is in fact affected by the readout noise, but this slight problem is overcome by the multiplication of such short exposures whose averaging is going to yield a negligible readout noise. Flat-Field measurements would lead in principle to the determination of ηij, that varies from pixel to pixel. This variation is of the order of 5 to 10 percent for thin CCDs and as low as the percent for thick ones (that are -one must pay for everything- more sensitive to cosmic rays and have a smaller blue quantum efficiency). One problem remains, that limitates the accuracy of the ηij map to 2 to 3 percent level: the Flat-Field source does not have the same color as the object studied so that their effective wavelengths are different.

To solve this, a good way would be to have available a large set of quasi monochromatic sources at different wavelengths and write color equations for each pixel. This, to my knowledge and the knowledge of some CCD experts, has never been tried. This may be tackled when an accuracy much higher than that necessary today will be crucial to have.

Another way is the use of techniques in which a point of the image corresponds to several pixels, so that the noise affected to each pixel is smoothed out (Figure 1). We shall see an instance in Section 1d. See also King (this workshop).

1c) Sky determination

Once the images are flat-fielded, the second step is to measure the sky contribution.

c1) Photographic Plates

A general difficulty when measuring the sky density on photographic plates is that we do not measure only the sky density on the emulsion, but in fact the combination of the sky density and the fog density. There is strictly no way to separate them from each other. The fog density is measured on the unexposed side of the plate and assumed to be constant throughout the emulsion: this is satisfactory only if the sky density is much higher than the fog density so that the variations of the fog density are quite negligible relative to sky density variations.

In addition, one is never sure that the interpolation of the sky densities in the innerfield is correct. The smaller the innerfield relative to the outerfield, the better, but small-scale inhomogeneities may have terrible effects. As far as I know, the only advisable remedy is plate compositing. Again, the more, the better.

c2) CCDs

For CCDs, this is quite a crucial problem for those relatively large objects that cover the whole field, e.g. bright galaxies that usually serve as standards and that are often observed for no other reason but checking the performances of the detector. A solution to this is to take several overlapping exposures so that the sky can be reached. Although a slight variation of the sky intensity from one exposure to another should not be too serious since the overlapping of the images allows for a correction (a procedure to use in desperate cases only !), this problem is certainly the more severe in CCD data reduction: how many times, for instance, have we all observed with a sky brightness affected by the (setting or rising) moon?

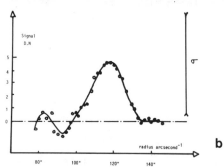

Figure 1: *Measurements of very faint shells around elliptical galaxies. Image of NGC 5018 obtained in the R band with an RCA CCD at the CFH Telescope (a). In NGC 3923, the signal has been integrated along the shell, following its curvature. It is then averaged on several pixels so that the profile of the shell can be provided with a high signal-to-noise ratio relative to the sum of the different noise contributions presented on the right of the shell profile (b). (Courtesy B. Fort).*

1d) An elegant solution to the problem: the scanning CCD

This overlapping procedure pushed to its limits is the scanning mode. It consists of letting the telescope track in the column direction at a speed relative to that of the sky equal to that of the charge transfer from one line to another. The whole exposure time depends on the speed at which the track is set. The final image will consist of an arbitrary long strip whose width is that of the CCD lines. So, large regions are covered and, again, "to the extent that the sky brightness variations can be interpolated spatially and temporally between the beginning and end of the frame" (Boroson et al., 1983), the sky can be reached and accurately subtracted.

The main advantage of the scanning mode for CCDs concerns in fact the Flat-Fielding Procedure. The intensity coming from each part on the sky (having the size of the pixel) and ending up on one single pixel of the final frame has been sampled by all the pixels of each row. Thus pixel-to-pixel variations are greatly reduced.

2. IMAGE DEGRADATION

2a) The characteristics of the problem

Atmospheric effects on optical images is a well-known phenomenon (e.g. Roddier, 1981). Three time scales can be considered in the degradation of the images by wavefront distorsions:
- i) very short exposures (a few thousandths of a second), shorter than the time scale of the large-scale atmospheric turbulence, yield a speckle structure;
- ii) short exposures, ranging from a few tenths of a second to a few seconds, correspond to the accumulation of the speckles so that the image becomes continuous, its structure being essentially affected by blurring.
- iii) Longer exposures (a few seconds or more) are affected by both variable blurring and the displacement of the centroid of the short-exposure images.

We shall concentrate our discussion on those short- and long-exposure images. These exposures depend essentially on one parameter (Fried, 1966), r_0, defined as the diameter of a zone where the quadratic mean of the phase displacement of the wavefront is less than one square radian. If the diameter of the telescope is larger than $4 \times r_0$, the resolution of a short exposure is limited by the atmosphere at $1.2 \times \lambda / r_0$, λ being the wavelength of the observations. In other words, r_0 is the diameter of a diffraction-limited telescope having a resolution equal to that of the observations. r_0 has been shown to depend on the observing wavelength as $\lambda^{6/5}$, which means for the resolution:

$$\rho \alpha \lambda^{-1/5},$$

illustrating the relevance of observations at long wavelengths for reaching a high ground-based resolution.

2b) Improving the image quality: what can we expect?

Following Fried (1966) and Hecquet and Coupinot (1984), the gain in correcting for image motion, namely the maximum gain of resolution of short exposures over long exposures is about 2 for a telescope diameter $D = 3.4 \times r_0$. Furthermore, the distribution of the resolutions of the short images, e.g. the resolution due to the only image blurring, shows that image selection criteria of short exposures would lead to a better resolution of the final image once the selected images are stacked. Again, the improvement depends only on D/r_0: for large telescopes, the higher the r_0, the better the improvement for a given selection criterion. The ideal D/r_0 ratio for this purpose is about 4 (Hecquet and Coupinot, 1984).

Values of r_0 at, typically, 5000 Å, are of 10cm at average observatories but reach very high values at good sites such as Pic du Midi or Mauna Kea, where r_0 as high as 40 cm have been measured. The above considerations mean that, at optical wavelengths, an ideal telescope on these sites should have a 1 to 1.5m diameter, and that, in particular, the CFH telescope is not adapted for image improvement of that kind.

In practice however, there are limitations to such considerations: guiding defects or mechanical problems in the telescope drive may seriously contribute to the image degradation. In addition, observations of faint objects requiring long exposure times (several tens of minutes or so) may be affected by variations of the seeing conditions, the r_0 value varying with time. It results from all this that, even with the 3.6m diameter of the CFH telescope, the final resolution of a long exposure is about twice as low as the "instant" resolution estimated by eye during the night (Nieto and Lelièvre, 1981; Racine, 1982).

2c) Improving image quality: how?

This subject is discussed at length by Roddier (1981) and in the 1981 Conference on the "Scientific Importance of High Angular Resolution at Infrared and Optical Wavelengths", that I refer to for further details. Speaking about surface photometry, I will restrict my talk to approaches in the domain of "active optics", e.g. that of low-frequency image corrections, that are valid on large fields (at least several tens of a second as opposed to "adaptative optics", which refers to high-frequency, but field-limited, corrections). For further discussion, see Smith (this workshop). Such active systems are envisaged with future telescopes, the MMT being the forerunner in this matter (initially for correcting misalignements, see Reed, 1978).

Several optical systems have been applied or are being applied to improve the image quality of astronomical data through a real-time control of optical wavefront, e.g. the "Image Stabilising Instrument System" (ISIS, Thompson and Ryerson, 1983), that has been applied successfully to surface photometry with a 0.6 arcsec resolution V CCD image of the radio galaxy Cygnus A (Thompson, 1984).

A posteriori techniques have been attempted at different levels of complexity, e.g. the classical method of deconvolution (that requires a good signal-to-noise ratio and always applicable on the final images, whatever the image improvement technique -if any; see for instance Lorre and Nieto, 1984 or Lauer, 1985). The simplest of all is the selection of the best image(s) in a sequence of exposures whose exposure time depends on the brightness of the object (Aurière, 1983 for globular clusters; Muller, 1983 for the solar surface; Nieto and Lelièvre, 1981 for extragalactic objects).

2d) CCDs in a cinematographic mode (or Cine-CCD)

The Cine-CCD (Fort et al., 1984) is especially designed to record short (e.g. a few tenths of a second) images on a frame whose size is intentionally limited in order to loose a minimum amount of time for data storage. The recentering of each individual image improves the resolution of the final integrated image. Unfortunately, this technique has been applied so far just once on the image of a double star whose resolution is so poor that the improvement in resolution is hardly visible. Because of the accumulation of (readout) noise, this technique can be applied to bright objects only, except if one uses an intensified CCD.

3. IMAGE QUALITY IMPROVEMENT WITH PHOTON-COUNTING-DETECTORS

Since they can record "instant" images with no readout noise, Photon-Counting-Detectors are perfect instruments for image improvement. This prompted us (Nieto et al., 1986) to develop algorithms devoted to an a posteriori computer image correction based on an image recentering at a typical frequency range between 0.1 Hz and 10 Hz and an image selection relying upon different criteria. These algorithms are based on the principle of correlating the individual images with the integrated image (or unsharp mask), the maximum of the correlation function giving the parameters of the correction to apply in the individual image. The algorithms were set up at the Centre Francais du Telescope Spatial (Laboratoire d'Astronomie Spatiale) in Marseilles on a VAX 780 with the MIDAS software developed at ESO on the same computer.

Experimental data were obtained with the European Space Agency Photon-Counting Detector in the Time Resolved Imaging Mode (di Serego and Perryman, 1986) at the Cassegrain Focus of the Canada-France-Hawaii Telescope. The image quality was excellent (FWHM about 0.5 arcsec on focus stars for 10 sec exposures at 3700 A, which means $<r0>$ about 18cm). We present the first experimental results obtained with the analysis of 30,000 frames, the frame exposure being 30ms. Because of the poor dynamical range of such detectors, these "instant" images had to be coadded until the number of photons to use for the determination of the image quality is significant. The field is the gravitational lens 2345+007 where two faint images of the same object, A and B, unresolved from the ground, are separated by 7.1 arcsec. The brightness ratio between the two objects is about 4. This configuration is ideal since it allows to check on the image of one object the effect of the algorithms applied to the image of the other object.

3a) Characteristics of the image degradation

Figures 2a and 2b show the evolution with time of the relative α and δ positions of the barycenter of the events from object A recorded in the 3x3 arcsec window for an adopted integration time of 2.5 sec. Periodic features appear in both coordinates, due to the imperfect telescope drive. The phenomenon has a 0.3 arcsec amplitude in α and 0.2 arcsec in δ. Otherwise, an amplitude of 0.2 arcsec is detectable between images separated by a few seconds. This effect is confirmed on similar figures made with longer exposures.

Figure 2c shows the evolution with time of the image "diameter" σ for a 2x2 arcsec window and an integration time of 10 sec. The face value of σ is only indicative since it depends on the size of the window of study. It appears that periods of good and bad seeing alternate, producing variations up to 25 to 30 percent.

3b) Influence of the number of photons

For a given integration time and given constant seeing conditions, the spread of the image characteristics should decrease with larger count numbers, following Poisson statistics. This determines the number of photons necessary to analyse the characteristics of the image degradation reliably, e.g. with a negligible contribution of the photon noise.

The analyses of the characteristics of the image degradation made on both objects A and B, i.e. with numbers of events differing by a factor of about 4 show that the minimum number of events for a reliable image correction is about 100-150 with the algorithm used. Below this number, the noise is predominant. This forced us to give a lower limit to our integration time of 10 seconds, seriously limiting the efficiency of the image improvement. For illustration, Table 1 shows the image quality obtained for recentering only and for recentering + selection.

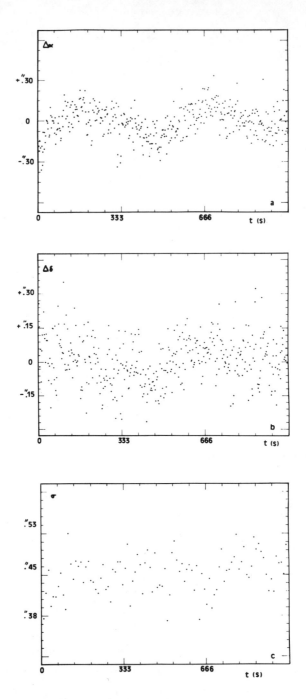

Figure 2: Evolution with time of the relative α (a) and δ (b) positions of the barycenter of the events (coming from object A of the gravitational lens 2345+007, for which each image was integrated for 2.5 sec) and of the image diameter σ for an integration time of 10 sec (c). The three figures show the improvement that we can expect from an image improvement procedure having these integration times for short-exposure images.

Table 1

Double QSO 2345+007

*Marginal distributions of the two images
A and B for 10 sec integration images corrected
with recentering and image selection algorithms
(applied to image A)*

	FWHM (A)		FWHM (B)	
	α	δ	α	δ
Integrated	0.66	0.60	0.70	0.73
Recentering	0.61	0.55	0.69	0.71
Rec. + Sel.				
20% Rej.	0.59	0.53	0.68	0.71
35% Rej.	0.58	0.51	0.67	0.68
60% Rej.	0.56	0.49	0.65	0.68

3c) Conclusion

In fact, our data are limited in two ways:
- The long integration time, that comes from the necessary use of a narrow filter. This is due to the poor dynamical range of Photon Detectors.
- The diameter size of the telescope relative to the Fried's parameter of our observations.

These algorithms should now be applied to a more adapted telescope, e.g. a 1 or 2m telescope for the same r0. Also, the exposure time can be reduced (and more adapted to the atmospheric perturbation time scale) by working at ... longer focal lengths: this will spread the signal on more pixels and lead to a shorter exposure time by adjusting the filter bandwidth to the saturation threshold of the instrument. A third improvement would be to work at longer wavelengths.

4. DISCUSSION AND PERSPECTIVES

It is very likely that, soon after its launch, the Space Telescope will make this paper obsolete, but it will also stimulate coordinated efforts from the ground to achieve the highest performances in both the detection of faint levels and the improvement of image quality.

For instance, observations requiring a 0.3-0.5 arcsec resolution will be the privilege of ground-based observations made with adaptative or active systems. Some astronomical questions that could be solved by this type of resolution are starting to arise now, stimulated by the high resolution recently attainable from the ground by simple direct imaging, and their number will probably increase with the first results from ST. Proposals for Space Telescope observations aimed at solving such questions will probably never go through Space Telescope Time Allocation Committee because they may not need the extreme resolution of ST or because they may be too time consuming. They will never come through especially if it has been proved that such resolutions are really achievable from the ground.

Therefore, we shall certainly see in the near future a systematic development of techniques similar to those described above that will require an always more and more sophisticated data analysis.

ACKNOWLEDGEMENTS

I am indebted to B. Fort for stimulating and enlightning discussions, and to E. Davoust and Ph. Prugniel for critical reading of the manuscript.

REFERENCES

Aurière,M., 1983, Thèse d'Etat, Université de Paris 6
Boroson,T.A., Thompson,I.B., Schectman, S.A., 1983, Astron.J., 88, 1707
Capaccioli, M., 1985, in "Data Analysis in Astronomy", Ettore Majorana
    International Science Series, V. di Gesù et al. Ed., p.363

Fort, B., Picat,J.-P., Cailloux, M., Mauron, N., Dreux, M., Fauconnier, Th., 1984, Astron. Astrophys., 135, 356

Fort, B., 1985, in "New Aspects of Galaxy Photometry", Springer Verlag Lectures in Physics Series, J.-L. Nieto Ed., p.3

Fried, D.L., 1966, J. Opt. Soc. America, 56, 1372

Hecquet, J., Coupinot, G., 1984, J. Optics, 15

Jones, W.B., Gallet, R.M., Obitts, D.L., de Vaucouleurs, G., 1967, "Astronomical Surface Photmetry by Numerical Mapping", Pub. Univ. Texas, Austin, Series II, vol I, n° 8

Lauer, T.R., 1985, Astrophys. J., 292, 104

Lorre, J.J., Nieto, J.-L., 1984, Astron. Astrophys., 130, 167

Muller, R., 1983, Solar Physics, 87, 243

Nieto, J.-L. Ed., 1985, "New Aspects of Galaxy Photometry", Springer Verlag Lectures in Physics Series

Nieto, J.-L., Lelièvre, G., 1981, in "Astronomical Photography 1981",CNRS-INAG, J.-L. Heudier and M.E. Sim Ed., p.189

Nieto, J.-L., Llebaria, A., di Serego Alighieri, S., Donas, R., Vernet, Ph., Lelièvre, G., Macchetto, F., Perryman, M., 1986, in preparation

Racine, R., 1982, private communication quoted in Fort et al.(1984).

Reed, M.A., 1977, in ESO Conference on "Optical Telescopes of the Future", F. Pacini et al. Ed., p. 209

Roddier, F., 1981, in Progress in Optics XIX, E. Woolf Ed., p. 281

di Serego Alighieri, S., Perryman, M.A.C., 1986, ESO Scientific Preprint n° 424

Thompson, L.A., 1984, Astrophys. J., 279, L47

Thompson, L.A., Ryerson,H.R., 1983, in Proc. SPIE, Instrumentation in Astronomy V

CLASSIFICATION OF LOW-RESOLUTION STELLAR SPECTRA VIA TEMPLATE MATCHING -
A SIMULATION STUDY

H.-M. Adorf

Space Telescope - European Coordinating Facility
European Southern Observatory
Karl-Schwarzschild-Str. 2
D-8046 Garching b. München, F.R. Germany

Abstract: The availability of wavelength- and intensity-calibrated low-resolution spectra directly from spectral survey plates and frames opens the possibility for quantitative spectral classification using modern pattern recognition methods. We investigate the classification power of a template matching technique, which cross-correlates each object spectrum with templates from an atlas of reference spectra. Numerical experiments show that this technique has a high noise-cheating capability and bears great potential for intragalactic as well as extragalactic spectral survey work.

INTRODUCTION

The established method for the two-dimensional classification of stellar spectra may be characterized as follows: In a tree-structured, multiple-decision process* a given spectrum is eventually classified using local spectral features, i.e. line ratios, line depths and widths, and continuum discontinuities (see Garrison, 1983, and references therein). This method is well suited for **individual**, medium- to high-resolution spectra which are observed long enough to guarantee a sufficiently high signal-to-noise ratio. It is required that the detector shows a smoothly varying sensitivity curve, whereas wavelength- and intensity-calibrations are not essential.

A line-oriented "visual" spectral classification procedure will almost certainly fail in the area of slitless spectroscopy, which must deal with medium to large sets of noise-contaminated low-resolution spectra. For instance, a single Schmidt prism plate may record up to ~50 000 spectra, most of which - except those of white dwarfs and perhaps A/F-stars - will not allow the detection and measurement of the rather weak small-scale spectral features required by an MK-like classification procedure. Particularly for the fainter objects, noise is capable of perfectly imitating non-existent narrow absorption lines or obscuring

---

* In the future "knowledge- and rule-based" expert systems, as they are now being used in the field of artificial intelligence, may well become important tools for this kind of approach.

existing ones. Attention therefore has to shift from **local** to increasingly more **global** spectral features, i.e. from narrow band colour indices (Christian, 1982), to broad band colour indices, to the continuum shape of the spectrum, as the only recognizable and possibly measurable quantities. An **automated** procedure* is favoured by the sheer amount of information present in the large sets of spectra from survey plates; making it **quantitative** leaves the possibility for stepwise improvements.

TEMPLATE MATCHING VIA CROSS-CORRELATION

From signal-to-noise considerations alone we are lead to conclude that a method which makes maximal use of the informational content of the **whole** spectrum would be best suited to cope with poor-signal spectra. Thus pattern recognition methods like "template matching" are suggested. In template matching the data being examined and the template being used are considered as vectors of a linear space; a metric (e.g. a Euclidean distance) is defined to measure the "similarity" between two vectors via their mutual distance.

We here investigate the use of template matching via cross-correlation, a technique which is well known in the field of pattern recognition (see e.g. Love & Simaan, 1985) and which has already been successfully used for the determination of radial velocities (Fehrenbach & Burnage, 1978). Using the usual Euclidean distance, the correlation coefficient ρ between two vectors s and t, both shifted to their mean, is defined as

$$\rho[s,t] := \frac{(s - \bar{s}, t - \bar{t})}{|s - \bar{s}| \, |t - \bar{t}|} := \frac{\sum_i (s_i - \bar{s})(t_i - \bar{t})}{\sqrt{\sum_j (s_j - \bar{s})^2} \sqrt{\sum_k (t_k - \bar{t})^2}} \quad . \tag{1}$$

As is well known, the correlation coefficient may be viewed geometrically as the cosine of the angle between the (shifted) data and the (shifted) template vector, and thus varies in the interval $[-1,+1]$, where +1 indicates a perfect match between the original and the template.

Computationally the method of "stellar spectral classification via template matching" consists of two steps: (1) the computation of the "measure of similarity", which in our case means to compute the correlation coefficient between a given spectrum and each reference spectrum, and (2) to determine the "best matching" template, which in our case means to compute the maximum** of the correlation function, the position of which indicates the searched spectral type. The correlation-coefficient measures the similarity in a global manner and as such is dominated by the shape of a given spectrum as a whole, i.e. its continuum-slope.

The cross-correlation technique not only determines the best match classification, but also specifies a "classification quality", measured by the peak value of the correlation function. In order to somehow control the number of misclassifications, a "quality threshold" level may be imposed on the classification: if the threshold requirement is not met, a data vector is left as "unclassifiable". A spectrum may fail the quality threshold either because the data is of low accuracy or because its template does not exist in the template set.

---

* For reviews on automated spectral classification see West (1973) and Schmidt-Kaler (1982).
** It is easy to show that <u>maximizing</u> the correlation coefficient over a set of suitably standardized templates is equivalent to <u>minimizing</u> the Euclidean distance.

On the astronomical part spectral classification via template matching requires firstly a suitable atlas of reference spectra, and secondly object spectra which are **wavelength- and intensity-calibrated**. Such spectra are now obtainable from a variety of observational records, including Schmidt-prism photographic plates (Adorf & Röser, 1984; Clowes, 1984, 1986; Hewett et al., 1984, 1985) and prime-focus grism photographic plates (Vaucher et al., 1982). The biggest problem for a method which uses the global shape of a spectrum for spectral classification is probably interstellar reddening.

For the sake of simplicity we will restrict ourselves to the one-dimensional classification of luminosity class V stellar spectra, for which the set of templates forms one-parameter grid of vectors, which vary smoothly with spectral type. Operating with these templates on a given object spectrum will result in a correlation function which also varies smoothly - a property which is fortunately preserved in the presence of noise.

NUMERICAL EXPERIMENTS

The method of template matching is best understood when looking at how templates classify themselves. We have used 13 main-sequence reference spectra of type O, B3, B5, A0, A5, F0, F5, G0, G5, K0, K5, M0, and M5, from the atlas of Straizys & Sviderskiene (1972), 12 of which are displayed in Fig. 1. The spectra have a resolution of 5 nm/pixel and we use the 37 pixels covering the wavelength range between 350 and 530 nm (which is approximately appropriate to simulate the classification of Schmidt-prism spectra recorded with a dispersion of ~140 nm/mm at Hγ on a Kodak IIIa-J photographic emulsion). All the templates were correlated among each other, including auto-correlating each template with itself.

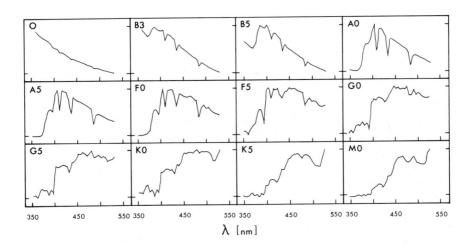

Fig. 1. Spectral intensity F_λ versus wavelength λ for twelwe out of thirteen luminousity class V reference spectra used for spectral classification via template matching.

Fig. 2 displays the resulting correlation functions, which consist of the correlation-coefficients between a fixed (template) spectrum and all the templates from the atlas. As has to be expected, each correlation function assumes a peak value of +1 (perfect match) at the position of its own spectral type.

The most remarkable property of the template matching method is its **high discriminating power,** which is preserved even in the presence of substantial noise. In order to quantitatively state the noise-cheating capability of the cross-correlation technique, a series of numerical experiments were conducted, which model the case where background shot noise is the dominant degradation effect. Artificially generated zero-mean Gaussian noise was added to the set of template spectra (see Fig. 3) with a predefined signal-to-noise ratio S/N of 100, 30, 10, and 3, where we have used a global S/N parameter defined via

$$S/N := \frac{\text{(Integrated signal of the spectrum)}}{\text{(RMS noise in the domain of the spectrum)}} := \frac{\sum_i s_i}{\sqrt{\sum_j n_j^2}} \quad . \quad (2)$$

This procedure was repeated 100 times resulting in a set of 13 x 100 x 4 = 5200 noise-contaminated spectra, which were subsequently classified. The simulations were executed - almost completely on the level of command procedures - within ESO's image processing system MIDAS (Banse et al., 1983; Crane et al., 1985; "The MIDAS Users Guide, 1986") on a DEC VAX 11/785, which needed about 2 min real time for the CPU-intensive correlation step.

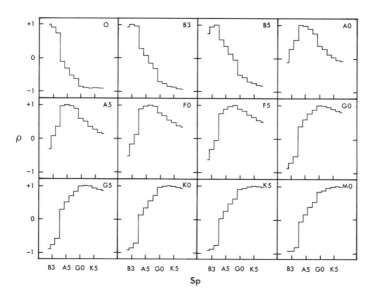

Fig. 2. Twelwe out of thirteen correlation functions resulting from correlating the templates with themselves. Each correlation function consists of the correlation coefficients between a fixed (template) spectrum and all the refererence spectra from the atlas. The location of the correlation peak indicates the "best match" spectral type; the peak value measures the achieved classification "quality".

RESULTS

As expected the quality of the classification, measured by the "best match" correlation coefficient, decreases with increased noise. Fig. 4 displays the mean correlation coefficient as a function of the S/N. As already mentioned this quality measure may be used to set a classification threshold T. An increased quality threshold decreases the number of misclassifications at the expense of more and more spectra rated as "unclassifiable".

Three quantities may be used to judge the results of the classification simulations: (1) the degree of classification "completeness" r_c, measured by the number of "classifiable" spectra divided by the total number of spectra, (2) the classification "success rate" r_s, measured by the number of correctly classified spectra divided by the number of classifiable spectra, and (3) the classification error σ_{Sp}, measured by the RMS dispersion of the assigned spectral types.

Table 1 displays some simulation results in terms of the observed degrees of "completeness", the "success rates" and the error values as a function of the signal-to-noise S/N and classification threshold T for all the spectral types. Looking at Fig. 4 we have chosen a threshold value T of 0.9, 0.7, 0.4 and 0.3 for the S/N levels of 100, 30, 10 and 3, respectively, and have obtained a mean degree of completeness $\bar{r}_c$ of 100, 86, 67 and 35 %, and a mean success rate $\bar{r}_s$ of 99, 80, 51, and 24 % among the classifiable spectra; spectral types could be assigned to classifiable spectra with a mean error $\bar{\sigma}_{Sp}$ of 0.0, 0.3, 0.7 and 1.3 spectral classes, respectively. Looking at Table 1c and 1d it appears, that the classification of early- and late-type stellar spectra works better than the classification of medium-type spectra, which may be explained by the more prominent continuum-signature of the former.

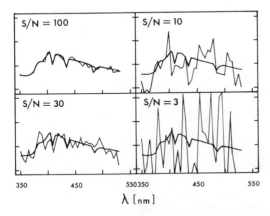
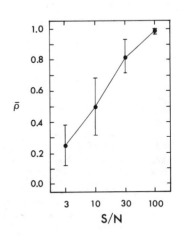

Fig. 3. Spectral intensity F_λ versus wavelength λ for a sequence of increasingly more noise-degraded A0-type stellar spectra. At S/N = 3 the spectrum as a whole is barely visible in the noise.

Fig. 4. The mean peak correlation coefficient $\bar{\rho}$, which measures the mean "quality" of spectral classification obtainable for a fixed signal-to-noise level S/N. As expected, increased noise (decreased S/N) makes classification more and more doubtful. The $\bar{\rho}$ value may be used to choose different classification thresholds T for different S/N levels.

Table 1. The "measured spectral type" as determined from the noise degraded spectra versus the "true spectral type" from the undisturbed original, displayed for various signal-to-noise ratios S/N and classification thresholds T: (a) S/N = 100, T = 0.9 (b) S/N = 30, T = 0.7 (c) S/N = 10, T = 0.4 (d) S/N = 3, T = 0.3. Each of the two-dimensional histograms shows the frequency of the observed "best matching" spectral types.

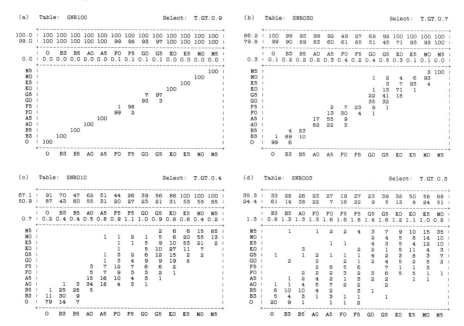

The top four rows of each subtable contain summary statistics for each spectral type: (1) the classification "completeness" r_c in %, (2) the classification "success rate" r_s in %, (3) the mean observed spectral type Sp, and (4) the RMS error of the assigned spectral types σ_{Sp} in units of spectral types. The three numbers on the left side of each subtable are from the top: The "mean completeness" $\bar{r}_c$ achieved, the "mean success rate" $\bar{r}_s$, and the "mean classification error" $\bar{\sigma}_{Sp}$ of the assigned spectral types, where the means are taken over all spectral types.

Under low noise conditions (S/N = 100, see Table 1a) the overall rate of misclassifications is on the 1% level; increasing noise (decreasing S/N) leads to a steadily increasing number of misclassifications. But even with extremely high noise-contaminations at the plate limit* (see Table 1d), where the individual spectrum can hardly be detected (Fig. 3d) by eye, a mean degree of completeness of $\bar{r}_c \sim 35\%$ and a mean success rate of $\bar{r}_s \sim 24\%$ among the classifiable spectra are still achievable, proving the power of the correlation technique; for early- and late-type spectra considerably higher degree of completeness and success rates may be obtained. (The remarkable classification success rate at the faint end would even sound better, if one allowed bigger spectral boxes than the "half-a-spectral-type"-sized ones we have been using.)

---

* In the absence of any established definition we have defined the "plate limit" somewhat arbitrarily to be at S/N = 3, below which the spectrum as a whole vanishes in the noise.

Varying the threshold value T affects both the degree of classification completeness and the success rate. Decreasing the demand by lowering T leads to an increased degree of completeness at the expense of a reduced success rate. At the S/N level around 10 an appropriate choice of T is critical for achieving a high degree of completeness.

The determination of the threshold value T depends on the classification objectives. Fig. 5 displays the mean success rate versus the mean degree of completeness obtainable for various S/N and T values. If maximum completeness is the goal, a rather low T value would be appropriate. On the other hand, if misclassifications must be avoided, a high T value would be the choice. Another possibility could be to maximize the number of correct classifications, which would lead to an intermediate value for T.

CONCLUSIONS

We have shown that "template matching" via cross-correlation provides a simple, yet powerful and robust method for the classification of large sets of highly noise-contaminated low-resolution spectra. Numerical experiments indicate that the classification works remarkably well even for spectra only 1.25 mag above the detection limit. Probably due to their unique continuum signature early- and late-type stellar spectra are somewhat easier to recognize than medium-type spectra. Although not time-optimized, the algorithms used are sufficiently fast to allow the classification of all the spectra on a deep Schmidt-prism plate.

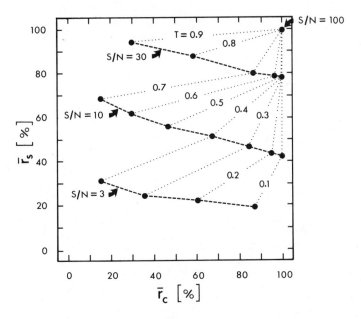

Fig. 5. The mean classification "success rate" $\bar{r}_s$ versus the mean degree of "completeness" $\bar{r}_c$. Each dashed line indicates how $\bar{r}_s$ and $\bar{r}_c$ vary with the quality threshold T for fixed signal-to-noise ratio S/N. At a fixed S/N the success rate may be increased by an increased T.

Spectral classification by template matching is not restricted to stellar spectra and lends itself to various quantitative spectral surveys. The most exciting results may even be expected in the area of extragalactic survey work, where the set of unclassifiable spectra may contain the objects of interest.

Note added in proof:
The cross-correlation technique has previously been used for automatic spectral classification by Kurtz (1982) in his Ph.D. thesis.

REFERENCES

Adorf, H.-M., Röser, H.-J., 1984, Reduction of Slitless Spectra - The Detection of Faint Emission Lines, in: "Astronomy with Schmidt-type Telescopes, IAU Colloq. No. 78, Asiago 1983", M. Capaccioli, ed., D. Reidel, Dordrecht, p. 423.
Banse, K., Crane, Ph., Ounnas, Ch., Ponz, D., 1983, MIDAS - ESO's Interactive Image Processing System, in: "Proc. Digital Equipment Computer Users Society", Zürich, p. 87.
Christian, C.A., 1982, Identification of Field Stars Contaminating the Color-Magnitude Diagram of the Open Cluster Be 21, Astrophys. J. Suppl. Ser., **49**:555.
Clowes, R.G., 1984, Automatic Processing of Objective-Prism Plates at the Royal Observatory, Edinburgh, in: "Astronomy with Schmidt-type Telescopes, IAU Colloq. No. 78, Asiago 1983", M. Capaccioli, ed., D. Reidel, Dordrecht, p. 107.
Clowes, R.G., 1986, Automated quasar detection in the SGP filed: a clustering study, Mon. Not. Royal Astron. Soc., **218**:139.
Crane, Ph., Banse, K., Grosbøl, P., Ounnas, Ch., Ponz, D., 1985, MIDAS, in: "Data Analysis in Astronomy", V. Di Gesù, L. Scarsi, P. Crane, J.H. Friedman, S. Levialdi, eds., Plenum Press, New York and London, p. 183.
Fehrenbach, C., Burnage, R., 1978, La mesure par corrélation des vitesses radiales au prisme objectif dans un champ stellaire situé à la latitude galactique $b_{II} = -30°$, Comptes Rendus Acad. Sci. Paris, **286**:289.
Garrison, R.F. ed., 1983, The MK Process and Stellar Classification, David Dunlap Obs., Univ. Toronto.
Hewett, P.C., Irwin, M.J., Bunclark, P., Bridgeland, M.T., Kibblewhite, E.J., McMahon, R., 1984, Automatic Analysis of Objective Prism Spectra, in: "Astronomy with Schmidt-type Telescopes, IAU Colloq. No. 78, Asiago 1983", M. Capaccioli, ed., D. Reidel, Dordrecht, p. 137.
Hewett, P.C., Irwin, M.J., Bunclark, P., Bridgeland, M.T. Kibblewhite, E.J., 1985, Automatic analysis of objective-prism spectra - I., Quasar Detection, Mon. Not. Royal Astron. Soc., **213**:971.
Kurtz, M.J., 1982, Automatic Spectral Classification, Ph.D. thesis, Dartmouth College, Hanover NH.
Love, P.L., Simaan, M., 1985, Segmentation of a Seismic Section Using Image Processing and Artificial Intelligence Techniques, Pattern Recogn., **18**:409.
"The MIDAS Users Guide, 1986, ESO Operating Manual No. 1", Image Processing Group, European Southern Observatory, Garching b. München, FRG.
Schmidt-Kaler, Th., Automated Spectral Classification - A Survey, Bull. Inform. CDS, **23**:26.
Straizys, V., Sviderskiene, Z., 1972, Catalogue of Energy Distribution Curves in Stellar Spectra of Different Spectral Types and Luminosities, Bull. Vilnius Astron. Obs., **35**.

Vaucher, B.G., Kreidl, T.J., Thomas, N.G., Hoag, A.A., 1982, Quantitative Measures of Slitless Spectra of QSOs, Astophys. J., **261**:18.
West, R.M., 1973, Automatic Classification of Objective Prism Spectra, in: "Spectral Classification and Multicolour Photometry, IAU Symp. No. 50", Ch. Fehrenbach, B.E. Westerlund, eds., D. Reidel, Dordrecht, p. 109.

AN APPROACH TO THE ASTRONOMICAL OPTICAL IMAGE COMPRESSION

L. Caponetti*, G.A. De Biase** and L. Distante***

*Istituto di Scienza dell'Informazione, University of Bari, Italy
**Istituto Asronomico, Università di Roma La Sapienza Italy
***Istituto di Elaborazione Segnali ed Immagini, CNR, Bari Italy

ABSTRACT

This paper analyses the problem of astronomical image compression. After an introduction to the peculiar nature of astronomical data a coding strategy, based on the cosine transform, is presented. The performance has been evaluated using statistical and photometric measures.

1. INTRODUCTION

Astronomical data are principally represented as two-dimensional images of energy maps. On the basis of the wavelength of observation and of the imaging system these images can be distinguished in three different classes:

- radio maps
- optical images
- x and gamma images

The astronomical images can contain a different number of pixels, going from the about 512 x 512 of the images coming either from orbiting observatories or from CCD detectors connected to ground-based telescopes, to the many millions of them obtained from the photographic plate sampling.

Consequently data compression becomes a fundamental problem in the management of astronomical data because of their relevant volume. The usefulness of the compression depends on the possibility to minimize both the image storage memory and the image transmission bandwidth.

2. OPTICAL IMAGES

Optical images consist of some point like objects enlarged by the Point Spread Function or of extended objects mixed to point ones (galaxies, stars, clusters, and so on).

The images include a background originated both by the sky brightness and by the imaging system. The background is usually characterized by gradients which vary at a lower rate than those of the objects included in the image.

Besides requiring algorithms apt to carry out morphological analysis and classification, image processing of optical data needs algorithms suitable to execute quantitative measures. Energy information (photometric measures) is most important for the astronomer and numerical methods have already shown to be superior to the traditional ones, owing to the higher accuracy, speed and objectivity.

3. COSINE TRANSFORM APPROACH

The basic idea of the transform coding process is that for most images the transform coefficients are relatively of low magnitude and the energy distribution in the transform domain is more suitable for coding than the pixel distribution in the spatial domain (Ref. 1).

In the transform image coding an image is divided into small blocks of pixels and a representation of each block is carried out by first taking a linear transformation of block pixels, next selecting the most significant transform coefficients and then quantizing the selected coefficients.

Figure 1 contains the block diagram of the transform coding strategy applied to some astronomical images.

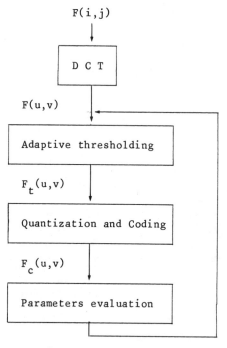

Fig. 1:
Block diagram of the coding method

The method consists of the following steps:

a) Discrete cosine transform
Through this step a discrete cosine transform (Refs 2,3,4) is applied to the input image in 16 x 16 pixel blocks

$$F(u,v) = 4C(u)\,C(v) \sum_{i=0}^{N-1} \sum_{j=0}^{N-1} \left[f(i,j) \cdot \cos\frac{2i+1}{2N}u\pi \right.$$

$$\left. \cos\frac{2i+1}{2N}v\pi \right]$$

for $u,v = 0,1,\ldots, N-1$, where

$$C(k) = \begin{cases} \frac{1}{2} & k = 0 \\ 1 & k = 0 \end{cases}$$

b) Adaptive thresholding

In this step an adaptive threshold T is selected for each block of the transform coefficients. The coefficients whose magnitudes are greater than the threshold are subtracted from the threshold using:

$$F_T(u,v) = \begin{cases} F(u,v) - T & \text{if } F(u,v) > T \\ 0 & \text{if } F(u,v) \leq T \end{cases}$$

c) Quantization and coding

The differences between the transform coefficients and the threshold are quantized in a fixed number of discrete levels. The number of levels for each difference is made proportional to its probability of occurence so that the spacing between output levels decreases in regions of high probability and increases in regions of low probability. The occurrence probability is estimated by means of the difference histogram. The output levels are coded using a runlength algorithm.

d) Parameter estimation

Some fidelity parameters must be evaluated to reach a desidered performance and the threshold is adaptively modified. These parameters estimate the compression goodness in terms of both the degradation of the reconstructed image (mean square error) and the compression rate (bit/pixel).

4. RESULTS AND CONCLUSION

This method has been applied to a picture (256 x 256 pixels) of the NGC 4438 image. This picture can be considered a good test since it includes both extended objects and point-like objects of various intensities.

Figure 2 contains the original image and the reconstructed images at different compression rate (2, 2.5, 3, 5, 10). It can be seen that the morphology degradation is very slow even in the case of maximal compression rate. Energy measures on four objects in the field has been carried out (see Fig. 3), both on the original image and on the reconstructed ones. The results of the analysis are reported in Table 1.

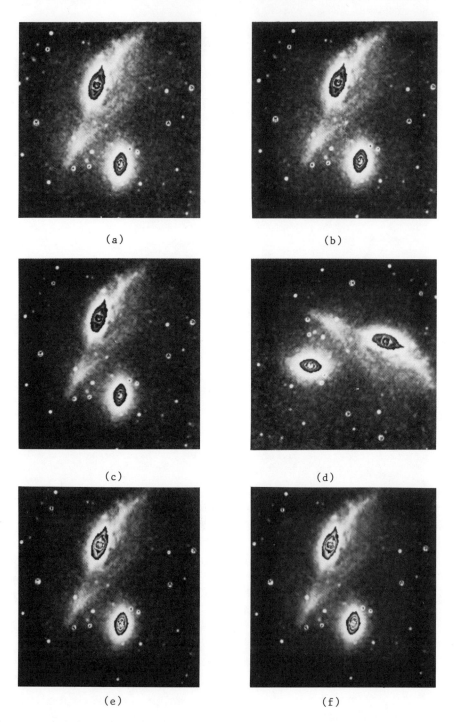

Fig. 2: Original image and reconstructed ones at different compression rates (a) Original image at 10 bit/pixel; (b) reconstructed image at 4 bit/pixel and e_{rms} = 0.26; (c) at 3.2 bit/pixel and e_{rms} = 0.50; (d) at 2.4 bit/pixel and e_{rms} = 0.82; (e) reconstructed image at 1.6 bit/pixel and e_{rms} = 1.54; (d) at 0.8 bit/ pixel and e_{rms} = 2.18.

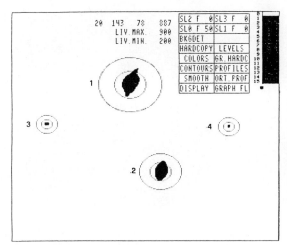

Fig. 3:

Map of the four selected objects: the inner loop describes the zone used in measuring the object intensity; the external one delimits the zone used in measuring the background intensity.

It must be noted that in spite of the different intensity of the analysed objects the relative error between the estimated energies on the original and on the reconstructed images is never greater than 1%.

The coding and decoding algorithms are implemented on a VAX 11/780 computer, running under a VMS operating system. For a picture of 256 x 256 pixels the execution time required for the coding or decoding process is about 60 sec.

The results obtained encourage the use of this approach for transmission in a computer network and eventually for the storage of astronomical images.

Table 1

Intensity normalized to the most intense object. The relative
error has been evaluated by the difference between the object
energy in the original image and the reconstructed one

Objects	Intensity	Relative errors at different compression rates (%)				
		2	2.5	3	5	10
1	1	0.6	0.9	0.8	0.08	0.5
2	0.95	0.6	0.9	0.8	0.2	0.4
3	0.06	0.6	0.9	0.9	0.09	0.7
4	0.02	0.8	0.7	0.9	0.5	0.2

REFERENCES

1. A.K. Jain, "Image data compression: A review", <u>Proc. IEEE 69,</u> pp. 349-389 (1981).
2. Ahmed, Natarajan and Rao, "Discrete cosine transform", IEEE Trans. on computer, pp. 90-93 (Jan. 1974).
3. Chen, Smith and Fralick, "A fast computational algorighm for the discrete cosine transform", IEEE Trans. Comm. Vol. COM-25n oo. 1004-1009 (Sept. 1977).
4. Chen and Pratt, "Scene adaptive coder", IEEE Trans. Comm. Vol. COM-32, pp. 225-232 (March 1984).

CODED APERTURE IMAGING

E. Caroli, J.B. Stephen, A. Spizzichino, G. Di Cocco, and L. Natalucci

Istituto TE.S.R.E./C.N.R., Bologna, Italy

INTRODUCTION

In comparison to other branches of astronomy, the production of sky images in the fields of X- and γ-rays is greatly hindered by the difficulty of focussing photons and the weakness of the fluxes with respect to the background counting rate. Interest has recently been shown in a class of telescopes based on multiplexing techniques for use in a range of astronomical applications. In these instruments the image formation consists of modulating the incident radiation by a suitable configuration of opaque and transparent elements and reconstructing the object from the detected events using a computer algorithm. These instruments may be conveniently divided into two subclasses. In the first of these the detector is not position-sensitive and the information on the object position is obtained by time multiplexing the signal by means of one or more collimators or grids[1]. The second class of telescopes utilizes the spatial multiplexing obtained with a regular set of transparent and opaque elements, whose shadow projected along the flux direction is intercepted by a position-sensitive detector; these instruments are known as "coded mask systems", and have already been used successfully in several fields: for monitoring of laser induced fusion[2], in nuclear medicine, particularly in tomography[3], and spectroscopy[4]. In the following sections we will discuss the principles and difficulties involved in designing a coded mask imaging system, with reference to a balloon borne low energy γ-ray telescope currently under construction[5].

IMAGE FORMATION AND DECODING

In coded aperture imaging systems a mask consisting of an array of opaque and transparent elements is set between the source fluxes and a position sensitive detection plane (PSD). Every source, or source element, within the field of view projects a shadow of the aperture on to the detection plane [fig. 1]. Thus for a single point source the detected two-dimensional distribution of events reproduces a mask pattern, while for a more complex source or distribution of point sources the recorded shadowgram is the sum of many such distributions. For each flux direction the part of the mask that contributes to coding on the position sensitive detector is called a "working zone". Of course, the general requirement is that for every direction the working zone has to be different so as to avoid ambiguities in the reconstruction of the source position. If we represent the transmission function $A(x,y)$ by a matrix $A(j,k)$, of dimensions nxm=N whose elements are equal to 1 and 0 respectively, corresponding to the transparent and opaque elements, the FOV may also be conveniently represented by a (pxq) array $S(j,k)$, the elements of which have dimensions governed by the solid angle subtended by one mask element. These angular dimensions define the geometric (or intrinsic) angular resolution of the instrument. Then the detected shadowgram is given by:

$$D(j,k) = \Sigma_i \Sigma_l A(j+i-1, k+l-1) S(i,l) + B(j,k) \qquad (1)$$

where the summations should be extended to infinity, even if they are limited by the physical dimensions of the detector and of the aperture A. The term $B(j,k)$ is a signal-independent noise term. This may also be expressed in the matrix form:

$$\underline{d} = W\underline{s} + \underline{b} \qquad (2)$$

where the matrix W and the vectors $\underline{d}$, $\underline{s}$ are obtained by lexicographical reordering of the matrices A, D and S respectively, following a method described by Andrews[6]. The elements of the matrix A are filled in the W array in such a way that each column k of W is the one-dimensional array representation of the coding pattern associated with the k-th element of the object $\underline{s}$, and each row j is the set of elements s_k which contribute to the counts in the pixel d_j of the detector. The W matrix has dimensions (pxq, pxq) and is also referred to as a "weighing design".

An estimate S' of the object S can be obtained by filtering the shadowgram D with a suitable decoding function. A direct general method of decoding can be written in the discrete case in the form of a cross-correlation with a matrix $G(j,k)$:

$$S'(j,k) = \Sigma_i \Sigma_l G(i+j-1, 1+k-1) D(i,l) = G*A*S + G*B \qquad (3)$$

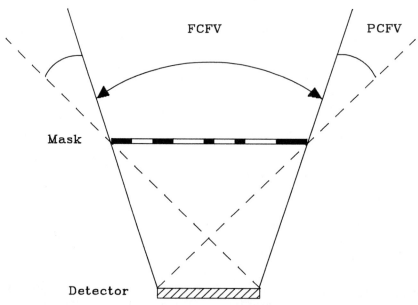

Fig. 1. The fully coded field of view (FCFV) is defined as comprising all directions in which the PSD records a completely coded image. The directions for which an incompletely coded shadow is recorded make up the partially coded field of view (PCFV).

The choice of the decoding matrix G must be such that G*A (System Point Spread Function - SPSF) is as close as possible to a delta function, in order to preserve the object features within the system resolution. Usually the design of the array G is based on the pattern of A^7. If the weighing design W of the mask is non singular, the formalism adopted in (2) suggests the use of the inverse W^{-1} as the deconvolution array G, whereupon the SPSF is exactly a delta function although large terms in W might cause noise amplification problems[8].

CODED APERTURE MASK DESIGN

Since the first proposal, by Mertz and Young in 1961[9], to use coded apertures in order to produce images of high energy radiation, a large number of mask designs have been conceived and employed with a view to optimising the quality of the images produced. Although advanced methods of deconvolution have recently been applied to this problem[10,11], resulting in the ability to produce high quality images from quite poor data sets, these methods are always very time consuming and require large amounts of computer power. On the other hand, relatively simple and fast methods of image reconstruction may be employed if the instrument design, in particular the choice of mask pattern, has been sufficiently well studied.

Most designs consist of aperture arrays based on multiple pinholes in an otherwise opaque plate. First introduced for use in X-ray astronomy, Dicke[12] and Ables[13] independently proposed apertures consisting of a large number of randomly spaced pinholes such that the overall transmission is 50%. In principle this is a logical extension of the simple pinhole camera, which itself has since been used successfully in the astronomical context[14]. The imaging properties of the simple pinhole camera are ideal, with the SPSF being of the form of a delta function, but it suffers greatly from a conflict between resolution and sensitivity. The angular resolution is proportional to the hole size, so for fine resolution a small hole is essential. The sensitivity, however, is determined by the amount of flux detected and so a large hole is preferable. These two conditions are clearly incompatible. The idea behind the random array is that the aperture may be increased by a factor equal to the number of pinholes, whilst retaining the angular resolution commensurate with the size of each hole. These Random Arrays allow reconstruction of the object distribution on a background of non-statistical noise determined by the pattern of pinholes.

A class of arrays which has found greater use in the field of high energy astronomy is that based on cyclic difference sets[15]. The SPSF obtained using these masks has only two values — one associated with the peak and the other with the background, so these designs do not suffer from systematic noise. These mask patterns may therefore be considered as being 'perfect'. The reason that the SPSF has only two values is that the number of times the vector spacing between a pair of 1's occurs is a constant regardless of that spacing (up to a certain limit). For this reason these arrays are also called Uniformly Redundant Arrays (URA's)[7].

Another class of masks was recently proposed[16] with optimum multiplexing properties. The algorithm to generate these masks consists of making the direct product of two PN sequences[17]. The resulting arrays are referred as a Pseudo Noise Product (PNP) and can be implemented in any geometry (pxq) for which PN sequences exist for each dimension separately. As each PN sequence is ~50% open the PNP arrays typically have a transparency of ~25%, which may be an advantage for use in medical imaging conditions. For this reason an alternative method of producing masks with less than 50% transmission has been suggested, resulting in a class known as Geometric coded apertures[18]. As the name suggests, they are characterised by their geometrical regularity, and hence by their ease of construction.

MASK PERFORMANCES

The performance of an imaging telescope depends on its ability to reconstruct faithfully the fluxes from the field of view both in terms of intensity and angular distribution. One parameter which may be employed to define the quality of an instrument is the Signal to Noise Ratio (SNR) which determines the minimum source strength that may be detected. In an image the SNR for a particular pixel is the ratio between the intensity in that pixel and a noise term:

$$SNR = S(j,k)/\sqrt{Var(B)+Var(S(j,k))+Var(S(j,k),S(l,m))} \qquad (4)$$

The first two contributions to the noise term depend only on the statistical fluctuations of the instrumental background and the source flux. The third contribution is due to the cross-talk between the different source elements and generally also contains a systematic contribution from the SPSF sidelobe structure. In the case of decoding by inversion of the weighing matrix W or by correlation, it is possible to obtain an analytical expression for the expected SNR in the reconstructed pixels by considering that the statistics of the fluxes and background are Poissonian and without presupposing the type of mask pattern[18]. Fenimore[8] has studied how the transparency of the mask affects the reconstructed SNR with a view to optimizing the aperture design with respect to a wide range of observational conditions (number of sources, source intensities, background count rate). On a basis of statistical considerations he has shown that in the typical conditions of high energy astronomy (few point source, high background) the optimum transparency is very close to 50% and furthermore that the loss in SNR encurred by the use of a mask transparency of 50% as opposed to of the optimum density is always less than 30% for all physical conditions.

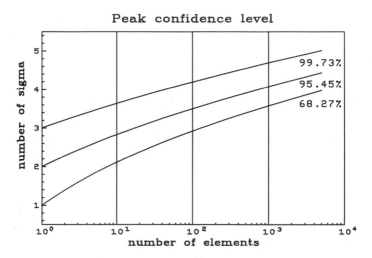

Fig. 2. The variation in confidence level for a particular SNR as the number of pixels in the mask is increased, in order to retain the same significance.

Care must be taken in interpreting the SNR in terms of significance. In coded mask systems it must be remembered that the observation of each element in the FOV consists of a number of simultaneous measurements of flux and background, the exact number depending on the corresponding working zone of the mask pattern. The larger the number of measurements, the more likely a certain value of sigma will be exceeded by chance and so the significance of a particular detected peak will be less. Figure 2 shows how the values of sigma have to increase as the number of measurements (pixels) in order to retain the same significance, assuming the counting rates in each pixel are mutually independent (i.e. no systematic effect in the reconstructed pixels)[19].

In order to provide a theoretical comparison between the different mask designs it is possible to analytically evaluate the response of a coded aperture telescope to a given source distribution. Fig. 3 shows the relative significance in the reconstructed image of a point source for systems, with the same detector area and intrinsic angular resolution, employing different mask designs. The evaluation assumed no unmodulated detection plane background, the object consisting of a point source superimposed on a uniform sky intensity. For situations of very low contrast, in which the background counting rate dominates the source counts, it can be seen that the pinhole camera provides the best response, but this situation is rapidly reversed as the relative source intensity increases. As each of the complex mask designs are 'perfect', the difference in reconstructed SNR depends mainly on the transmission of the mask pattern, and hence the number of counts detected. For the URA and PNP however, the structure of the modulated background has no effect on the predicted SNR, the variance of the background being given by the square root of the total detected counts, whereas for the geometric pattern the weighted terms in the decoding array lead to a dependence on this parameter[18]. Figure 4 shows the reconstructed images obtained from simulated observations of two point sources with four different mask designs. A uniform detection plane background was assumed in each case, and mask designs of similar geometries were adopted to facilitate comparison.

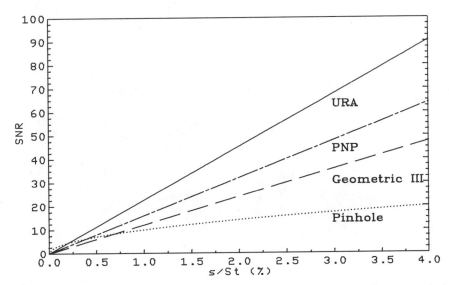

Fig. 3. A theoretical comparison between the expected performances of a 50% transparent 31 x 33 URA; a 25% transparent 31 x 31 PNP; a 12% transparent 31 x 31 geometric mask and a single pinhole mask (~0.1%) (from ref. 34).

SOME DESIGN CONSTRAINTS ON CODED APERTURE SYSTEMS

In the following sections some criteria which have important consequences on the efficiency of a coded aperture imaging device are outlined. They fall into three main categories: the detection plane, the mask, and the conditions in which the device is to be used.

The Position Sensitive Detection Plane

The finite positional resolution of the PSD is the basic limiting factor of both the angular resolution and source location accuracy of the final instrument. In the latter case, the ability to position a point source in the field of view is determined by the possibility to detect the shadows of the edges of the mask elements and so, for a given statistical quality of the data, is purely dependent on the detector performance. The angular resolution, on the other hand, is dependent on the angle subtended by a mask element at the PSD, and hence on the mask element dimensions (given a fixed mask/PSD separation). However, for a particular value of positional resolution on the PSD, the smaller the mask elements are, the more effect that random errors in the positioning of detected events will have. Figure 5 shows the reduction in SNR which occurs in one pixel for a single point source as the mask element dimensions are changed. From this figure it is apparent that, in order to lose less than 50% sensitivity from the ideal case, the mask element dimensions must be at least twice the positional resolution (σ) of the PSD.

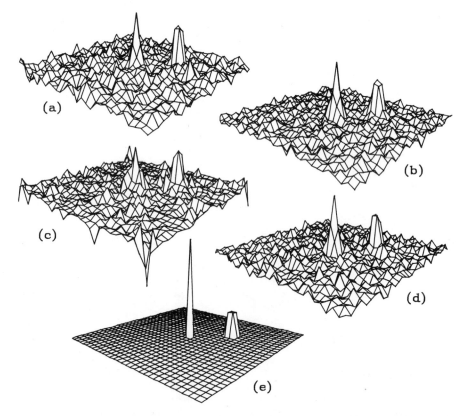

Fig. 4. Simulated images of the two source distribution (e) by: a 31 x 31 random mask (a); a 31 x 33 URA (b); a 31 x 31 geometric mask (c) and a 31 x 31 PNP (d). The peak heights have been scaled to the same value so as to emphasise background structure.

The Coded Aperture Mask

In a coded aperture system which is designed to produce images over a wide range of photon energies, the opaque mask elements vary in efficiency with respect to photon energy, passing through a minimum at about 2-3 MeV. The increased transmission of unwanted photons will inevitably lead to a degradation in image quality. It is not always possible, however, to increase the mask element thickness until the desired opacity is accomplished over the entire energy range of interest, both due to the large weight of a thick mask and to collimation by the mask elements. For non-perpendicular angles of incidence some of the photons which would pass through an open element will be stopped by a neighbouring closed element, thus limiting the effective field of view of the instrument. To reduce this effect the profiles of each element may be rounded, the only problem being that, for sources close to the centre of the field of view, this can lead to a loss of definition at the element boundaries — information which is vital for point source location purposes.

A similar problem occurs when a practical mask has to be constructed. The vast majority of mask patterns have isolated elements within the structure and so require some means of support. In general this requires the presence of material in the position of the open mask elements.

For a URA both the above cases may be expressed together in terms of the expected SNR in an image of a point source, giving[20]

$$SNR_{ij} = \frac{MS_{ij}(T_o+T_c) + S_{ij}T_c}{[MS_{ij}(T_o+T_c)-T_cS_{ij} + (M(T_o+T_c)-T_c)S_t + NB]^{\frac{1}{2}}} \quad (5)$$

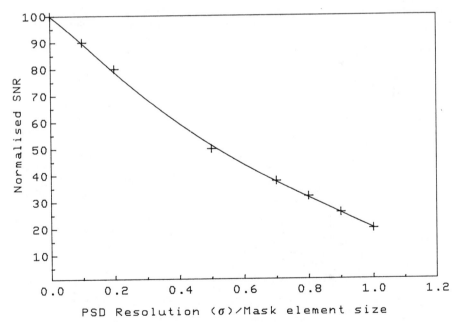

Fig. 5. The decrease in SNR from misallocation of photons to neighbouring pixels due to the positional resolution of the PSD, as a function of mask element dimensions. In order for the loss to be less than 50%, the mask element has to be at least twice the positional resolution (σ).

where the mask has N pixels, M of which are transparent, the source strength is S counts/mask element, the background is B counts/pixel, T_o and T_c are the transmissions of the open and closed elements respectively and S_t^c is the integrated intensity of all other sources in the field of view. It is interesting to consider two limiting conditions for (5): For the case of perfect opacity of the closed elements, no background or background sources it reduces to $SNR = (MT_o S)^{1/2}$, which is equivalent to imaging a source of reduced strength ($T_o S_{ij}$) with a perfect mask. Where the open elements transmit 100% of the incident radiation, under the same conditions we get

$$SNR_{ij} = \frac{[N - (N-1) T_c] S_{ij}^{1/2}}{[N + (N-1) T_c]^{1/2}} \qquad (6)$$

The term incorporating T_c in the numerator is responsible for a reduction in peak height, whilst the equivalent term in the denominator represents the increased variation in the background.

A further degradation in image quality may occur due to the effect of having a massive mask nearby the detection plane. Butler et al.[21] have shown by means of Monte-Carlo simulation that, although the absolute background count rate is not significantly altered by neutron activation of the mask structure, there may be significant modulation introduced when the aperture and PSD are in close proximity (see figure 6).

Instrument surroundings

The presence of a non-uniform background distribution across the PSD can lead to large fluctuations in the deconvolved image. The nature and intensity of these fluctuations are dependent not only on the background counting rate and distribution but also on the mask pattern and type of deconvolution employed (see fig. 7 for a one dimensional example). For instruments which may be employed in such conditions, sufficient shielding is essential, or background subtraction must be undertaken at the data analysis stage, thus reducing the benefits of the spatial multiplexing inherent in the imaging technique.

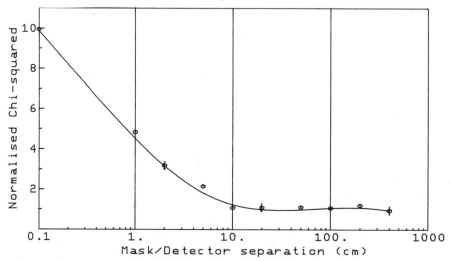

Fig. 6. The modulation of the background counting rate across the PSD introduced by the interactions of an isotropic photon flux within the mask elements, measured by the chi-squared between the resulting distribution and a flat background.

Another limitation depends on the distribution of the sources to be imaged. In the system employing a replicated mask in order to give a wide, fully coded field of view, those sources which cast an incomplete shadow of the mask onto the PSD will be reconstructed in the wrong position with a false intensity, and will also give rise to highly structured background effects. In the X-ray regime this may be obviated by means of a slat collimator over the detection plane restricting the field of view to that coded by the mask, and in the high energy γ-ray regime these events may be rejected as the PSD's in use allow the reconstruction of some positional information. For low energy γ-ray imaging, however, neither of these two techniques are possible, and the data analysis must be able to handle such situations. At present such techniques rely on either using advanced non-linear methods of performing the entire analysis[10,11] or on decoding each image with suitably altered decoding arrays in order to identify possibly vignetted sources[22]. The latter has the advantage of speed, whereas the former is superior in sensitivity but requires much computing time.

FINAL REMARKS

The technique of coded aperture imaging has been shown to be an efficient method for the production of images in astronomy in the entire range from X-rays to very high energy γ-rays. The wide range of mask patterns allows flexibility in instrument design, and relative ease of manufacture over other imaging methods such as grazing incidence or Compton telescopes. In particular, the angular resolution obtained with such devices is, in principle, limited only by diffraction and it may be forseen that in the future very large mask/detector separations will be employed in order to realise this potential. Already astronomical results from coded aperture instruments have been obtained[23,25] and many more experiments are planned for the near future[26,32]. The theory of coded apertures, as presented above, is of importance as the analysis of the data obtained from these instruments will include a large amount of digital image processing and restoration, which can only be correctly applied once a detailed and accurate understanding of the encoding/decoding steps is achieved.

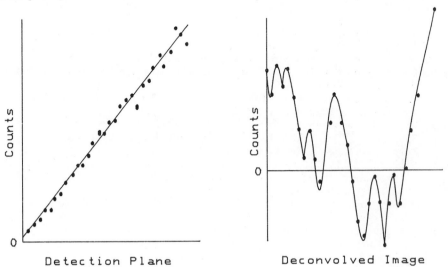

Fig. 7. The transformation of a non-uniform detection plane background by the decoding process, in the case of a monodimensional Hadamard sequence mask.

REFERENCES

1. M. Oda, Appl Opt 4:143 (1965).
2. C. Yamanaka, M. Yamanaka, H. Niki, A. Yamada, Y. Yamamoto and T. Yamanaka, IEEE Trans NS 31:490 (1984).
3. J. S. Fleming and B. A. Goddard, Med Biol Eng Comp 20:7 (1982).
4. E. D. Nelson and M. L. Fredman, J Opt Soc Am 60:1664 (1970).
5. R. E. Baker, P. M. Charalambous, A. J. Dean M. Drane, A. Gil, J. B. Stephen, L. Barbareschi, G. Boella, A. Bussini, F. Perotti, G. Villa, R. C. Butler, E. Caroli, G. Di Cocco, E. Morelli, A. Spizzichino, M. Badiali, A. Bazzano, C. LaPadula, F. Polcaro, P. Ubertini, Proc 18th ICRC 8:11 (1983).
6. H. C. Andrews and B. R. Hunt, "Digital Image Restoration", New York (1977).
7. E. E. Fenimore and T. M. Cannon, Appl Opt 17:337 (1978).
8. E. E. Fenimore, Appl Opt 17:3562 (1978).
9. L. Mertz and N. U. Young, Proc International Conference on Optical Instrumentation, London p. 305 (1965).
10. G. Ducros and R. Ducros, Nucl Instrum Meth Phys Res A 221:49 (1984).
11. R. Willingale, M. R. Sims and M. J. L. Turner, Nucl Instrum Meth Phys Res A 221:60 (1984).
12. R. H. Dicke, Astrophys J Letters 153:1101 (1968).
13. J. G. Ables, Proc Astron Soc Aust 4:172 (1968).
14. S. S. Holt, Astrophys Space Sci 42:123 (1976).
15. L. D. Baumert, "Cyclic Difference Sets" Lecture Notes in Maths Springer-Verlag, Berlin(1971).
16. S. R. Gottesman and E. J. Schneid, IEEE Trans NS 33:745 (1986).
17. F. J. MacWilliams and N. J. A. Sloane, Proc IEEE 64:1715 (1976).
18. A. R. Gourlay and J. B. Stephen, Appl Opt 22:4042 (1983).
19. E. Caroli, R. C. Butler, G. Di Cocco, P. P. Maggioli, L. Natalucci and A. Spizzichino, Nuovo Cimento 7:786 (1984).
20. P. M. Charalambous, A. J. Dean, J. B. Stephen and N. G. S. Young, Appl Opt 23:4118 (1984).
21. R. C. Butler, E. Caroli, G. Di Cocco, P. P. Maggioli, A. Spizzichino, P. M. Charalambous, A. J. Dean, M. Drane, A. Gil, J. B. Stephen, F. Perotti, G. Villa, M. Badiali, C. LaPadula, F. Polcaro and P. Ubertini, Nucl Instrum Meth Phys Res A 221:41 (1984).
22. P. P. Maggioli, E. Caroli, A. Spizzichino and M. Badiali, Nucl Instrum Meth Phys Res A 221:86 (1982).
23. R. J. Proctor, G. K. Skinner and A. P. Willmore, Mon Not R Astr Soc 185:745 (1978).
24. U. Graser and V. Schonfelder, Astrophys J 263:677 (1982).
25. M. Badiali, D. Cardini, A. Emanuele, M. Ranieri and E. Soggiu, Astron Astrophys 127:169 (1983).
26. A. P. Willmore, G. K. Skinner, C. J. Eyles and B. Ramsey, Nucl Instrum Meth Phys Res A 221:284 (1984).
27. S. Miyamoto, Space Sci Instrum 3:473 (1977).
28. G. Spada, in "Non thermal and very high temperature phenomenon in X-ray Astronomy", Rome (1983).
29. N. Gehrels, T. L. Cline, A. F. Huters, M. Leventhal, C. J. MacCallum, J. D. Reber, P. D. Stang, B. J. Teegarden and J. Tueller, Proc 19th ICRC 3:303 (1985).
30. M. R. Garcia, J. E. Grindlay, R. Burg, S. S. Murray and J. Flanagan, IEEE Trans NS 33: 735 (1986).
31. P. Mandrou, Adv Space Res 3:525 (1984).
32. W. E. Althouse, W. R. Cook, A. C. Cummings, M. H. Finger, T. A. Prince, S. M. Schindler, C. H. Starr and E. C. Stone, Proc 19th ICRC 3:299 (1985).

"STUDY OF PULSAR LIGHT CURVES BY CLUSTER ANALYSIS"

V.Di Gesù[1,2], R.Buccheri[2], and B.Sacco[2]

[1] Dipartimento di Matematica e Applicazioni
Università di Palermo
Via Archirafi, 34 - 90123 Palermo, Italy
[2] Istituto di Fisica Cosmica ed Applicazioni
dell'Informatica/CNR
Via Mariano Stabile, 172 - 90139 Palermo, Italy

Summary

The distribution of the phase numbers, corresponding to the arrival times of the gamma-ray photons detected by the COS-B satellite from the directions of the Crab and Vela pulsars, is analyzed by a clustering technique with the aim to detect possible microstructures in the pulsed emission. The method is found to be promising especially in view of the future gamma-ray experiments where better photon counting statistics is expected.

1. Introduction

The study of the pulsed emission of pulsars is usually done by the use of the "phase histograms" obtained by folding the arrival times of the photons coming from the pulsar direction, modulo the pulsar period. When applied to the gamma-ray range where the counting statistics is generally low, this method has the disadvantage that the intrinsec time resolution cannot be fully exploited because the width of the histogram bins has to be made large enough to include a consistent number of counts. As a consequence small scale structures of the light curves can be lost in the binning process.
Clustering methods applied to the phase numbers corresponding to the gamma-ray photon arrival times, can provide a powerful tool for the study of microstructures present in the gamma-ray light curves of pulsars. In this paper is presented the application of a clustering technique to the case of the pulsed gamma-ray emission from the Crab and Vela pulsars as observed by the COS-B satellite (Scarsi et al., 1977).

2. Data base and analysis method used

The data used here were derived from three observations of the COS-B satellite performed in 1975 and 1979 and aimed to study the pulsed gamma-ray emission from the Crab and Vela pulsars. Table I shows the parameters of these observations.

TABLE I

Pulsar	Start/End of COS-B Observation	Photons	Reference
Vela	Oct.20 - Nov. 8, 1975	537	Kanbach et al., 1980
Crab	Aug.17 - Sep.17, 1975	478	Wills et al., 1982
Crab	Feb.22 - Apr. 3, 1979	320	Wills et al., 1982

For each of the observations, a set of N photon arrival times was selected using an energy-dependent acceptance cone such to optimize the signal-to-noise ratio (Buccheri et al., 1983). The corresponding phase numbers have been obtained by folding the photon arrival times through the formula:

$$\alpha_i = \text{fractional part of } ((t_i - t0)/p - \dot{p}(t_i - t0)/2p) \qquad (1)$$

where t_i is the i-th arrival time of the gamma-ray photon coming from the pulsar direction (i=1 to N), p and $\dot{p}$ are the pulsar period and its time derivative and t0 is a reference epoch.

To the obtained set of α's we applied a single-link clustering technique (Zahn, 1971) using the natural distance as the interpoint relation.

Under the hypothesis of a uniform distribution of the phases in their interval of definition (0,1), the number of clusters expected after cutting the edges with weights above a threshold ϕ (Di Gesù and Sacco, 1983), is given by

$$Nc(\phi) = 1 + (N-1) \ast \exp(-N\phi) \qquad (2)$$

and the probability to have an edge with weight greater than ϕ is $P(\phi > \phi_0) = \exp(-N\phi_0)$; it follows that the probability to detect Nm-1 clusters at a threshold ϕ is

$$P(Nm:\phi) = \binom{N-1}{N_c-1} \ast \exp(N\phi)^{N_c-1} \ast (1-\exp(-N\phi))^{N-N_c} \qquad (3)$$

For values of the threshold near to $(\ln 2)/N$ the probability distribution of eq.3 approaches the normal distribution; in this case it is easy to derive analytically an estimate for the expected number of points per cluster, x=N/Nm, and its variance. These are given by

$$E[x] = N/N_c + N*Var[N_m]/N_c^3$$

and

$$Var[x] = N^2 * Var[N_m]/N_c^4 \quad (4)$$

where Nm is the measured number of clusters, Nc is given by eq.2 and $Var[N_m] = (N-1)*\exp(-N\phi)*(1-\exp(-N\phi))$.

Our analysis procedure consists on the following steps:

a- find the set of α's by the use of eq. 1 and pulsar parameters p and ṗ as supplied by radio observations. Here we used the same pulsar parameteres as in Kanbach et al., 1980 and Wills et al., 1982.

b- apply the Minimal Spanning Tree procedure to the set of α's and find the distribution $N_m(\phi)$ i.e. the experimental distribution of clusters at all cutting thresholds ϕ (see Maccarone et al., 1986).

c- Compare $N_c(\phi)$ with $N_m(\phi)$. If they agree, the set of α's are uniformly distributed in the interval (0,1): no structures are present in the pulsed emission.

d- If there is a consistent difference between the two curves, important structures are present; in order to measure the parameters of these structures we compute, for all clusters detected at any threshold ϕ, the quantity

$$(N_{ph} - E[x])/\sqrt{Var[x]} \quad (5)$$

where Nph is the number of α's found in the cluster and E[x] and Var[x] are given by eqs.4. The results presented in this paper refer to threshold values near to (ln2)/N such that eq.4 can be used. We will consider as "significant" the clusters where the quantity of eq. 5 is greater than 3: this choice is made on euristic basis and assures a confidence level of the order of 1% in our case studies. The results will be presented as phase histograms in which the phase intervals with no "significant" clusters are set to zero.

Results and discussion

Fig.1 shows the gamma-ray light curve of the Vela pulsar as measured by the COS-B satellite in a one-month observation between October and November 1975 (Kanbach et al., 1980). Due to the low counting statistics, the allowed no. of bins of the histogram was limited to 50 corresponding to a bin width of 1.8 ms, sensibly far from the COS-B experimental time resolution of 0.5 ms. The result of the application of the cluster technique is shown in the figures 2 and 3. Fig.2 shows the experimental distribution of clusters as compared to that expected from uniformly distributed phases in the interval (0,1). Fig.3 shows

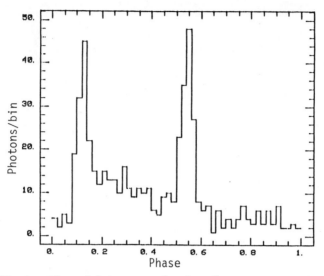

Fig.1 - Phase histogram showing the structure of the pulsed gamma-ray emission from the Vela pulsar as observed by COS-B from Oct. 20 to Nov. 8, 1975.

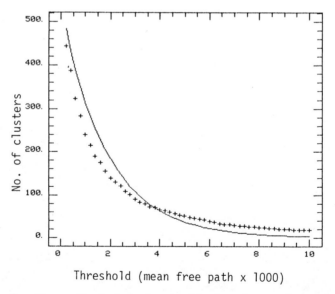

Fig.2 - Theoretical (continuous line) and experimental (crosses) cluster distributions for the Vela pulsar case.

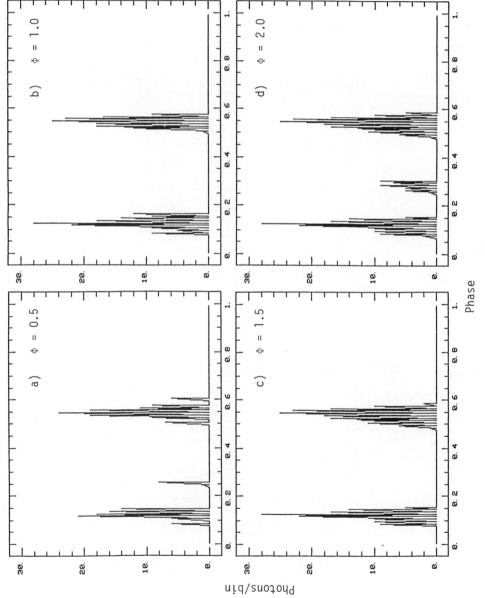

Fig.3 - Pulsed gamma-ray emission from the Vela pulsar. Structures found by the clustering technique at four values of the threshold Φ.

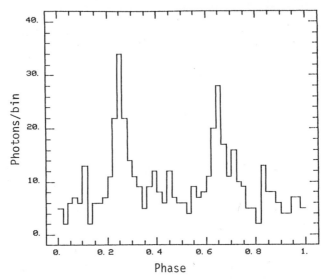

Fig.4 - Phase histogram showing the gamma-ray pulsed emission from the Crab pulsar (1975 COS-B observation).

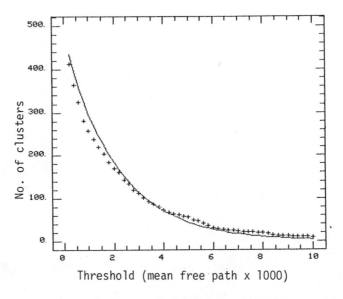

Fig.5 - Same as fig.2 for the Crab case (1975 observation).

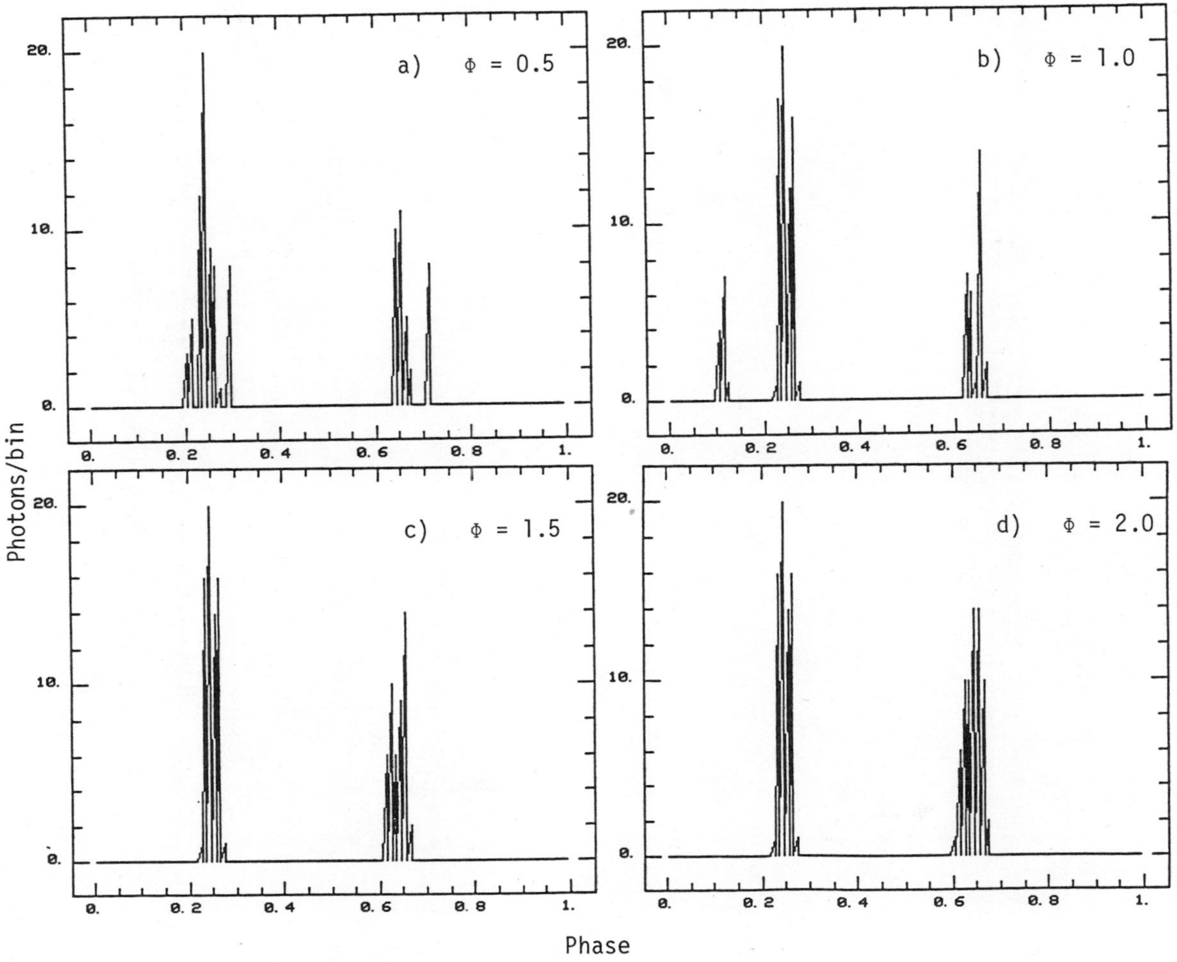

Fig. 6 - Structures found by the clustering technique in the pulsed gamma-ray emission of the Crab pulsar (1975 observation). The splitting of the main pulses into several narrower peaks has to be noticed.

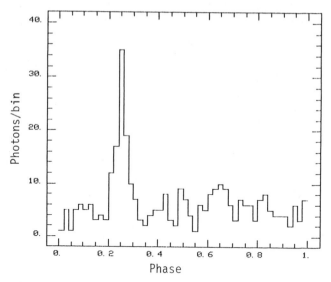

Fig.7 – Same as fig.4 but referred to the 1979 observation by COS-B of the Crab pulsar.

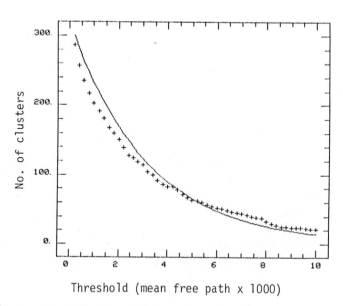

Fig.8 – Clusters distributions for the case of the Crab pulsar (1979 COS-B observation).

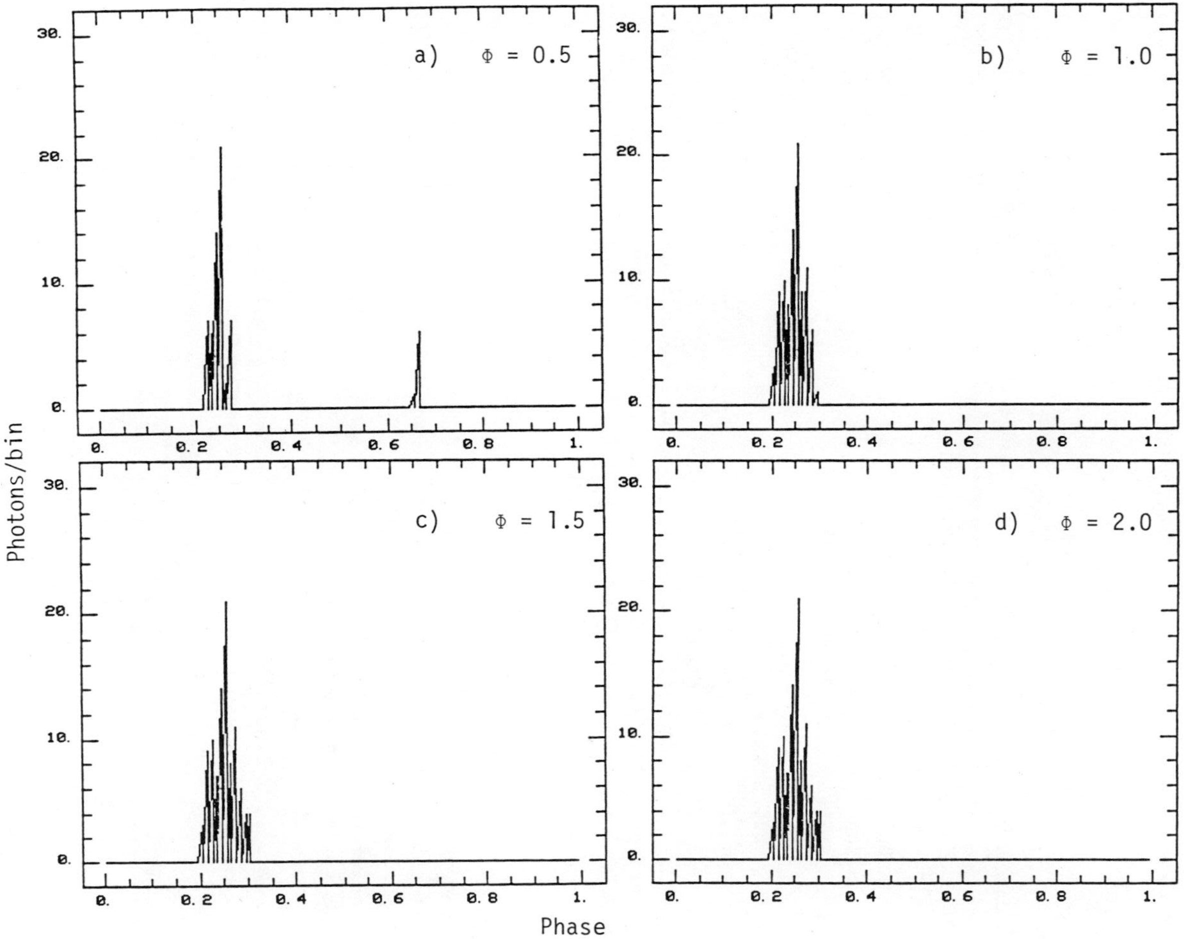

Fig.9 - *Structures found in the pulsed gamma-ray emission of the Crab pulsar (1979 COS-B observation).*

the "significant" clusters found. Apart the two main peaks, also showing up with the classical histogram (fig.1), a new feature is observed at approximately $\alpha=0.25$ consisting in a peak 0.05 wide (see fig.2d) enclosing a narrower peak (fig.2a). The histogram of fig.1 shows in the same phase interval a not resolved broad structure (interregion between the two peaks).

Figs.4 to 9 show the results of the same analysis for the case of the Crab pulsar as seen by COS-B in the 1975 and 1979 observations. Here again, while the conclusions reported by COS-B about the general structure of the Crab pulsed emission (see figs.4 and 7) are confirmed, the appearance of smaller scale, not yet known, structures is reported (see for example fig.5 relative to the 1975 observation).

It is planned to perform more detailed applications to all the available COS-B data aiming to confirm the present results.

References

R.Buccheri, K.Bennett, G.F.Bignami, J.B.G.M.Bloemen, V.Boriakoff, P.A.Caraveo, W.Hermsen, G.Kanbach, R.N.Manchester, J.L.Masnou, H.A.Mayer-Hasselwader, M.E.Oezel, J.A.Paul, B.Sacco, L.Scarsi, A.W.Strong
Astron.Astrophys., 128, 245, 1983

V.Di Gesù, B.Sacco
Journal of Pattern Recognition, Vol.16, N.5, pp 525-531, 1983

G.Kanbach, K.Bennett, G.F.Bignami, R.Buccheri, P.Caraveo, N.D'Amico, W.Hermsen, G.G.Lichti, J.L.Masnou, H.A.Mayer-Hasselwader, J.A.Paul, B.Sacco, B.N.Swanenburg, R.D.Wills
Astron.Astrophys., 90, 163, 1980

M.C.Maccarone, R.Buccheri, V.Di Gesù, this volume

L.Scarsi, K.Bennett, G,F,Bignami, G.Boella, R.Buccheri, W.Hermsen, L.Koch, H.A.Mayer-Hasselwader, J.A.Paul, E.Pfeffermann, R.Stiglitz, B.N.Swanenburg, B.G.Taylor, R.D.Wills
Proc.of the 12th ESLAB Symp. "Recent Advances in Gamma-Ray Astronomy", ESA SP-124, p.3, 1977

R.D.Wills, K.Bennett, G.F.Bignami, R.Buccheri, P.A.Caraveo, W.Hermsen, G.Kanbach, J.L.Masnou, H.A.Mayer-Hasselwander, J.A.Paul, B.Sacco
Nature, vol.296, no,5859, 723-726, 1982

C.T.Zahn, IEEE Trans Comput. c-20, pp.68-86, 1971

"MULTIVARIATE CLUSTER ANALYSIS OF RADIO PULSAR DATA"

M.C. Maccarone[1], R. Buccheri[1], and V. Di Gesu[1,2]

[1] Istituto di Fisica Cosmica ed Applicazioni dell'Informatica del C.N.R., Palermo, Italy
[2] Dipartimento di Matematica e Applicazioni, Università di Palermo Palermo, Italy

SUMMARY

Data relating to 336 radio pulsars are analyzed by a cluster method aiming to study the relevant features of the sample.
As a first result, few known selection effects are retrieved by the method thus encouraging more detailed applications in the future.

1. Introduction

Statistical studies of pulsar data have been undertaken by many authors aiming to infer the physical characteristics of the pulsar population. Such studies (see for example Lyne et al., 1985 and references therein) have been done using the pulsar data collected in the framework of few large surveys made by radiotelescopes located in different sites (Arecibo-USA, Jodrell Bank-UK, Molonglo-Australia,...) and having different sensitivity chacacteristics in relation to the various measurement parameters (flux, pulsation period, distance...). The inference on the intrinsec characteristics of the parent pulsar population was done by using the a priori knowledge about the surveys (range of periods and fluxes and regions of galaxy investigated).
The retrieval of informations related to the classification of the objects and the description of such classes has not been so far an objective of statistical studies of pulsar data. We plan to investigate the possibility to recover such information by the use of non parametric methods; we believe in particular that the multivariate cluster analysis can be a powerful tool for the study of the features of the available sample of pulsar data.
In this paper we give the results of the application of a clustering technique to a sample of 336 radio pulsars aiming to test its potentiality to select and describe different classes of objects in the data sample. We found encouraging that some known selection effects, characteristic of the surveys from which the data derive, have been detected by the method. We plan to perform more extensive applications

oriented to detect possible fine structures in the sample as resulting from the overlapping of different classes of objects.

2. Description of the method

Given a sample of N points (nodes) in a d-dimensional space, the Clustering methods aim to select the structures present in the sample by using a "distance" function defined on the parameter space. The technique here adopted is based on a multi-dimensional single-link algorithm: the data are seen as a random undirected weighted graph where the weight is the euclidean distance between the nodes. These are linked together by a Minimum Spanning Tree (MST) i.e. the spanning tree with minimum weight (the sum of the weights of the edges in the tree, see C.T.Zahn, 1971). The clusters are then derived by cutting all the edges of the MST which weight is greater than a given threshold. Finally the cluster distribution is compared with that expected from uniformly distributed data (Di Gesù and Sacco, 1983, 1986).
The expected number of clusters NC is a decreasing function of the threshold; in the case of a Poisson uniform distribution of the nodes, this function is given by

$$NC = 1 + (N-1) \times \exp(-\lambda V_d(\phi)) \qquad (1)$$

where λ^{-1} is the mean free path and $V_d(\phi)$ is the equivalent hypervolume at the threshold ϕ in the d-dimensional parameter space.
The comparison between the shape expected in the case of uniformly distributed data, as described by equation (1), and the corresponding experimental cluster distribution found by the procedure, allows to detect the presence of non-statistical structures in the input data base. These structures are subsequently described by identifying the elements of the clusters obtained at various thresholds; the use of small thresholds will give informations on fine structures while large scale disuniformities will be visible at high thresholds.
The visual representation of the detected classes is only possible when the number of parameters is smaller than four. In other cases the description can be done by the use of dendrograms. These are rooted trees where the clusters detected at the i-th level (nodes) are partitions of those detected at the (i-1)-th level. The value of the threshold at which the partition occurs is an indicator of the "peculiarity" of the object with respect to the average properties of the sample (see Murtagh,1984)

3. The data base used

The data sample used in the present application derive from a compilation of pulsar data obtained from few surveys devoted to pulsar discoveries and was made available to us by the courtesy of R.N.Manchester.

Such a "catalog" (the latest version was published by Manchester and Taylor, 1981) contains the parameters relative to 336 pulsars. The values of some parameters are always present in the catalog because their measurement is inherent to the discovery process: this is true for coordinates, pulsar period, dispersion measure and, in dependence of the radio observing frequency (generally 400 and 1400 MHz), flux and pulse width. A value for the period derivative is not always present since it derives from several other observations after the discovery. Other parameters (distances from the sun and from the galactic centre, height above the galactic plane, age, luminosity, braking power and surface mangetic field) are calculated on the basis of the current models about the pulsar evolution and the distribution of electrons in the galaxy.

For the present analysis we have used only the parameters obtained during the observations: the main reasons for this choice are i) to reduce as much as possible the bias connected with the use of theoretical models and ii) to have a fast answer on the potentiality of the method by using a restricted number of parameters. Concerning the period p and the period derivative $\dot{p}$, the range of observed values covers several orders of magnitudes and therefore is too large for a meaningful analysis. We have in the following used the decimal logarithm of p and $\dot{p}$ thus limiting our attention to orders of magnitudes instead of values. The results must be confronted with this choice.

It must be stressed that the data come from many different experiments aimed to look at different regions of the galaxy with different sensitivities; we therefore expect severe asymmetries in the data sample, which must be taken into account when drawing the conclusions about the results. We expect, for example, biases against the pulsar period due to the particular data sampling used in the discovery process and against the dispersion measure DM because of the interstellar scattering which reduces the visibility at high distances. Furthermore we expect strong asymmetries in the spatial distribution of the pulsars of the present catalog due to the different coverage of the sky by the different surveys.

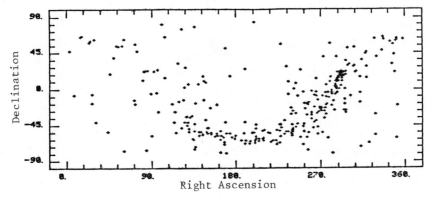

Fig.1 - *The distribution of observed pulsars in celestial coordinates*

4. Results and conclusions

Fig.1 shows the spatial distribution of the pulsars used in the present analysis. When the clustering algorithm is applied to the 2-dimensional space "Right Ascension-Declination"(RA-D), we obtain the distribution of clusters shown in fig.2.

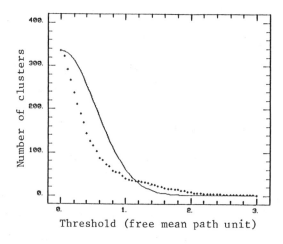

Fig.2 - Theoretical and experimental cluster distributions relative to the parameter space RA - D.

This distribution (dots in the figure) does not agree with that expected from uniformly distributed data (continuous line in the figure); this may be due to the asymmetries of the sample discussed in sect.3.

Fig.3 shows the clusters with more than 5 pulsars at four thresholds; the choice of 5 objects as a limiting threshold to show the most relevant clusters has been done on heuristic basis and has not a statistical meaning. The appearance of structures at smaller and smaller scale is visible when the threshold is reduced. In particular the galactic plane structure is visible in fig.3b while the cluster shown in fig.3d corresponds to the intense observing activity of the Arecibo radiotelescope.

Figs.4 to 6 show the results of the same analysis but in the 2-dimensional space "Dispersion measure-Period"(DM-p). Selection effects due to the reduced sensitivity at too low and too high periods and too high DM values is evident.

Figs.7 to 9 refer to the 2-dimensional space p-$\dot{p}$. The non uniformity of the sample is evident also in this case where structures at various scales are visible when different thresholds are chosen to cut the MST. The estimate of the significance of these structures must be done using the proper statistical tools and their interpretation in the framework of the pulsar evolution. It is not our aim to discuss these items; it is, however, interesting to notice the microstructures appearing at low thresholds, where significantly more clusters were expected from uniformly distributed data (see fig.8).

Fig. 10 is the result of the application of our technique to a 5-dimensional space defined by the coordinates, the period, the dispersion measure and the flux measurement at 400 MHz. The faster-than-expected clustering of the objects at low thresholds is due to the presence of many microstructures in the data, perhaps due to the

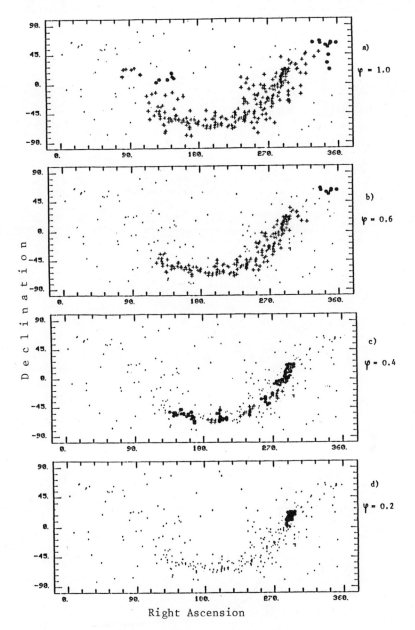

Fig.3 - Relevant clusters found in the RA-D space at various thresholds ϕ

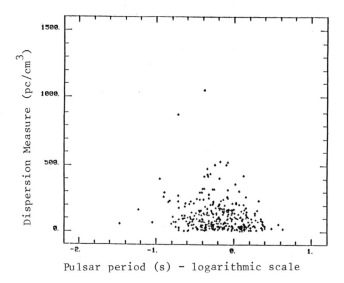

Fig.4 – *Period – Dispersion Measure plot.*

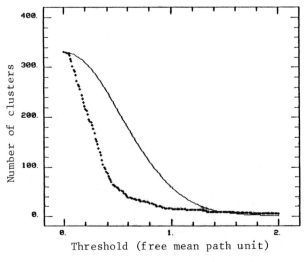

Fig.5 – *Clusters distributions relative to the parameter space P – DM*

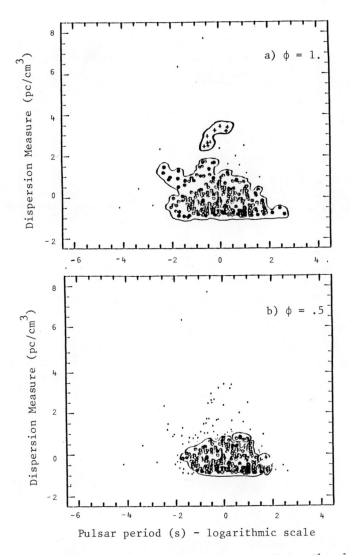

Fig.6 – Relevant clusters found in the DM-P space at two threshold values. For representation purposes the coordinate scale is built using parameters values referred to their mean and with unit standard deviation. In fig.6a the crosses refer to a cluster of pulsars observed in the course of deep surveys.

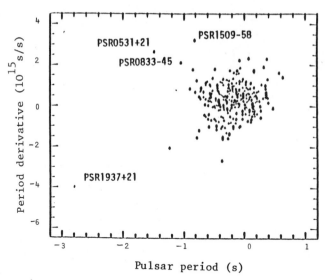

Fig.7 - P-Ṗ plot (logarithmic scale); some peculiar objects are indicated.

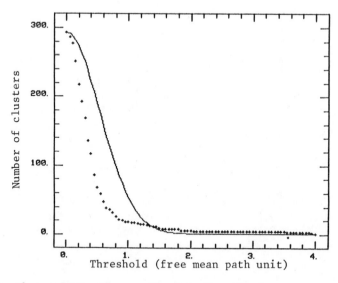

Fig.8 - Theoretical and experimental distributions of clusters in the parameter space P-Ṗ (few isolated points have been excluded from the analysis).

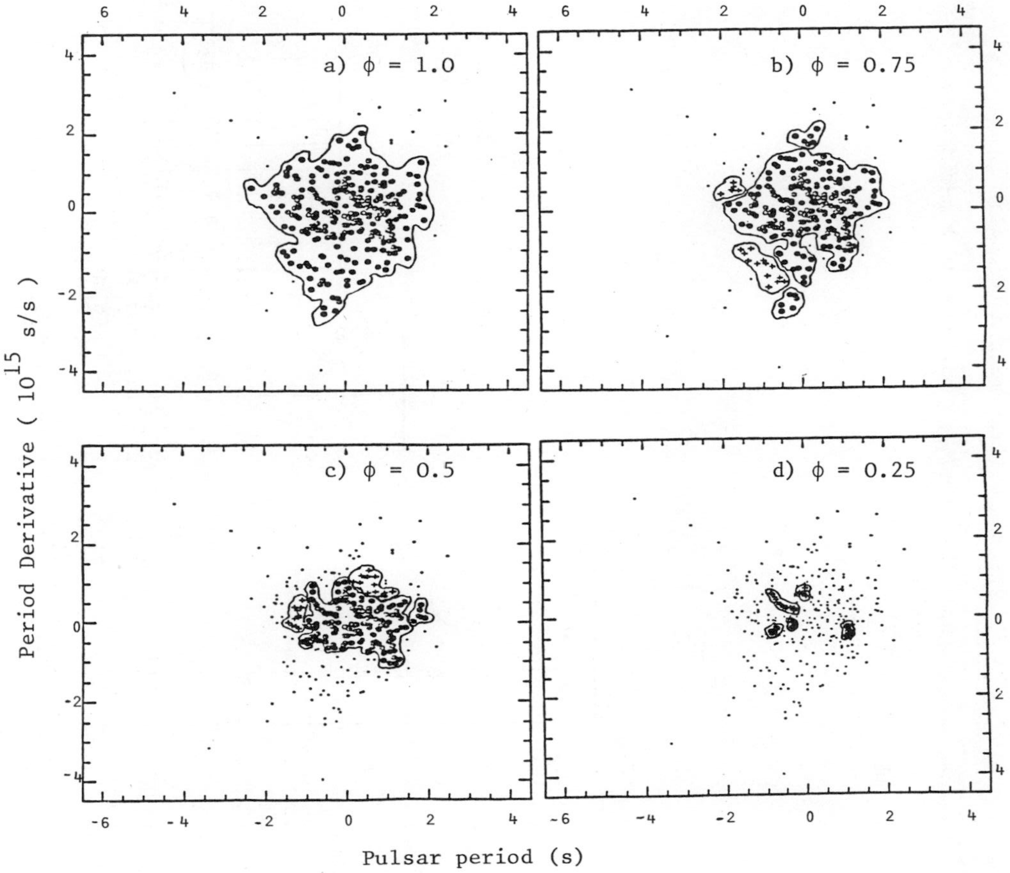

Fig.9 - P-Ṗ parameter space; structures revealed by the method at four threshold values.

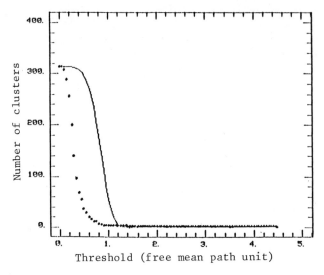

Fig.10 - Cluster distributions relative to a 5-dimensions parameter space (RA, D, DM, P, flux).

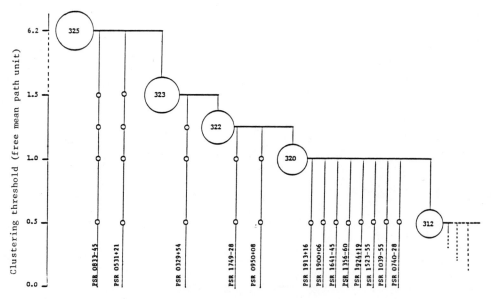

Fig.11 - Upper part of the dendrogram relative to the clusters detected in the case of fig. 10.

non-coherence of the data and possibly to the presence of different physical classes of pulsars in the sample.

Fig.11 shows the upper part of the dendrogram relative to this case. It is interesting to notice that pulsars like the Crab (PSR0531+21) and Vela (PSR0833-45), known to be the only two "young" objects of the catalog, are immediately isolated from the sample. When the threshold is further reduced, other known interesting pulsars are isolated (for example PSR0950+08, a very near-by pulsar, PSR1913+16, a binary system including two neutron stars, PSR1641-45, a glitching pulsar, etc...) thus revealing the potentiality of the method to select different classes of pulsars from the sample catalog.

We plan to perform in the next future more detailed applications and to define the suitable statistical tools such to detect and describe the clusters of pulsars that significantly deviate from the expectation. This would give us indications of the presence in the catalog of underlying structures compatible with the presence of physically different classes of pulsars.

References

V.Di Gesù, B.Sacco
Pattern Rec., Vol.16, n.5, p.525-531, 1983

V.Di Gesù, B.Sacco
Pattern Rec., submitted, 1986

V.Di Gesù, M.C.Maccarone
Pattern Rec., vol.19, n.1, p.63-72, 1986

A.G.Lyne, R.N.Manchester, J.H.Taylor
Mon.Not. R. astr.Soc., vol.213, p.613-639, 1985

R.N.Manchester, J.H.Taylor
Astron.J., vol.86, no. 12, p.1953, 1981

F.Murtagh
Discrete Applied Math., 7, 191-199, 1984

C.T.Zahn
IEEE Trans. Comput. C-20, p.68-86, 1971

AUTOMATIC PROCESSING OF VERY LOW-DISPERSION SPECTRA

P. Schuecker, H. Horstmann, and C.C. Volkmer
Astronomical Institute of Muenster University

The microdensitometer PDS 2020GM of the Muenster Astronomical Institute is part of ADAS (Astronomical Data Analyzing System) managed by ADAM (Astronomical Data Analyzing Monitor, Teuber 1984). The PDS 2020GM is a fast machine, able to measure large photographic plates (500mm x 500mm) with high accuracy, while the ADAS-software performs all functions, including on-line processing under the supervision of ADAM. It is feasible to obtain information on large numbers of astronomical objects within reasonable times. The bulk of data resulting from the digitization of Schmidt plates (a few GBytes per plate), requires that all procedures must work without supervision by the astronomer. This leads to fully automated working procedures, simulating human skill in visual perception. In this context, pattern recognition, expert systems and rule-based systems are among the most powerful tools. Programs relying on these techniques provide much flexibility and are widely applicable. In order to produce the appropriate software we introduced the theory of trainable classifiers into astronomy. Trainablility offers a way to increase program performance without the need for higher-level computer language like LISP or PROLOG. Several procedures developed for automated processing of very low-resolution spectra are illustrated with examples taken from the application to the classification of M-type stars and redshift measurements of normal galaxies.

BASIC PROBLEMS

Pattern recognition provides a general scheme for processing digital images (Pavlidis 1977). For astronomical applications this scheme had to be modified. The modules "object filter" and "quality filter" were instituted in addition to the commonly used modules quantization, segmentation (object isolation), pre-processing (scene analysis - calibration, filtering etc.), shape description (feature extraction) and object description (classification).
The object filter separates astronomically classifiable objects (single spectra) from astronomically not classifiable objects (plate flaws, artefacts, multiple spectra etc.). The quality filter separates over-, well- and under-exposed spectra. This is necessary, because over- and under-exposed spectra contain less information. The complete scheme is given in Fig. 1.

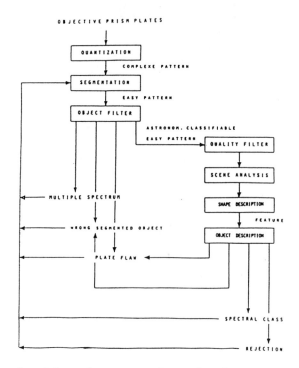

Fig. 1. Scheme for processing objective prisme plates

SEGMENTATION

The first step in processing digitized Schmidt plates is segmentation or object isolation. The results of this procedure are

- information on local background density and plate noise
- coordinates of all isolated objects
- various moments for all objects (measures of brightness, shape etc.)
- segments (frames), each including one object for further processing.

In most cases the general size and shape of an object to be isolated on direct or objective prism plates are known. This prior knowledge can be used for the segmentation process (top-down segmentation). Sometimes the classification of objects into diffuse (galaxies) and not diffuse objects (stars) is necessary. This can be done in a two-dimensional feature space with the axes "central-object density" and "effective radius". For more details see Horstmann (1986).

Fig. 2 is a plot of all stars and galaxies found automatically on a direct SRCJ-Atlas plate in a 11'x 11' field centered at RA=0^h54^m3, Decl.=-28°46'. The presentation is in form of ellipses whose sizes are measures of object brightness and whose orientations correspond to the position angles of the major axes of the galaxies. Fig. 3 gives spectral density profiles from the corresponding section of a UK-Schmidt objective prism plate, found by the same program as that employed for the search of stars and galaxies on direct plates. Spectra marked G belong to objects automatically classified as galaxies on the direct plate.

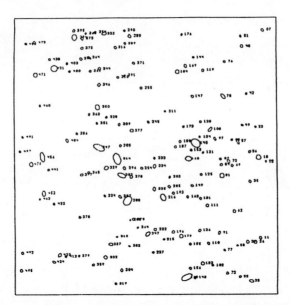

Fig. 2. Identified stars and galaxies in the field RA=$0^h54^m.3$, Decl.=$-28°46'$

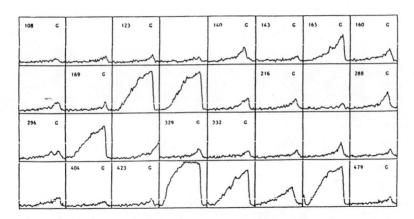

Fig. 3. Spectral density profiles from a UK-Schmidt objective prism plate (emulsion: IIIa-J, dispersion: 246 nm/mm at H_γ)

OBJECT FILTER FOR VERY LOW-DISPERSION SPECTRA

The object filter separates astronomically classifiable and astronomically not classifiable spectra. For this purpose the object frame is subdivided into five cells. Cells Z^1 to Z^4 are peripheral cells, Z^5 is the central cell. E(Z), S(Z) and R are functions characterizing the cells:

E(Z) = 1 : cells including (parts of) the object
S(Z) < 1 : artefacts
R   < 1 : large deviations from Gauss fit to marginal density distribution (perpendicular to dispersion) inside frame.

Only objects with E(Z)=0 for Z^1 to Z^4, $S(Z^5)>1$ and R>1 are astronomically classifiable spectra (Fig. 4).

	$\epsilon(2^h)$	$\epsilon(2^h)$	$\epsilon(2^h)$	$\epsilon(2^h)$	$s(2^h)$	R
	1	0	0	0	>1	>1
	0	1	0	0	>1	>1
	0	0	1	0	>1	>1
	0	0	0	1	>1	>1
	1	0	0	1	>1	<1
	0	1	0	1	>1	<1
	0	0	0	0	>1	<1
	1	1	0	0	>1	>1
	0	0	0	0	>1	>1
	0	0	0	0	<1	>1
	1	1	0	0	<1	<1
	0	0	1	1	>1	>1

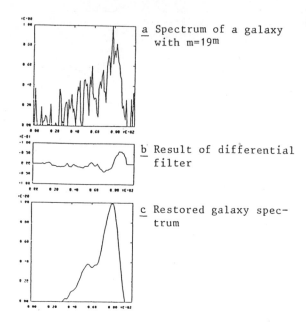

Fig. 5. Pre-processing of faint galaxy

a Spectrum of a galaxy with m=19m

b Result of differential filter

c Restored galaxy spectrum

Fig. 4. Features of object filter

PRE-PROCESSING OF SPECTRA

The main purpose of pre-processing is to enhance the relevant features. In the case of objective prism spectra, intensity calibration, filtering, normalization, rectification of the continuum etc. are often used. An example is given in Fig. 5 where a galaxy spectrum in three different states of processing is shown:

<u>a</u> normalized density profile of a galaxy with m=19^m
<u>b</u> average slope in the vicinity of each point in <u>a</u>, obtained by differentail filtering
<u>c</u> galaxy spectrum restored by numerical integration of <u>b</u>.

This technique of pre-processing was used for redshift measurements from galaxy spectra (see below). The importance of pre-processing is evident in the case of noisy objects.

TRAINABLE CLASSIFIER

Trainable classifiers are often used in pattern recognition. They are decision-making algorithms which are able to learn: with increasing information input they improve their performance. In the case of Bayesian classifiers (optimal statistical classifiers), training is defined as the procedure of estimating probability density functions (pdf) from a representative sample of prototypes with known class membership (Sklansky et al. 1981). Once the pdfs are known, the system works as a maximum likelihood classifier, assigning the object to the group with the highest probability. The probability that the observed feature value c is associated with the group w_i is

$$p(w_i;c) = p(c;w_i) p(w_i) . \qquad (1)$$

The joint probability $p(w_i;c)$ is the product of the "a posteriori" (constituent) probability $p(c;w_i)$ and the "a priori" probability $p(w_i)$. $p(c;w_i)$ can be estimated by a training procedure, for $p(w_i)$, in most cases, only rough estimates are possible. Therefore, operating characteristics are needed which describe the classification error under varying $p(w_i)$. The classification is then performed in two steps:

- estimate the joint probability $p(w_i;c)$ for all groups
- choose the group which maximizes $p(w_i;c)$.

Two kinds of training are possible:

- parametric training (PT)
  the pdfs are described by a family of functions (normal distributions etc.)
- non-parametric training (NPT)
  the pfds are described by the prototypes themselves (histograms etc.).

PT depends on a statistical model, NPT requires no assumptions. One possiblity of NPT to estimate the pdfs is:

$$p(c;w_i) = \frac{m_i \, n \, \Gamma(n/2)}{2 \, r^n \, \pi^{n/2}} \quad , \tag{2}$$

$$m_i = \sqrt{N_i} / N_i \tag{3}$$

N_i = number of members in group w_i of the sample
n = dimension of feature space
r = radius of a hypersphere in feature space where m_i objects are found.

This procedure leads to the required smooth distributions (eq.2: see also Loftgaarden et al. 1965; eq.3: Niemann 1983).

The a posteriori probabilities are also useful for finding the best feature. For example, the Bhattacharyya metric (Kailath 1967)

$$G_{ij}^B = - \ln \left\{ \int (p(c;w_i) p(c;w_j))^{1/2} dc \right\} \tag{4}$$

is the distance between classes w_i and w_j in feature space. The best feature is that, which maximizes (4); (see also Fig. 7).

APPLICATION 1: CLASSIFICATION OF M - TYPE STARS IN STELLAR AGGREGATES

Fig. 6 illustrates the classification problem with density profiles of low-resolution objective prism spectra taken in the near infrared (610-890nm, dispersion 400nm/mm at the atmospheric A-band). The rows show spectra of different spectral types (from left to right: late M - type stars to early-type stars) but nearly equal densities. The columns show spectra of equal spectral types but different densities. The molecular absorption dips, caused by TiO and VO, influence the shapes of the spectra in a systematic way: the later the spectral type the steeper the profile. From Fig.6 it is also seen that there is a tendency to classify under-exposed spectra (rows 6 and 7) systematically too late, over-exposed spectra (rows 1 and 2) too early.

To avoid these classification errors, one must use either

- intensity instead of density profiles or
- restrict the comparision between standard and unkown spectra to objects of nearly equal densities.

The low dispersion together with the high noise limits the number of classification groups to three:

- MK (early-type stars to early M-type stars)
- M (intermediate M-type stars)
- ML (late M-type stars).

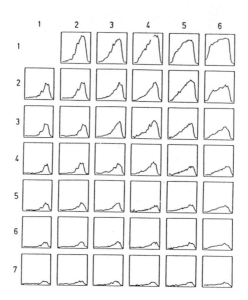

Fig. 6. Spectral density profiles from Rozhen (Bulgaria)-Schmidt objective prism plates (emulsion: I-N; dispersion: 400nm/mm at the A-band)

Two feature are tested:

- spectral ratios and
- cross correlation

between standard and unkown spectra. The resulting features are vectors in a multi-dimensional feature space. Because the components of these vectors form almost everywhere monotonic functions, the integrals over the vectors in the wavelength range 650-810nm were calculated. This reduces the feature space to one dimension.

With a training set of 100 prototypes the training procedure described above was applied to both features. The results are illustrated in Fig.7. It is clearly seen that the ratio (c_{sv}) is a better classification criterion than the cross correlation (c_{kk}).

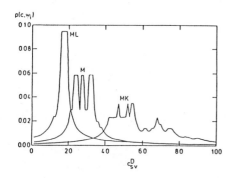

Fig. 7a: Feature space for the statistical classification c=spectral ratio

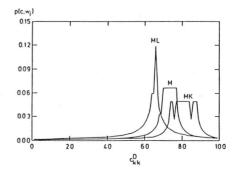

Fig 7b: Feature space for the statistical classification c=cross correlation

The spectral ratio was used for the classification of 3300 objects in the regions of NGC 7000 and M42. 253 M-stars identified in these areas are listed in a catalogue and atlas (Schuecker 1986), 219 M-stars was found in the field of NGC 7000 and 34 in the field of M42.

114

APPLICATION 2: INTERACTIVE REDSHIFT MEASUREMENTS OF NORMAL GALAXIES

Spectra like those given in Fig.3 were used for the determination of galaxy redshifts. They cover the wavelength range 320-540nm and have a dispersion of 246 nm/mm at H_γ. Simulations were carried out with a model atmosphere of T_{eff} = 5500K, g = g_o and solar metal abundances (Kurucz 1979). After application of the relevant dispersion curve and atmospheric extinction values, it is seen that the following absorption dips are useful for redshift measurements:

- CaII 393.4nm, 396.8nm or
- MGII 279.5nm, 280.2nm or
- FeII 258.6nm, 263.1nm.

The emulsion cutoff at 538nm of the IIIa-J plate serves as the reference wavelength. Measureable changes occur up to at least a redshift of z=1, which is the value reached for galaxies of M = -24 at the plate limit 19^m and H_o = 100 km $sec^{-1} Mpc^{-1}$, q_o = 0.5 .

The position of the red cutoff is defined through the half maximum density point on the cutoff slope. The positional accuracy is of the order of 3nm on the wavelength scale (Nandy et al. 1977), corresponding to dz = 0.015 . Astrometrically obtained positions from direct plates with a potential accuracy of dz = 0.001 will be used in the future. The spectrum shown in Fig. 5 is that of a galaxy with z = 0.13 . The measuring error is uncertain, due to the lack of slit spectra for comparison. A safe, though exaggerated margin, is dz = 0.03 . The measuring process will be automated in the near future.

The method illustrated above was used for the determination of redshifts of faint galaxies in the test field Fig. 2 . For 15 of the galaxies found on the direct plate, spectra were detected on the objective prism plate. 13 of these had measurable redshifts. The results are given in the accompaning Table of Fig. 8 . All galaxies are so faint that comparison with published data is not possible. Slit spectra of faint galaxies obtained with large telescopes are needed.

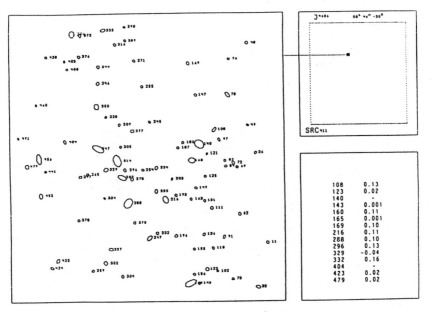

Fig. 8. Identified galaxies of the field RA=$0^h 54^m.3$, Decl.=-28°46'

REFERENCES

Horstmann, H.: 1986, Mitt. astron. Ges., in press
Kailath, T.: 1967, IEEE Trans. Communication 15, 52
Kurucz, R.L.: 1979, Astr-phys.J. Supp. 40, 1
Loftgaarden, D.O.; Quesenbury, G.P.: 1965, Ann. Math. Stat. 36, 1049
Nandy, K.; Reddish, V.C.; Tritton, K.P.; Cooke, J.A.; Emerson, D.: 1977, Mon. Not. R. astron. Soc. 178, 63P
Niemann, H.: 1983, "Klassifikation von Mustern", Springer-Verlag, Berlin
Pavlidis, T.: 1977, "Structural Pattern Recognition", Springer-Verlag, Berlin
Schuecker, P.: 1986, Diplomarbeit, Astron. Inst. Univ. Muenster
Sklansky, J.; Wassel, G.N.: 1981, "Pattern Classifiers and Trainable Machines", Springer-Verlag, New-York
Teuber, D.: 1984, Proc. of the International Workshop on Data Analysis in Astronomy, (eds.: De Gesue, V.; Scarsi, L.; Crane, P.; Friedman, J.H.; Levialdi, F.), Erice (Italy), 253

DATA HANDLING AND SYSTEMS DEDICATED TO LARGE EXPERIMENTS

Chairman : P. Crane

VLA: METHODOLOGICAL AND

COMPUTATIONAL REQUIREMENTS

> Pat Moore
>
> National Radio Astronomy Observatory*
> P.O. Box O
> Socorro, New Mexico 87801   USA

The Very Large Array (VLA) is an aperture synthesis radio telescope consisting of 27 antennas, each of 25m diameter. See, for example Napier et. al. [1] for a detailed description of the instrument. An instrument of this power and versatility is capable of an extremely high rate of data acquisition as will be described. Worse still, the data collected are not in the form of images (the form required by astronomers), but their Fourier transform. Details of this Fourier transform relationship is given by Bracewell [2], and Fomalont and Wright [3]. These two circumstances conspire to produce a unique data processing problem of considerable magnitude. This paper first describes the computing problems posed by the VLA, then goes on to discuss the methodologies that have been used to tackle them. Finally there is an outline of unresolved problems, and possible paths open in the future to help solve them.

Data Acquisition at the VLA

In order to appreciate the magnitude of the computing requirements of the VLA, it is necessary to understand some basics about the functioning of the instrument. The sensitivity of an interferometer increases as the observing bandwidth increases, so the hardware is designed to maximize this value. Each antenna produces 4 independent channels each with 50 MHz bandwidth. These analogue signals are all fed into the central electronics room, making a total bandwidth from all antennas of 5.4 GHz. These signals are immediately digitized at the Nyquist rate using 3 level (1.5 bit) samplers. These sampling rates and levels are a compromise between performance and cost. The total resultant digital data rate is 2 Gigabytes/sec.

---

*The National Radio Astronomy Observatory is operated by Associated Universities, Inc. under contract with the National Science Foundation.

These digitized signals are fed into the correlator, which forms the heart of the interferometer. This correlator combines signals from every pair of antennas (baseline) by performing a complex multiply followed by a time average. This correlation step first increases the data rate (27 antennas can produce 351 baselines), and then decreases it again with the time average. The choice of this time average period (the integration time) is again a compromise between performance and cost. Short integration times are necessarily for large fields of view and transient phenomena, but necessarily increase the data rate. The minimum integration time practical with the existing correlator hardware is 3.3 sec. The exact data rate after correlation depends on the details of the observing mode and correlator configuration, which are rather flexible. Two common configurations are typical of continuum and spectral line modes:

Continuum mode

    351+27   baselines
    4   polarizations
    2   spectral channels
    4   bytes (16 bit real and imaginary parts)

Spectral line mode

    351+27   baselines
    1   polarization
    512   spectral channels
    4   bytes (16 bit real and imaginary parts)

The extra 27 baselines are the results of antenna autocorrelations, which are required for accurate calibration. The resultant data rates are 12 and 774 kilobytes per integration respectively. With the minimum integration time these correspond to 3.6 and 232 kilobytes/sec.

The data is read out from the correlator by the online computer system. This imposes two major limitations. The minimum integration time is 10 seconds, and the number of correlation products is limited. This latter limit has no effect in continuum mode, but severely restricts spectral line mode. A typical spectral line configuration which runs at the maximum data rate allowed by the online computer system is as follows:

    300+25   baselines
    1   polarization
    32   spectral channels
    4   bytes (16 bit real and imaginary parts)

The resulting data rate is limited to about 4.2 kilobytes/sec. At this stage the data are archived on tape. The maximum data rate corresponds to a rate of about 15 1600 bpi tapes per day. Work is now underway to upgrade the online computer system to remove this limitation and allow the full bandwidth of the correlator to be realized.

Post Processing

After archiving, the data are first calibrated and edited. This is performed on a DEC 10 computer. Most of the steps required in this process are relatively straightforward, and involve few passes through the data. Current plans are to incorporate these calibration steps into both the on and offline computer systems directly, making this step unnecessary. The plan is to roughly calibrate the data online, and to keep a record of any changes made. The offline system will then be able to refine the calibration as necessary.

When the VLA was first proposed, it was expected that a simple Fourier transform would be the final step in the processing to produce an image. The quality of an image produced in this simple manner is degraded by two major factors. First, the aperture plane is rather poorly sampled by 27 antennas, and second, there are variations in atmospheric refractivity over each antenna. The maximum resulting dynamic range achievable is thus limited to about 100:1. More recently, however, techniques have been developed to counter both of these problems.

The effect of poor sampling of the aperture plane is to multiply the true observed visibility by a weighting function. This weighting function is simply the tracks made by the various baselines in the aperture plane as the Earth rotates. When this is Fourier transformed to the image plane it corresponds to convolving the true image with the Fourier transform of this weighting function, commonly called the dirty beam of the interferometer. This dirty beam has a central primary response, and fairly large and extended sidelobes. The configuration of the 27 antennas at the VLA has been expressly designed to achieve low sidelobes on this dirty beam, but they are still at a significantly high level. Fortunately, this dirty beam can be very accurately determined as it solely a function of array geometry. Consequently, deconvolution algorithms have been developed which work fairly effectively. The first of these is the CLEAN algorithm (Hogbom, [4]). This attempts to deconvolve the image into a number of point components by simply subtracting dirty beams centered on any flux visible in the image. It is an iterative procedure that works well for sources that are not too extended. The second is the maximum entropy algorithm. This states from the outset what it attempts to do, unlike CLEAN, and it is possible to make some changes to this "prior information". There are at least two variants of the algorithm in common use, Gull et al. [5] and Cornwell and Evans [6]. The basic idea of such algorithms is to maximize the entropy of the final image (i.e. extract the least contentious information) subject to the constraint that the image be compatible with the observed data. These algorithms also work very well, and are complementary to CLEAN in that they are most effective for very extended sources.

The effect of atmospheric refractivity is much easier to correct at radio rather than optical wavelengths. This is because the size scale on the sky for atmospheric irregularities (the isoplanatic patch) is larger than the antenna primary beam size for all existing observing frequencies. Consequently, the effect of these irregularities is to multiply the gain of each antenna by a single complex number. The amplitude of this represents the atmospheric attenuation, and the phase the signal phase delay. For each integration with the VLA there are 27 unknown complex gains, and 351 measured complex visibilities. There is some redundancy of information which means that given a rough model of the image and sufficient source flux to be above the noise level, it is possible to make a least-squares solution for the gains (Schwab [7] and Cornwell and Wilkinson [8]). The visibilities may then be recalibrated and a refined image produced. This refined image can in turn be used for more accurate self-calibration. It turns out that this procedure is remarkably insensitive to the initial model used, and reliable results are regularly obtained.

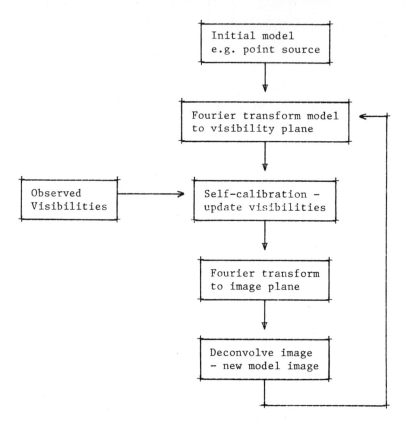

Fig 1 - Typical continuum mode post processing.

Fig 1 summarizes the sequence of operations typically followed to produce deconvolved amd self-calibrated images at the VLA. Note that this diagram shows two nested iterative procedures - the loop shown here, together with the iterative scheme implicit in the deconvolution procedure. This results in an enormous increase in the computing required over the simple Fourier transform imaging available when the VLA was first proposed. The computer system is not only acting as the "lens" of the VLA, but also removing its two most serious defects. The improvements possible with schemes such as the above are spectacular. Dynamic ranges of 10,000:1 are routinely achieved, and with care they can exceed 100,000:1. The increase in quality of the images has an enormous impact on their scientific value. As an example, Fig. 2 shows the improvement achieved for the radio galaxy Cygnus A (Perley et al. [9]).

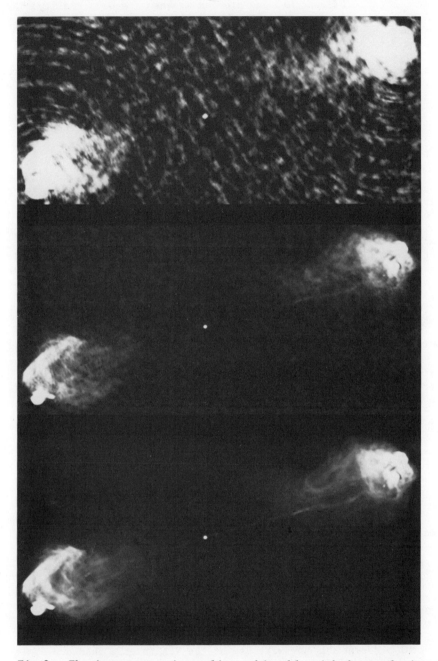

Fig 2 - The improvement in quality achievable with deconvolution and self-calibration. The top image is the raw Fourier transform image. The middle image is after deconvolution. The bottom image is after self-calibration.

An important feature of the algorithms used in radio astronomical imaging is that they require some interaction. The convergence must be monitored and there are several refinements, not shown in the diagram, that need careful control. One problem is ensuring data quality. The VLA hardware automatically flags bad data under many circumstances, but there are some events that defeat this. For example, interference can be very difficult to detect. In some cases it is not possible to fully judge the data quality without performing several iterations of self-calibration. The data must then be inspected, and perhaps edited to remove bad points. This interactivity has one very important effect - the algorithms must run sufficiently quickly so that human interaction is feasible.

Offline Computing Facilities

The offline computing facilities at the VLA fall into two categories. First are fairly conventional VAX computers with attached array processors. The VAX handles routine computations and manages the I/O, while the array processor handles some of the compute intensive algorithms much more efficiently. The software running on these systems, the Astronomical Image Processing System (AIPS), is portable to various hardware and operating system configurations. This system enables us to preserve our investment in algorithm development, as the algorithms typically outlast the system on which they run. The advantage of this portability was recently demonstrated by the purchase of a new CONVEX computer, which was very quickly running AIPS.

The second offline computing facility is a purpose built assembly of PDP 11 computers, array processors and memory. The various units have been assembled into a "pipeline" for processing VLA data that is particularly efficient when many spectral channels are present. This is achieved by having each part of the hardware dedicated to performing a small part of the computation in parallel with the others. Consequently this pipeline is much faster than the more conventional computers. It does, however, have two major drawbacks - the code developed for it is difficult to port to more modern hardware, and it is difficult to implement new algorithms. As an example, the self-calibration technique described earlier was developed after the pipeline was designed. It has not been possible to implement this algorithm on the pipeline.

Recent Developments

The offline computing requirements have grown far beyond the original concept of the VLA. The deconvolution and self-calibration algorithms together result in at least a two orders of magnitude increase in computational power over what is required for simple Fourier transforms. A number of recent developments threaten to increase this still further.

Several observations now make use of the VLA in more than one of its four configurations. These different configurations are different arrangements of the antennas to achieve different resolution and brightness sensitivity (Napier et. al. [1]). Combining these observations allows imaging over a large field of view with unprecedented detail. Images with 4096 * 4096 pixels have now been made. Existing computer facilities at the VLA are totally unable to handle these observations. As an example, the supernova remnant Casseopea A has been observed in this mode. The total observing time was 26 hours, resulting in 5 million complex

visibilities - a total of 340 MBytes. Initial calibration of this
database took several days, and initial I, Q and U polarization images
at limited 2048 * 2048 pixels resolution took 3 days on a VAX 11/780
and array processor. In order to achieve full resolution, time was
rented on a Cray XMP supercomputer, and AIPS ported to it. An I
polarization image at full 4096 * 4096 resolution took 10 hours of
time on this Cray XMP. The resulting image has 0.7 arcsec resolution
over an object 6 arcmin in size.

Spectral line observations can also saturate the available computing
facilities. Modifications to the on-line computers are in hand that will
allow a full 512 spectral channels to be recorded on all 351 baselines.
The previously two dimensional images now expand in a third dimension.
Fortunately, the data reduction problem does not scale linearly with
the number of channels for two reasons. The sensitivity in a single
narrow channel is reduced, and any self-calibration gains will usually
be constant across the band. Conversely, there are additional
calibrations to be performed to achieve a flat spectral response. In
addition, for sources with complex spatial and frequency structure it
may be necessary to perform a full three-dimensional deconvolution to
resolve ambiguities.

The VLA is now being equipped with receivers at 327 MHz, and 75 MHz is
being investigated. Moving to these lower frequencies raises an additional
set of problems. The first is caused by the increase of the size of the
antenna primary beam coupled with the increase in radio source brightness
at low frequencies. These two effects cause confusing sources to appear on
the images. This would not be harmful were it not for the typical
dirty beam for the VLA which has very extensive sidelobes. A confusing
source may be away from the region of interest, but its sidelobes will
not be. The only way to solve this problem is to map the whole antenna
primary beam to full resolution. The magnitude of this approach can be
minimized by first making a low resolution image to locate the regions
of emission, and then to only image these small regions to full
resolution, This works because the radio sky frequently consists of a
number of discrete sources with negligible emission between.

There are, however, additional problems with this large field imaging.
When the field is extended to cover the several degree primary beam
size of the antennas, it is not possible to make the approximation
that the sky is flat. This means that the simple two-dimensional
Fourier transform relationship breaks down. It is necessary to break
the image into a number of small fields, each sufficiently small that
the approximation is good. A second problem is that for these large
images, variations in atmospheric refraction become severe. The
isoplanatic patch becomes smaller than the image size, and antenna
calibration errors can no longer be represented by a single complex
gain. Again, breaking the image into small fields is probably the way
to solve this problem, making different self-calibration solutions for
each field. The magnitude of these effects on the VLA computer load is
still being investigated. It is clear that there will be a substantial
increase in the amount of computation required.

The Future

In order to try to make a detailed estimate of the offline computing
needs of the VLA, the following question was asked: "How would VLA
observing time be used if there were no offline computer limitations,
and what offline computing resources would this require ?" No changes
to the VLA hardware are envisaged, other than the online computer

upgrade to allow 512 spectral channels now being implemented. The details of this investigation are included in [10]. The major result of this is that an achieved throughput of 60 Megaflops is required. Note that this is "achieved" throughput, many manufacturers quote "maximum" throughput which is typically several times greater. This puts the requirements well into the regime of class VI supercomputers.

There are several possible ways to achieve this kind of throughput. A simple approach would be to duplicate existing computer systems many times. This could indeed achieve the required throughput, however, it denies the possibility of human interaction on large problems. As described earlier, this is an essential feature of some algorithms. Another possibility would be to build a "super-pipeline" tailored to these algorithms. This would, undoubtedly, be the cheapest in terms of the hardware costs. It does, however, have major disadvantages. The software costs to build such a machine are very large and require highly skilled individuals. Also, when the machine finally ends its useful life most of this software effort is lost with it.

The approach most favoured to solve computing problems at the VLA is to purchase an integrated supercomputer. This would have the performance required together with a simple architecture allowing algorithms to be easily coded in high level languages. It must have sufficient speed to allow true interaction, and languages suitable for easy migration to other machines when it becomes obsolete.

REFERENCES

[1] P. J. Napier, A. R. Thompson and R. D. Ekers, "The Very Large Array: Design and Performance of a Modern Synthesis Radio Telescope", Proc IEEE, vol. 71, no 11, page 1295, 1983.

[2] R. N. Bracewell, "Radio interferometry of discrete sources", Proc. IRE, vol. 46, page 97, 1958.

[3] E. B. Fomalont and C. H. Wright, "Interferometry and aperture synthesis", in Galactic and Extragalactic Radio Astronomy, G. L. Verschuur and K. I. Kellermann, Eds. New York: Springer, 1974, ch. 10, page 256.

[4] J. Hogbom, "Aperture synthesis with a non-regular distribution of interferometer baselines", Ap J Supp, vol. 15, page 417, 1974.

[5] S. F. Gull and E. Daniell, "Image reconstruction from noisy and incomplete data," Nature, vol. 272, page 686, 1978.

[6] T. J. Cornwell and K. F. Evans, "A simple Maximum Entropy deconvolution algorithm", Astron. Astrophys, Vol. 143, page 77, 1985.

[7] F. Schwab, "Adaptive calibration", Proc. S.P.I.E., vol. 231, page 18, 1980.

[8] T. J. Cornwell and P. N. Wilkinson, "A new method for making maps with unstable radio interferometers", Mon. Not. R. Astr. Soc, vol. 196, page, 1067, 1981.

[9] R. A. Perley, J. W. Dreher and J. J. Cowan, "The jet and filament in Cygnus A", Ap. J. vol. 285, page L34, 1984.

[10] NRAO staff, "A supercomputer for radio astronomical imaging", 1985.

PERSPECTIVES ON DATA ANALYSIS FOR THE SPACE TELESCOPE

Peter M. B. Shames

Space Telescope Science Institute
3700 San Martin Drive
Baltimore, MD 21218   U.S.A.

INTRODUCTION

The systems that are being created to provide data management and analysis facilties for the Hubble Space Telescope are similar to many that preceeded them, but also employ some techniques that are relatively new to astronomical data processing. These systems will be briefly reviewed, both from an overall view and from the user view. The various elements that have been used in the design and construction of the systems and the more important technological features will be discussed in some detail.

The user's point of view must be considered by a system designer from the beginning, yet it is one of the most changeable aspects of any system. The environment must provide users with access to all of the services and facilities that are available: the data catalogue and archive, calibration reference files and algorithms, and various hardcopy output devices. The analysis environment must be easy to use for a preliminary scan of the data, and accomodate the demands of sophisticated analysis scenarios. The system should ideally permit the user to customize the analysis and data reduction functions to match his understanding of the process.

The architectural elements that make up the system: the processors, networks, database and data archive sub-systems, display systems, and software modules are also important to consider. There are three key elements: distributed processors dedicated to specific functions; database and archive data management facilities; and portable software and data. Networks, both local area and wide-area, are the glue that holds such distributed systems together. Systems of distributed processors can be well matched to the tasks at hand, and clean interfaces allow these elements to be updated as dictated by changing processing requirements or technological advances.

It is significant to note that actual hardware elements may change rather quickly, driven by technological advances, but the software systems and data must gracefully survive these changes. The data, which is the reason for these projects, and the software systems, which are such a major investment in any large project, must

persist. The use of system-independent data formats and portable
software to protect this investment is well established in principle,
if not yet fully in practice. Portability of data and programs must
be considered in two different contexts: within a site as a way of
surviving changes over time; and between sites as programs and data
are distributed to the community.

OVERVIEW OF PROCESSING FUNCTIONS AT ST ScI

The systems at the Institute can be loosely grouped into three
separate sub-systems: the dedicated ground support systems; the
archive and output product systems; and the science support systems.
The primary ground support system which is called SOGS, for Science
Operations Ground System, was developed by TRW for NASA and contains
the planning and scheduling, real-time observation, pipeline
processing and calibration functions. The Proposal Entry Processor
(PEP) and Guide Star Selection System (GSSS) may also be considered as
a part of the ground support system though they are not really part of
SOGS.

The archive and output product systems provide services both to
the ground support system and the users, and in some sense are the
interface between these two worlds. These systems are the Data
Management Facility (DMF) and the film processor (FILM). The rest of
the systems at the Institute perform science support functions,
including data analysis, document processing, software development and
maintenance and related activites. The system that supports remote
proposal submission (RPSS) will also be briefly discussed.

The SOGS Ground Support System

The SOGS system is actually distributed between the Goddard Space
Flight Center and the Institute, but most of the Goddard functions are
duplicated at STScI and only the local ones will be discussed. The
Science Planning and Scheduling System (SPSS) is the first SOGS
function that science proposals come in contact with. There are
actually two prior systems, outside of SOGS, that proposals pass
through: the Remote Proposal Submission System (RPSS) and the Proposal
Entry Processor (PEP). These pre-processors were built at the
Institute to provide a more user-friendly front-end to the SOGS
system.

Once proposals have been submitted and passed through review by
the Time Allocation Committee (TAC), they are entered into SPSS where
they can be scheduled as actual observations. The schedule of
observations are passed to the Payload Operations Control Center
(POCC) for transmission to the spacecraft. The Observation Support
Subsystem (OSS) is the only one of the SOGS systems that deals with
the spacecraft in realtime. It can generate command requests that are
checked and relayed by the POCC. OSS also provides facilities for
supporting the small-angle maneuvers required by some instrument modes
and apertures, and for monitoring instrument performance.

Pipeline processing of data is done in the Post-Observation Data
Processing System (PODPS). PODPS is an automated processing system
that provides standard pipeline calibration and instrument signature
removal for all data that is taken by the system. Calibration
reference data, processing parameters and pointers to flat fields are
stored in a Calibration Data Base. The Calibration Data-Base System
(CDBS)[1] is responsible for processing calibration observations and
creating the reference files that are used in the pipeline.

Much of the data that is processed by PODPS will have hardcopy output products made, graphical or image output on film. These requests are generated by PODPS and passed to the FILM processor for handling. All of the data that is generated, both raw and calibrated, is archived in the Data Management Facility (DMF). Also archived are the reference files, calibration parameters, engineering data, and copies of the software so that the configuration of the system during data acquisition and processing can be re-created as desired.

The Data Management Facility

The DMF[2] system has several functions: archiving the science, calibration and engineering data that flows through the SOGS ground system; providing an on-line catalogue of this data and other data sets; and making these data available to researchers and other users. The DMF also serves as the access port between the archive data and both SOGS and non-SOGS components. It has the dual functions of providing user access to required data while preserving the integrity of the ground system.

Archive data will be preserved in two forms, on magnetic tape and on current technology optical disks. Although the data that is used within SOGS is stored in a "waivered FITS" form on disk, and VAX floating point is permitted for processing efficiency in the pipeline, the Flexible Image Transport System (FITS)[3] will be used as the standard data structuring technique in the archive, so that data portability may be ensured.

Users will be able to browse the on-line observation index and then issue requests for archived data. Any data that is available to the user (personal or public domain) will be returned to the user in a disk file accessible to the analysis system currently in use. Further discussion of data analysis system features will be presented from the user point of view in the next section.

The DMF provides local user access to the HST data archived at the Institute, a full copy of the data will be available on optical disk at the European Coordinating Facility in Garching FDR, and other archive copies are being considered elsewhere. Remote access to the data will be a consideration in the future but brings its own set of challenges because of limited wide area network bandwidths. Controlling access, providing sufficient network throughput (adequate bandwidth and data compression), and good remote user services are topics that have been identified for future investigation, but are not yet resolved.

The Remote Proposal Submission System

RPSS provides a set of functions to support remote submission of observing proposals, entered via a Telenet packet switched network link. Telenet was chosen because it is available via local phone call in most parts of the US and can be accessed from many foriegn countries using gateways from the local PTTs. The RPSS itself is implemented on a dedicated uVAX II workstation using captive accounts and a limited set of functions. It is a public service and appears to be a secure way of providing general access to a wide community.

Proposals that are to be submitted via RPSS are typed into a file in a fairly straightforward way and are then shipped to the system over the network link. Kermit[4] or a simple file dump procedure may be used. Proposals are run through a verification step, and once they pass, may be submitted for consideration. A signed cover page must be mailed to complete the transaction. RPSS also supports a simple information service and may be expanded at some point to offer other public services.

Science Support Systems

A Science Cluster, that presently contains two 8600's and a 780 is the heart of the science support facilities, augmented by a small but growing number of workstations. These systems handle all of the data analysis, image processing, software development and maintenance, document preparation, and other activities required by a typical research organization. While these systems are a substantial computer resource, each system typically has from 50-70 users logged in at any one time, performing a variety of different tasks. The 780, of course, only handles 15-20 users at any given time. These systems will also support HST Observers at the Institute.

The primary data analysis system for HST will be SDAS/IRAF (see below). In addition to the standard SDAS/IRAF system, a variety of other software packages are in use, with AIPS, IDL and MIDAS being the most common. The tasks performed range from normal scientific computing, through calibration and evaluation tasks, to functional work in preparation for operational activities. Since there are four software development groups and two maintenance groups at work (managing 800K lines of code), there is a large mix of compiles and links in progress as well. Several automated tools are in use to help manage all of this code (CMS, MMS and DTM), but while these make the developers' job easier they represent a system load in their own right. A broad mix of other tasks include modelling, TeX and Runoff processing, spread-sheet use, library catalogue management and all of the other typical functions.

A limited number of image display stations are now available, to be augmented in the near term with more standard image displays or with imaging workstations. Workstations are only just now starting to be used at the Institute, linked into the other systems via an Ethernet link running both DECnet and TCP/IP protocols. These systems are well suited for certain tasks, as will be discussed later, and are expected to see wider use in the future.

USER'S VIEW - THE ANALYSIS ENVIRONMENT

The Institute has developed a set of data analysis tools (Science Data Analysis System - SDAS)[5] which are embedded in an interactive analysis environment (Image Reduction Analysis Facility - IRAF)[6] that was developed at Kitt Peak. These two software packages together form the users' data analysis environment, both for online and offline data analysis. A variety of analysis packages are provided in the system, both instrument-specific tools and generalized utilities for image processing, analysis and display. The system provides standard analysis functions, and permits users to customize the tools to suit their own view of the analysis process.

The system can be used in command mode; new commands may also be created to perform a user specified processing action, the canonical "doit" routine. Beyond that it is possible for a user to bring existing code into the environment, either executed as a direct command, or fully integrated into IRAF. Both options are available, the latter being slightly harder to accomplish, but also more efficient. We have brought in subroutine packages from other systems, and over time we expect that other outside software will be added to the system to enhance its functionality.

In addition to the increasing number of standard SDAS algorithms, the analysis environment also supports other packages. All of the NOAO analysis routines supported within IRAF are available. Although some of these are quite specific to instruments at Kitt Peak, many are general purpose image and spectral processing packages of more general interest. Some routines from the MIDAS system are also provided and the SDAS packages use MIDAS table format interfaces to pass parameters among themselves; a table editor is also available.

The CDBS system, which runs mainly within SOGS, also uses IRAF and the SDAS packages as part of its environment. CDBS routines will be made available to the science user, along with calibration reference files, so that re-calibration may be performed as desired. Either the standard reference files or user derived files may be used for this operation. The standard reference files and parameters, archived science or engineering data, and other datasets may be requested from the DMF by invoking its User Interface (UIF) from within IRAF.

The analysis environment must support all of the functions that a user requires to perform data reduction tasks. This includes the analysis tasks themselves, access to archive data, handling of output product or image hardcopy requests, and managing a local catalogue of the data that the user is processing. All of these functions are to be accessible from within the IRAF/SDAS environment, but some of them (like the user catalogue) are now still in development. Because of the extensible design of the environment, a variety of interfaces and services can be made available while development of the full system progresses.

We plan on exporting copies of the software systems for a number of reasons, not the least of which being inadequate resources at the Institute to support all analyses of all of the HST data. SDAS is now available for distribution to sites running VMS, and IRAF may be requested from the development group at NOAO or from us. Periodic revisions, updates and enhancements are planned, as is a newsletter and clearinghouse for related information. All of the systems developed at the Institute are in the public domain, but only SDAS is expected to receive this level of support for public distribution.

DISCUSSION OF ARCHITECTURAL ELEMENTS

While many of the elements used to create the HST systems are familiar from other contexts, the use of this combination of them appears to be unique to the Institute. In this section four of the most significant elements are described and some of the considerations that enter into their use in our environment or in others are discussed.

Data Management Sub-Systems

Data management has become increasingly important as the amount of data increases, and as the cost of large-scale scientific experiments continues to climb, because the value of each bit of data is significant. Even on modest projects, there is an increasing awareness of the value of preserving copies of the actual data, observing modes and conditions, and other information, not just for the original researcher, but for others who may seek to do archival research. For the Hubble Space Telescope, a rather complete data management system is being prepared, and archival research is expected to be an important component of the scientific output of the facility.

DBMS and archival storage. The two essential elements of the HST data management sub-systems are database management systems (DBMS) and high density archival storage. Data-base management systems are used to maintain spacecraft configuration files, operational history, proposal tracking information, calibration reference data, observation indices, and many other status indicators. During all handling of observing proposals, data indicating the current status are maintained on-line in a relational database. All steps in processing are traced, from initial proposal entry and selection to scheduling, data acquisition, pipeline calibration, archival storage and subsequent use. An observation catalogue will be maintained on-line for public access, with descriptions of sources observed, instrument modes and other pertinent information.

Database management systems, whether hardware based as in our case, or based upon shared software systems, are essential to good data management practices. Modern database technology provides very good solutions for local database management and remote access, and now even distributed access systems are becoming available. A variety of topics relating to direct control over the database, protection of proprietary rights in data, wide access to the data and support of a public information facility deserve a broader discussion than is possible here.

In the past, archival storage has meant stacks of photographic plates or racks of tapes, but these media have not always been well labelled as to source or content and are typically (with rare exceptions) not accessible to or not very useful to any but the originator. For the HST data, the data itself will be preserved in an optical disk archive and complete records will be preserved describing the generation and processing of the data. A portion of this data will be on-line at all times (present plans call for 200 gigabytes to be on-line) with other disks available for mounting as required.

Optical disks offer some significant advantages over tapes in that they may be accessed randomly and have much higher storage densities, a factor of about 20 to 1 over magnetic tape. While the technology is still quite new, data storage lifetimes are now quoted at a conservative ten years, but longer storage periods are clearly to be expected. Data density and transfer speeds are also expected to increase above the current levels of 2 GB per double-sided platter and 200K Bytes/second. The doubling time is anticipated to be on order two years. Media costs are now roughly $250/GB, but these are also expected to drop in the near future as more production facilities come on-line.

Jukebox (robotic disk handler) technology can now provide random access to any of 100 or more platters in ten seconds or so, and the costs are quite reasonable. Use of a juke box minimizes manual handling of the data. Here too, developments are expected, but progress will be more rapid in units that can handle the smaller media than in large-scale archival storage.

Probably the most rapid area of change will be in smaller 5 1/4" devices, rather than the 12" devices we will use for our archives. These smaller devices now only store 200MB on their Write Once Read Many times (WORM) media, but they will develop rapidly because of the wider market for office level products. Such units are very well matched to work-stations and should find a use in many analysis environments. The small CDROM disks will also be of considerable interest as a way of "publishing" data catalogues and selections of interesting datasets.

Data access and distribution. Providing wide access to the data produced by the HST is part of the charter of the Institute. This access will be achieved both by supporting on-site researchers and by distributing data to other groups and individuals. To permit the broadest access to the data set, it is essential that the data (and the programs that analyze it) be widely distributed, so archive copies will be made available to other qualified sites.

Data distribution will take at least three forms: shipment of tapes and optical disks; remote access via networks; and creation of other archive data centers. There is now a European center for support of HST data; the Space Telescope - European Coordinating Facility (ST-ECF) is an ESA funded facility located at ESO in Garching FDR, that will provide access to the full HST archive set for European scientists. A full copy of the archive is to be available at that facility, and software to access it is being jointly developed at the Institute and at the ST-ECF. Other archive locations are being discussed.

A very important part of this data archive and distribution activity is the consideration given to portable data formats. The FITS image transport format has had wide acceptance inside (and outside) the astronomy community because it provides a system-independent means of storing and describing image data on tape. FITS has also grown to include techniques for storing data types (group data, tables) that were not envisioned in the original specs and has proven itself as a reliable data transport technique. All of the HST archive data will be distributed on tape in FITS format, to ensure data portability.

Optical disks, however, are not magnetic tapes, and yet a similar level of data portability is to be sought for this data medium too. We have defined a data-structuring technique for optical disks that provides many of the same features that FITS does, in a similarly portable way. This disk format supports a directory structure to identify disk contents so that advantage may be taken of the large-volume random access nature of the disk media. The disk directory is in ASCII for portability, the format is self-descriptive and extensible, and is not restricted to a single size or type of disk platter. If efficient access to the directory is required it will be necessary to create a copy on magnetic disk, but this is a small price to pay for portability. We encourage evaluation and use of this format by other groups as a way of providing data portability on this new data media.

Portable Software and Systems

Along with portable data, portable software is also of great importance. Hardware changes, systems evolve, users want to take analysis programs home to their own institutions; portable software can make this possible. The AIPS package developed at NRAO was the first major astronomical analysis system that actually achieved a reasonable level of portability, and has shown the way to other developers with similar needs. The Institute needed an analysis environment for the SDAS software it was developing, and, after a review of available systems chose the IRAF package that was then being developed at Kitt Peak National Observatory (now NOAO).

The SDAS software is a layered design, where the applications routines are separated from the underlying environment by a set of high level get/put interfaces that provide isolation from changes in system structure. These routines have been further layered, and a common set is now in use both in SDAS and the MIDAS system from ESO.

<u>The IRAF analysis system.</u> The IRAF system can be thought of as a third generation analysis environment, where the first generation is the classical lash-up of bits and pieces and the second is represented by carefully done, but non-portable systems. IRAF was designed to be fully portable from the beginning. It is coded in a pre-processed dialect of RATFOR that translates into FORTRAN-66, a widely supported language. Parts of the command language, but not the support software and applications, are written in C, a language well suited to systems level programming.

IRAF was designed to be highly portable, and is now operational on several computer systems including VAX VMS and UNIX, SUN UNIX and DG AOS. It uses the concept of a virtual operating system (VOS)[7] to provide isolation from underlying system details, and has proven to be quite easy to port to new systems. SDAS itself has been coded in standard FORTRAN-77 and uses a well defined set of interfaces to layer the programs from the operating environment. There are elements of VMS system structure which were permitted in SDAS, which will be excised during the conversion to UNIX that is now being contemplated.

Both the IRAF approach that uses a rather low-level (open, close, read, write) set of interfaces to a virtual operating system, and the SDAS approach that has higher functionality (get_image, put_keyword), permit a clean layer of isolation between the applications program and the system. As long as such interfaces are uniformly used, and system-specific language extensions are forbidden, portability can be achieved. In fact, the rules for generation of truly portable software are rather more complex and are properly the subject of another paper.

<u>ST and IDI interfaces.</u> The interfaces that are used by the SDAS applications programs were defined to provide a layer of isolation between the applications and the environment that they were to run in. At the time that SDAS was designed the environment was not well defined and this layer of isolation was crucial. The migration of SDAS from its original testbed environment to IRAF was quite easy as a result of this design, and there have been other benefits of software sharing.

The MIDAS group working at ESO had defined a similar layer of interfaces, and there was considerable interest in making the SDAS algorithms available within the MIDAS environment. A jointly agreed upon set of routines, called the STandard interfaces (ST for short)[8] were defined for use in SDAS and MIDAS. These are now fully implemented on both systems and have permitted the transport of applications codes between both systems with a small amount of effort. Because these two systems are in some sense complementary, the ability to share this wide body of applications routines will enrich both environments.

The ST routines are not the end of the story however, for they are somewhat VMS dependent, deal only with program parameters and image and table data, not other types, and do not provide a complete layer between an applications program and the underlying operating system. It is intended that the ST routines will be further developed to accomodate these additional requirements in an upward compatible way. Completion of the system level interface, name bindings that are compatible with VMS and UNIX, and removal of VMS dependencies are topics for immediate discussion.

There is also the opportunity, during this review, for defining other data structures (for radio, X-ray, and other data forms) and other interfaces (terminals, graphics and image displays, system services) so that a more complete interface exists. The benefit to be derived from use of such interfaces has been shown to outweigh the cost and is highly encouraged. Collaboration in this effort with other interested groups is actively sought.

Another recent effort along these lines that is already bearing fruit is the definition of a set of device-independent Image Display Interfaces (IDI)[9]. This activity began by gathering together several individuals with extensive experience in the design and creation of interfaces for a variety of image displays. We defined both a user model of interactions with an image device and a programmer model of a virtual device that served those needs. From this, a set of image display interfaces was defined that would be relatively easy to code for new devices. A draft of this interface specification is now being circulated, and three trial implementations of the draft version have been made.

While the IDI specifications are still in a state of flux, the results are encouraging and it appears that the intended functionality and ease of development has been achieved; the three versions took an average of two man-weeks each to code and test. The specifications will receive further review and refinement before being published and widely distributed. Image devices are still quite costly and rather idiosyncratic, but are essential to much astronomical data analysis. The ability to port applications code to new devices in a clean way, or to interface several applications programs to a new device by building one new interface, is very attractive.

Distributed processing functions

Distributed processing, the allocation of specific resources to specific functions, is a key element in the design of the HST systems. By taking the approach of using distributed processors, systems may be selected on the basis of proper fit to the problem and to the processing load. Networks or some sort of high speed interconnect must be used in a distributed environment. Current LAN technology provides moderately high speed data transfers (~100K bytes/sec), distributed system interconnections, and access controls which may be used to advantage in isolating dedicated production facilities from other less controlled functions like software development. Another advantage of distributed systems is that upgrades tend to be less costly and individual bottlenecks can be dealt with as they are identified. Careful selection of network interfaces, transport layers and user services is essential if a variety of hardware architectures are to be eligible for deployment.

One of the problems of using distributed systems is the increased management load represented by many small systems rather than one large one. The DEC cluster concept has been a tremendous boon in this regard, in that management of disk storage and system functions has been centralized on the cluster. Although this technique is not directly applicable to other systems that cannot support this kind of "cluster", use of suitable network protocols can provide a quite workable solution. A distributed file system, such as represented by the SUN NFS[10] system and the "Yellow Pages" directory services, offer many of the same benefits as a cluster, but supports a heterogenous mix of systems. Hardware solutions that should offer the benefits of distributed system clusters are being built by several vendors.

Network interconnections

Network links are important for connecting distributed systems, but local network throughputs are not yet as fast as disk transfers, nor even of fast tape drives. Current maximum network throughputs and the problems of managing distributed systems (file backup, installations, upgrades, more hardware pieces to manage) are areas of some concern that will receive increasing attention in the near future. Because these concerns are widespread, we anticipate solutions from the developers of such systems in the course of the next year or two.

Local site networks are just part of a hierarchical network view that must include network connections to other sites, and internet connections among networks. Astronomers are just starting to seriously investigate wide area networks, and there are some major benefits, as well as some problems, to be dealt with. One of the more significant issues is the choice of network protocols, both in terms of the kinds of systems that can be connected, and of the services that are offered. Agreement on a standard set of interconnect protocols would allow creation of an astronomy domain[11] that would provide a logical network for support of science.

Vendor specific protocols (DECnet, IBM SNA) are facile choices when only one type of equipment is to be connected, but present serious problems for a heterogenous mix of systems. Proprietary protocols do tend to perform better than the open system alternatives because they are carefully tailored for the hardware and operating systems on which they run. The problem is that such protocols typically only run on one kind of system.

Where a diversity of systems must be accomodated (the typical situation in a wide area network) then some "open system" approach must be taken. At present there are two widely available open system networks, TCP/IP and CCITT X.25. There is a third suite of protocols, the ISO OSI (International Standards Organization - Open System Interconnect) that is emerging as a viable alternative, but the suite is not yet fully defined and partial implementations are only just now becoming available.

The TCP/IP suite[12], which was defined for use in the ARPA Internet, has seen wide use outside that community and is the network of choice on most UNIX systems. Implementations exist for all major vendor systems across a broad range of performance levels. It is a true open system and supports communication among the different vendor systems with a common set of lower-level and user-level protocols. Included in the basic set are remote login, file transfer and mail services, as well as the control functions that make it all work. These protocols can be used to create dedicated networks, but more importantly, can connect such networks together into an "internet".

The X.25 protocols are derived from early ARPA packet switching experiments, and provide a basic level of network transport services. X.25 is a connection oriented network service, and some higher level user services have been defined upon it, but there is not a complete standard set. Remote login is well supported as is process-process communication, but there is not a single, widely supported, suite of other user services. With the full development of the ISO OSI suite this will all change, but full deployment of such network services is expected to take several years. Migration from TCP/IP (and other proprietary protocols) has been committed to by ARPA and by several vendors, but the timetable is still uncertain.

FUTURE DIRECTIONS

Some of the projects at the Institute are just now completing a period of intensive development and are phasing into a "maintenance and enhancement mode"; others are continuing active development or beginning new development. Some of the projects that may be of wide community interest will be briefly discussed.

Network Developments

The Institute has started a wide area network project with NASA backing to provide high speed connections among an initial group of eleven astronomy and astrophysics sites. The initial participants are widely distributed geographically, cover much of the electromagnetic spectrum (radio, ir, optical, uv, x-ray) and include user observing facilities, significant data archives, and development groups with direct interest in distributed access. There are two central themes for the project: 1) that the various centers which support active data catalogues and archives will want to make these research resources widely available; and 2) that appropriate selection of network protocols and services can provide the vehicle for this to take place.

The plan is to help create a backbone network that will support inter-center communication, and provide access to these centers by the community. It is expected that other major centers will be added as the project continues and that wide-spread access by the community will be possible. A mail communication facility, and access techniques for remote database, image browse, and computer access (the Portal suite) are the most obvious projects to be tackled. The network will use the TCP/IP protocol suite for backbone communications, but mail relays will also support Bitnet, DECnet, and uucp mail services. The ARPA Internet, including ARPA, NSF, and other science networks[13], will be used until the NASA supplied lines are in place. The proposed NASA Science Network, which is to eventually host the project, will also be TCP/IP based.

Workstations

There have already been some low-level explorations of workstation technology at the Institute, but much more activity is expected in this area in the coming years. Super-micro workstations already approach or exceed VAX 11/780 power levels and are well suited for dedication to a single researcher in support of data analysis tasks. We see at least three distinct areas where suitably configured workstations can be usefully deployed: image analysis; support for AI projects; and remote access to data and computational resources. Do note that portable software and high speed network connections play an important role in permitting optimum use of these workstations.

The canonical last generation image processing "workstation" was a VAX 11/750 (or other dedicated minicomputer) with some sort of image display device. Because developments in display technology, CAD/CAM workstations, and related areas with broad commercial markets have lowered system costs, it is now possible to buy workstations suitable for imaging for $50-80K. These prices will continue to drop with time. Low-end stations, with less capability, can now be had for $10-20K or less. Clusters of mid-range workstations that share tape and disk drives, and high speed communications lines, perhaps linked to high-end VAX-class machines, can provide dedicated computational and display services while sharing otherwise expensive peripherals.

Remote access to distributed databases and archives, preparation and analysis of supercomputer runs at remote centers, and remote observing runs on distant telescope facilities are all nicely supported by super- micro workstations. While it is true that many of these functions could be handled by using a portion of some central computer, the ability to dedicate a system to these functions does have significant benefits. Use of a workstation permits development and use of interaction modes that would be costly or difficult to do on the typical central system. Another advantage of a dedicated system is that the user can "use it all up" without affecting other users of a central resource.

Exploration of AI Applications

There have been a few applications of AI technology at the Institute, and more areas will be explored in the near future. Use has been made of a production rule-based language (OPS5) in support of the proposal transformation process. LISP has also been used in a decision-support environment where natural language processing techniques were usefully employed.

The implementation of HST observations requires much detail about observing modes and spacecraft operation. While the proposal forms themselves are quite complicated, due to the several instruments and options, the actual commands that must be sent to the spacecraft are more complicated still. Transformation from the proposal form to a form that is suitable for detailed spacecraft scheduling involves a hundred-fold or greater expansion in level of detail. Now that this transformation process is well understood (by the humans who did it initially), their knowledge has been expressed as a set of rules in OPS5, and the process is now largely automated.

The Time Allocation Committee (TAC) that must review and help select proposals needed tools that would assist in this review process and in the analysis of the resulting observing load. A set of tools has been developed in LISP that provides part of this decision support function and uses tabular representation of various aspects of the problem. The impact of additions or subtractions on the overall observing schedule of proposals can easily be seen.

LISP was employed to provide a natural language front-end, so that the TAC would not have to learn a cumbersome new syntax to operate the tools. The goal was to supply tools quite easily accessible to untrained users which will help the TAC deal with an extremely large number of proposals in a finite amount of time.

There are other areas, such as the planning and scheduling of detailed spacecraft observing sequences, where AI techniques can be well applied, and these will be under study during the coming years.

One lesson learned from these early activities however, is that AI tools tend to be rather computationally intensive. There is an obvious trade-off between human and computer resources, but the costs are justified. Because of the resource load that these tools represent, they are ideally suited for running on workstations, where the load cannot affect other users, and where the graphical representation functions and clean user interactions can be readily provided.

UNIX Port of SDAS

Although the IRAF analysis environment is portable to other than VAX VMS systems, SDAS is not yet. We feel quite strongly that SDAS should be ported to other environments, both for our own long-range plans and in support of users who run other than VMS systems. Because of its wide-spread use on a broad range of hardware architectures, UNIX is the obvious choice for the next system to support. UNIX now runs on all flavors of processors from micros to super-computers, and on both classical and parallel architectures. The result of any port project must be software that can run on both VMS and UNIX systems without modification.

During the coming year we plan to do a port of SDAS to a UNIX system. The port will probably be done on a Motorola 68020 based system like a SUN or on a uVAX II Ultrix system, since either can provide a modest cost 1-2 user workstation for image processing. Future ports may be made to other systems such as one of the new mini-supercomputer class processors (Alliant or Convex), but this will probably be done at some other facility with our help. Since all of these systems use UNIX, once the initial port is done the others should be quite straight-forward.

CONCLUSION

Any project the size of the Hubble Space Telescope is subject to major development costs, long lead times, inevitable (and unforseen) delays, and much long and short range planning. In the specific areas of data management, computational support, and analysis these considerations are compounded by the periodic obsolescence of equipment and the need to move forward with the times. Part of the planning effort must be spent in "crystal balling" the future, trying to discern long term trends so that the systems that are being developed will be viable for significant periods of time. Portable software systems, well designed and cleanly layered, are an important part of any long term strategy.

Use of "standards" in coding, data formats, system interfaces, network connections, and languages are also part of this long term strategy. Some of these standards are industry wide and easy to identify, modular programming being an obvious one. Others are not so easy, or are driven by the needs of the particular kind of data processing that is being done. The FITS standard is a case in point, developed to meet the needs of astronomy in transporting image data among a divers collection of computers. Interfaces like the ST and IDI set (and GKS) are similar, but aimed at porting applications codes from one environment to another.

Many of the astronomy specific standards efforts require cooperation among researchers and developers at different institutions. Such efforts should be encouraged, for the obvious advantages that accrue from sharing technological developments and resources in a synergistic way. The network project we have helped begin should similarly be a positive benefit, by improving communications as well as providing new opportunities for interesting collaborative and correlative research.

It is clear that developments by talented individuals will continue to be a source of much of the most innovative work that is done, but that the cooperative efforts of many others at different sites are required to create, enhance and maintain these systems. Cooperative efforts among different organizations are also required to develop and maintain the networks that can connect these sites into a domain for support of astronomy.

REFERENCES

1. CDBS--Calibration Data Base System, SO-11, Vol. 2 Requirements; ST ScI, 15 Feb 85.
   Calibration Data Base Data Design; V1.2; Lezon, Botway; ST ScI; 6 Feb 86.
2. DMF--Data Management Facility--Design Guide; McGlynn, Russo, Richmond, Shames, Ochsenbein; ST ScI; 9 Oct 85.
   ST Science Data Archive, Starcat User's Guide--Preliminary; Richmond, Russo; ST-ECF 0-02; 15 Jan 86.
3. FITS--Flexible Image Transport System; June 1981; Wells, Greisen, Harten; Astron. and Astrophys. Suppl. Series 44, (1981) pp 363-370.
   An Extension of FITS for Groups of Small Arrays of Data; Greisen, Harten; NRAO.
4. KERMIT--A Simple File Transfer Protocol for Microcomputers and Mainframes; Cruz, Catchings; Columbia University Center for Computing Activities; May 1983.
5. SDAS--Science Data Analysis System; SO-03, Vol. 2 Requirements; ST ScI; 13 Aug 82.
   SDAS Applications Prog Guide; SO-03, Vol. 4; ST ScI; May 1986.
6. IRAF--Image Reduction and Analysis Facility; The Role of the Preprocessor; Tody, NOAO; Dec 1981.
   A User's Introduction to IRAF, V2.0; Shames and Tody; ST ScI; Feb 1986.
   Programmer's Crib Sheet for the IRAF Program Interface; Tody, NOAO; Sept 1983.
7. A Reference Manual for the IRAF System Interface; Tody, NOAO; May 1984.
8. ST Interfaces--Software Interface Definition--Preliminary; ST-ECF 0-11; 12 Sept 84.
9. IDI--An Image Display Interface for Astronomy, V.04, DRAFT; Terrett, Shames; ST ScI; 3 Feb 86.
10. NFS--Networking on the SUN Workstation; Sun Microsystems; No 800-1177.01; 15 May 85.
    See also RPC, YP
11. Domain--Domain Names--Concepts and Facilities; RFC 882; Mockapetris; ISI; November 1983.
    Mail Routing and the Domain System; RFC 974; Partridge; BBN; January 1986.
12. TCP/IP--Official ARPA Internet Protocols; RFC 961; Reynolds, Postel; ISI; December 1985.
    See also MIL-STD-1777; Internet Protocol Transition Workbook; March 1982; IP-RFC 791; ICMP-RFC 792; TCP-RFC 768; and many others.
13. SCINET--Computer Networking for Scientists; Jennings, Landweber, Fuchs, Farber, Adrion; Science, Vol. 231, pp 943-950; 23 Feb 86.

PRESENT AND PLANNED LARGE GROUNDBASED TELESCOPES:
AN OVERVIEW OF SOME COMPUTER AND DATA ANALYSIS APPLICATIONS
ASSOCIATED WITH THEIR USE

Harlan J. Smith

McDonald Observatory
University of Texas at Austin
Austin, TX   78712

I. INTRODUCTION

As a non-expert in computers and their uses, my aims in this paper are not to explore any subject in detail, but rather to survey the subject of large ground-based telescopes along with some of their present and future demands on computing and data-analysis facilities, and to try to stimulate discussion from experts as to how these needs can be met.

Hundreds of telescopes in the 0.4- to 2-meter class now exist (nearly 200 in the US alone). Many are well maintained, adequately instrumented, and in good enough locations to encourage useful research. Most of these are involved with imagery or photometry, and are being equipped with mini- or microcomputers. Some of these relatively modest telescopes will eventually draw on some of the more advanced topics discussed at this meeting. But because of expense and complexity, most of the sophisticated new computer-oriented developments will be made in connection with large or very large telescopes, which I arbitrarily define here as having apertures of 2 meters or greater. More than 30 such telescopes are already in existence, 13 of them in the 3- to 6-meter class, giving wide opportunities for participation in the massive computer revolution in groundbased optical observational astronomy. At least 20 more are under construction or being planned, half of them with primary-mirror units in the 8- to 10-meter class, giving a much-needed tripling of presently available light-gathering power by the end of the century.

II. SOME COMPUTER-ORIENTED PROBLEMS COMMON TO MOST
    OBSERVATORIES

1. Systems Approach and Strategy

Though the problems discussed below stress, for particular applications, the desired output and some of what is needed to achieve this output, the entire subject should be embedded in a systems approach. Simple, reliable interfaces and high-bandwidth communication links, e.g. from instruments to control rooms, are almost as important components of modern observatory equipment as the computers themselves. Specialized hardware is usually much more expensive and harder to maintain than commercially available, vendor-supported equipment. And software is now normally much more expensive than hardware, especially if it has to be developed locally rather than bought. For many higher level functions it may be a major economy of both money and time to avoid either the very latest or the very cheapest computers, and rather to invest in well-developed and strongly supported units (currently such as MicroVAX II's or SUN 3's). For very large-scale projects, these kinds of units along with microprocessors for specialized functions can be integrated into highly cost-effective and reliable systems. As an example of this approach, Kelton (1984) outlined a system for the proposed University of Texas 300-inch telescope, featuring many real-time control microprocessors governed by about half a dozen higher-level control processors, all networked and provided with a high-bandwidth link to Austin 700 km from the observatory.

A second general point may be thought of as strategy. The staggeringly large flow of data coming from some present and many future telescope/detector combinations increasingly requires that the computer systems include stages of data compaction and reduction as close to the telescope as possible. Appropriate microprocessors should get to work on the data stream immediately as it emerges from the detector, so that it has become at least somewhat manageable by the time it is recorded, transmitted, and analyzed for significance.

A final general point concerns the analysis of the data. Even that which is almost entirely automatically reduced can overwhelm a mere human being. In an increasing number of research areas attention must be given to semi-intelligent programs which derive the desired results or spot the interesting anomalies in the reduced data.

2. Telescope Control

The computer revolution is quite young. Only 20 years ago, while the Texas 2.7-meter telescope was in final design, we undertook the then-revolutionary decision to omit setting circles altogether. It thus became the first major optical telescope ever to be equipped with accurate coordinate encoders and digital readouts coupled to a "powerful" computer -- an IBM 1800 with all of 4k 16-bit words of core memory, plus hard disk drives. A major part of the programming which enabled this system, so primitive by modern standards, to work so well for some 15 years was done by a graduate student, Don Wells, now a key figure both at this conference and in the world of astronomical computing.

Large telescopes are expensive to build and operate, ($1,000 per observing hour), so any significant time wasted in their use is undesirable. With modern optical and IR detectors some of the objects to be observed are invisible to the eye, and the astronomer can lose a lot of time simply in finding and centering. Ideally a telescope should be able to point automatically to any specified object with a precision of about an arcsecond. With the marginal exception of the MMT no optical telescope yet does this, although precisions of a few arcseconds are becoming commoner. The tasks to be solved are straightforward: given the desired coordinates, calculate and correct for precession, nutation, refraction, mount misalignment, and flexure. The problems lie with asymmetric mount flexures and slippages of optical elements especially in older telescopes, requiring complex mount models which may change appreciably with every change of auxiliary instrument. The best solution is a generalized mount model with built-in learning curve, the model being continually refined and updated with every successful centering on an object of accurately known coordinates. A few observatories have implemented such programs, which require only minimal minicomputer facilities.

Once the object is in the field, automatic centering and guiding are desirable. For large telescopes dealing with faint or even invisible objects, full-scale pattern recognition represents a generalized approach to finding and guiding. This is now becoming feasible with moderate-cost CCD's and large-scale fast-access memories associated with quite substantial computers. Without some form of compaction, a single grey-scale 512 x 342 CCD field requires about 200 kilobytes. A few megabytes of memory could thus store, for immediate access, the image fields for a subset of targets which might be desired in a given observing program. It is more efficient, however, to convert the target-field information into a database, for example by scanning each field in a helical pattern emanating from the target object. The computer can then tabulate the radial coordinates plus additional image-characterizing information about every other object in the field at a typical cost of only about a kilobyte of memory per field. This technique also puts the information in a form immediately useful for pattern recognition. For any trial acquisition field it is only necessary to ask the computer to run the helical pattern separately for every object in the field (the whole operation requiring about a second on a typical minicomputer), and to compare each of the resulting patterns with the reference table for the object whose acquisition is desired. If the target proves to lie in the telescope field, the computer then also knows its offset from the center, and can command the telescope drive to correct the positioning. Such a system has been designed and built at McDonald Observatory, using a slightly modified IBM PC computer, by B. Hine and R. E. Nather (unpub.), at a total equipment cost of about $4000.

Once the target is located and centered, guiding corrections can be made as rapidly as necessary, typically at about 1 or 2 Hz with ordinary CCDs, but up to 20 Hz or more is practical if an intensified CCD is used. Comparing the current field with the standard field and commanding telescope motions to correct for drifts can thus be done in real time. Any anomalies in the field, caused by novae, asteroids, or stars of high proper motion, should also be called to the observer's attention by the comparison program. The finding and guiding field for any telescope equipped in this way becomes useful in the manner of the serendipity mode of the "wide-field" camera of ST. All of these functions including storage of a large number of target fields can be handled with particular ease by one of the next-generation workstations with 2-4 MIPS speed, 2+ megabytes of primary memory and 30-40 megabytes of hard-disk secondary storage, due to appear within a year or so at only $5-6000 (Crecine1986).

3. Available Data Bases

Next is the question of the acquisition and interpretation of data-bases readily available to the working astronomer. The NASA Goddard Astronomical Data Center (NSSDC) and their counterpart European group at the Strasbourg Stellar Data Center (CDS) for years have been putting astronomical catalogs onto tape (Hauck and Sedmak 1985). By now about 450 of the most commonly used catalogs are in machine-readable form. Intermediate-sized catalogs typically run less than a megabyte, while large ones exceed ten megabytes. As of this date the total storage is 454 megabytes.

The rapidly developing technology of optical disks currently offers low-cost 550-megabyte 5 1/4-inch ROM disks, and reasonable-cost gigabyte 11- and 12-inch disks. Thus before long observatories may be able to keep effectively on line essentially the complete file of important recorded astronomical data, for the immediate real time convenience of the observer or data analyst. Even the Palomar Sky Survey (all five colors) is planned to be scanned by the U.S. Naval Observatory for digital storage. Each color might fit on one of the Mark III high-density tapes with its hundred-gigabyte capacity, although possibly (and certainly more conveniently) one might wish them available on future ultra high-density optical disks. However, only the 5 1/4-inch disks have so far been standardized by the industry, and most observatories may wish to wait on larger disks until complete and standard subsytems are available which offer appropriate software with vendor service and support.

III. SOME COMPUTER-RELATED PROBLEMS SPECIFIC TO MOST FUTURE LARGE TELESCOPES

1. Design

For any given telescope design, the cost of the telescope increases roughly with the 2.6 power of the aperture. This is primarily, although not exclusively, a function of mass. Thus to first order keeping down the cost of ever-larger telescopes becomes a problem of seeking simpler and lighter-weight designs. Older telescopes were made with large and excessive safety factors in strength. Modern computer-aided design programs and analysis tools enable engineers to find optimized structural design solutions, giving next-generation telescopes a minimum of weight for the required stiffness, thus getting away from the old battleship-like construction and helping to hold costs within attainable limits.

2. Primary Mirror Support

An old rule of thumb which worked well for small- and intermediate-size telescopes specified that the aspect (thickness-to-diameter) ratio of primary mirrors should be about one to six, in order to achieve adequate stiffness and supportability. With increasing size of telescopes, the resulting cube factor in mirror volume led to very serious consequences. These include the great cost and enormous weight of the glass, the excessive weight and cost of the telescope tube and mount required to support the mirror, and the intolerably long thermal time-constant of so thick a mirror. The extreme dead-end example of this evolutionary track is represented by the Soviet 6-meter telescope with its succession, over the years, of replacement solid-glass mirrors each more than a meter thick, each weighing some 50 tons.

Affordable telescopes in the five-meter or larger class clearly require light-weighted primary mirrors. Three principal approaches are currently under development: honeycomb, thin monolithic, or segmented construction. In addition, second-of-arc performance was acceptable in most of the older telescopes. The new designs usually specify something closer to the quarter-arc-second limiting performance set by the atmosphere under truly excellent conditions. This desired final result must be divided among many elements of the error budget, with the primary mirror figure in actual operation being held to about 0.1 arcsecond. Achieving such performance requires not only extraordinarily good initial optical figuring and support, but also the retention of this figure at any inclination of the telescope and under thermal and wind stresses.

Honeycomb construction, successfully pioneered in the 1930's by Corning for the 200-inch telescope, seems the simplest of the large primary mirror designs. Its deep ribbing can give great stiffness along with relatively light weight. However, at the new high levels of performance specification, the calculation of thermal effects and of rib structure and the placement of axial and radial mirror supports become crucial (Ray and Chang, 1986). Large computers and quite sophisticated programs are required for the finite element analysis of the models.

Thin-meniscus monolithic mirrors are relatively simpler to analyze, but require many more supports working to a higher level of precision in order to maintain the required mirror figure. Analysis of the natural modes of bending of the mirror allows selection of an optimized set of support locations able to control the significant amplitude contributions from virtually all modes (Ray and Chung, 1985).

The 10-meter Keck telescope, with its 36 separate hexagonal mirrors, requires a means of holding each in exact reference to its neighbors, in order to generate in effect a single large mirror. Mast and Nelson (1982) have tackled this problem by inventing a sensitive capacitive position indicator to be placed between each pair of mirror edges. Working from these readings, a computer algorithm continuously supplies commands to actuators which adjust the relative heights and tilts of adjacent mirror surfaces. The computational facilities required for this operation consist of a mirror control computer operating through a bidirectional bus and segment multiplexers synchronized with a master oscillator.

3. Active Optics

In principle an edge-referenced segmented primary mirror, once adjusted, can "fly blind", relying on its edge-sensors and computer-controlled actuators to demand a perfect total mirror surface under all conditions. Likewise, for many purposes a honeycomb mirror may be stiff enough to hold an adequate figure using only passive mirror supports, although large borosilicate mirrors will require elaborate active thermal control. But the thin meniscus requires, and all large telescopes can probably benefit (Hardy 1982) from, "active optics" -- some place in the optical train where the average effective figure of the mirror can be improved as necessary in the course of observing (Wilson et al. 1984). Since primary mirror supports are crucial to maintaining the original figure of the mirror, small adjustments in their forces are easily able to warp the shape of the mirror over a wide range. This property of active optics can be used to correct for thermal or gravitational bendings, even to improve the figure over that originally supplied by the optician (Noethe, et al. 1986).

The information needed to correct the shape of the primary mirror can be supplied by what amounts to a continual Hartmann test. The simplest form of the test is probably the Shack-Hartmann device (Wilson 1984) wherein the converging cone of light from some star in the field is intercepted by an array of "lenslets" near an exit pupil. Each lenslet produces an image of the star from the small bundle of rays coming from the portion of the primary illuminating that portion of the lenslet array. The resulting pattern of star images is registered on an imaging detector. Any deviation of some portion of the primary mirror from the desired figure displaces the corresponding spot or spots on the detector, supplying the partial derivative information needed to derive corrections for the mirror support actuators. Since the mirror will react slowly to most disturbances, these corrections normally need to be made only infrequently. A single-board microcomputer with enhanced floating point operations suffices for this method of low-bandwidth figure correction (Noethe 1986).

4. Adaptive Optics -- Partial Correction for Seeing

Telescopes in the 5- to 10-meter class have theoretical resolving powers approaching 0.01 arcsecond in visible light. Unfortunately plane wave fronts entering the earth's atmosphere suffer phase distortions in passing through blobs of air of different density, leading to the phenomenon of "seeing" -- the blurring of resultant images formed by the telescope (Roddier 1981). In principle one could correct for seeing, if there were a method of sensing the error in the wavefront from the source being studied (or from a very nearby reference source), and if this information could be relayed to an active optical element able to correct rapidly enough all the phases in the wavefront. This approach, known as adaptive optics (e.g. Woolf 1984), is in effect a rapid real-time use of active optics. It will surely represent one of the more important developments in observational astronomy over the next several decades, greatly enhancing the efficiency of groundbased telescopes especially in the infrared.

Although the principle of adaptive optics is simple, realistic applications are difficult to achieve for several reasons. First, the spectrum of atmospheric seeing noise has significant frequency components going up to several hundred Hz. It is obviously impossible to readjust elements of the primary mirror at that rate. Accordingly an extremely flexible optical element -- a "rubber mirror" -- must be in a projected pupil near the focus where the beam is small enough to keep the mass and cost of this element reasonable. Next, one must have a means of detecting and analyzing the entire wavefront at several hundred Hz, and the readout rate of ordinary CCD's is far too slow for this. Finally, the rubber mirror needs as many actuators as the number of elemental sections of plane wavefront that are present in the wave trains.

Fried (1966) has defined a seeing parameter, r_o, as the diameter of the largest telescope which would produce a diffraction-limited image equal to the diameter of the seeing image. This diameter is just the size of the plane wavefront sections referred to above. To make this argument more graphic, recall that a telescope of 10-cm aperture has a 1-arcsecond visual diffraction limit. Thus if r_o is 10 cm, telescopes of 10 cm or larger aperture will report 1 arcsecond seeing disks. The larger telescopes will of course put more light into the seeing disk of each star, also creating a more complex speckle pattern (see below) in the image. In order to clean up the entire wavefront being imaged by the telescope, and to produce an image as small as the diffraction limit of the telescope itself, it is necessary to have separate actuators able to bring each small element of wavefront back into phase with the others. The required number of actuators to attempt full correction of the wavefront is thus equal to the area of the primary mirror divided by the area of r_o. For a 10-m telescope in 1-arcsecond seeing, this number is about $1000^2/10^2 = 10^4$, an unfeasible requirement at least with present fast mechanical actuators costing nearly $\$10^4$ each. But suppose the seeing were so good that $r_o = 50$ cm. The number of actuators required for a theoretical full correction would then be only 400, a somewhat more reasonable number. This example brings out the extreme importance of starting with excellent seeing if one hopes ever to achieve something approaching full theoretical resolution of a large telescope working in visible light.

There are two further important points to note regarding the theory of adaptive optics. First is the fact that seeing correction works only for the light coming from a small solid angle. Wavefronts from sources a few seconds of arc (or under excellent conditions, a few minutes of arc) away from the reference source have had a quite different history, such that the phase corrections put into the mirror surface for the benefit of the reference source will actually further distort the wavefronts for sources outside this small isoplanatic patch. In other words, adaptive optics is doomed to remain a technique good only for improving ground-based telescopic resolution over very small fields of view. This however can be important for imaging small fields or for putting concentrated starlight through the very narrow slit of a high resolution spectrograph.

The second major point, however, is much more encouraging. The Fried parameter is directly proportional to the 6/5 power of the wavelength being used. Thus if r_o is 10 cm at visible wavelengths of around 0.5 micrometers, it will be nearly 400 cm at a wavelength of 10 micrometers in the infrared, requiring only 8 or 10 actuators on a 10-meter mirror for nominally full correction of the wavefront. In conditions of outstanding visual seeing, with r_o of 25 cm, the 10-meter telescope will be fully diffraction-limited when working in the infrared at 10 micrometers or longer.

The computer systems required for realistic adaptive optics systems in the coming decade will thus need to interpret several hundred elemental wavefront errors at a rate of a few hundred Hz, calculating and instructing the set of actuator motions which will minimize at each moment the phase distortions in the total wavefront approaching focus. Ultimate designs will probably involve an array processor in the solution of the large linear system of equations used to model the flexible mirror. An early version of an adaptive optics system is currently under development at NOAO under the direction of J. Beckers, using a Sun workstation.

IV. SOME EXAMPLES OF MAJOR LARGE-TELESCOPE USES REQUIRING EXTENSIVE COMPUTER AND/OR DATA ANALYSIS FACILITIES

Most of the uses to which large telescopes are put fall into the categories of photometry, spectrometry, imagery or interferometry. Virtually every specific use has interesting computer or data problems or applications, some representative examples of which are noted here.

1. Photometry

The new large telescopes will be heavily involved in measuring accurate magnitudes and colors of objects comparable with or fainter than the sky background. Standard photoelectric photometry, using photomultipliers behind small apertures, can give acuracies approaching .001 mag (0.1% intensity) on relatively bright sources. But the standard approach fails for very faint objects because the total skylight admitted through the aperture swamps the signal from the desired object. Also, extraneous objects with brightness below the observed threshhold may often be included in the aperture, which must necessarily be much larger than the star image being measured in order to avoid loss of light at the diaphragm edges caused by vagaries in guiding and seeing. In principle the solution lies in improved use of area detectors such as CCD's, which detect with high quantum efficiency the photons impinging on each pixel. It should then be possible to integrate the total number of counts in all the pixels included in the image of the star or other object being measured, and to subtract an average sky count per pixel determined from clear sky areas very close to the image.

A number of practical difficulties currently limit the utility and accuracy of this approach. Most important is the fact that no present area detector has truly uniform pixel sensitivity. CCD's in particular are susceptible also to loss of entire lines or columns, to cosmic ray spots, and to incomplete charge transfer during readout. In addition the limited dynamic range of most current area detectors causes the central core of a bright image to saturate before surrounding pixels accumulate enough signal to override readout noise. As a result, CCD photometry even of only 0.01 mag (1% intensity) accuracy has been very hard to achieve. Solutions must include better-quality chips supplied from manufacturers, improved techniques of flat-fielding to determine pixel sensitivities, and improved algorithms to analyze the rich information content of any good area-detector frame.

A second development in photometry over the past decade (Nather 1973) has been the availability of dual fast photometers able to record an essentially uninterrupted stream of photons from objects whose brightness may vary significantly in only fractions of a second. Pulsating white dwarfs (ZZ Ceti stars), cataclysmic variables, and optical pulsars are the leading examples of such objects, which have opened up an exciting branch of stellar seismology.

The pulsating white dwarfs are all multiply periodic, and need observations spanning many uninterrupted hours or even days to unravel the complex light curves and extract the individual periodicities. Since such extended coverage is impossible from a single observatory, R. E. Nather and D. E. Winget are building a number of portable ("brief-case") photometers with matched portable computers for data acquisition and analysis at many observatories around the globe. This is a new form of "stellar seismology" that makes use of time series analysis by microcomputer (using non-linear least-squares fitting, periodogram analysis, cross- and autocorrelations, fast Fourier transforms, etc.) to extract information about the early universe by examining denizens of the stellar graveyard. Such studies have already yielded an independent measure of the age of the oldest white dwarfs, and hence a limit to the age of our galaxy.

2. Spectroscopy

The first major advance in optical spectroscopy over the past dozen years came from the advent of Reticons -- linear solid-state detectors -- initially applied at McDonald Observatory (Tull et al. 1975; Vogt et al. 1978). These record up to 1872 spectral elements with photoelectric accuracy. The hardware and software needed to handle the output from such systems are now relatively standard. The principal opportunity for further computer enhancement of their use in ordinary spectral analysis probably lies in the development of expert systems with limited artificial intelligence capacity, to determine continuum levels, to identify lines and blends, and to measure equivalent widths. A prototype of such a system has been set up at McDonald Observatory by C. Sneden.

Multiple-object spectroscopy replaces the Reticon with an area detector, normally a CCD, and lines up many dozen images of objects at a long spectrograph slit, usually with the aid of individual optical fibers. This approach can increase by one to two orders of magnitude a telescope's efficiency in gathering spectral data when many desired objects such as faint galaxies or quasars lie in the same one or few square degrees of sky which some telescopes are able to image. Other than multiplicity, nothing new is introduced in the recording and reduction of such spectral data. The novel feature lies in devising mechanical systems and their appropriate computer control to place as automatically as possible the receiving ends of each fiber precisely at the point in the telescope image plane corresponding to the celestial coordinates of each desired object, and to keep this "medusa" of actuators and fibers from tangling.

The next level of complexity comes with imaging spectroscopy, or with echelle formats where a great many spectral orders are stacked on the face of the chip. Here the task involves not only the usual problems with CCD flatfielding, etc, but in addition the problems of straightening out the shapes of the curved spectral segments and correcting for vignetting of the field -- the goal of course being to produce a spectrophotometrically accurate one-dimensional spectrum from the rather messy two-dimensional array of data.

The new detectors open up previously impossible lines of research. One interesting example is the determination of radial velocity variations of stars with a precision of a few meters per second. A promising approach (Cochran et al. 1982) uses an ultra-stable Fabry-Perot etalon in absorption in front of the spectrograph slit to impose on the entering starlight a regular array of uniformly spaced absorption lines. The system is being developed around the extraordinarily stable Tull coude spectrograph at the McDonald 2.7-meter telescope, using Tull's new Octicon detector -- a system of eight co-linear Reticons each with 1872 detector elements. In principle, for any star the program begins by obtaining an original stellar spectrum through the Fabry-Perot absorption system. Each subsequent exposure involves removal of a plain spectrum of the star from the new image and cross-correlation of the residual Fabry-Perot pattern with the original one, to calibrate any shift of wavelengths which may have occurred in the system. Next, the standard Fabry-Perot pattern is removed from the same new spectrum, and the residual stellar spectrum is cross-correlated with the original stellar spectrum to measure any apparent average wavelength shift which may have occurred since the initial spectrum was obtained. Subtraction of any spectrograph shift measured by the above Fabry-Perot cross correlation leaves as the desired residual the doppler shift which the star has manifested since the standard exposure was made. Sub-programs can readily be set up to correlate shifts of only certain sets of stellar lines such as those of different excitation or ionization potential, in order to measure motions at different levels in the stellar atmosphere.

The above data form the raw material for stellar seismology based on whole-disk radial velocities. Given long enough time series and sufficient accuracy in the measurements, the first several l modes of oscillation should be detectable in a number of stars, with implications on interior structure, especially the depth of the convection zone. The detailed theory of the process is complex, but once the intrinsic variations of each star are well enough understood, residual variations over long time scales may then be interpreted as barycentric motions caused by very low-mass stellar or high-mass planetary companions.

3. Imagery

Some astronomically useful 800 x 800 CCDs have been produced, and 2048 x 2048 units should soon be commercially available from Tektronix. Still larger CCDs are unlikely to appear for many years, because of the great cost of their production coupled with the very small market for them, and because of the problems of handling so large a volume of data. Even the present large chips put out between 10^7 and 10^8 bits per frame. Also the readout time can be quite long -- up to a few minutes -- when low-noise readout is required.

Nevertheless, certain astronomical problems require larger image area than can be served by the biggest chips. A number of groups are approaching this problem by using arrays of CCDs. Apart from any area losses arising from incomplete overlap of the chips, this solution is good. The cost of the mass-produced smaller chips is much lower, and their parallel readout is relatively fast compared with the larger CCDs.

Among the many virtues of CCDs, one of the greatest is the relative ease of continuing to add digitally the photon counts from successive exposures, in order to achieve exceedingly deep penetration against sky. Signal from a faint source increases linearly with exposure time, while sky noise grows only with the square root of the time. Accordingly ground-based telescopes can reach almost arbitrarily faint objects provided one is willing to invest the necessary telescope time for the successive long exposures. A. Tyson has shown the potential of this approach by bringing out galaxies as faint as 27th magnitude using about 20 summed deep RCA CCD exposures from the prime focus of the Cerro Tololo 4-meter telescope. To gain the field necessary for efficient survey work he is constructing a camera with four 2048-square CCDs. To solve the data-handling problem in real time he plans to feed the output on line into a MicroVAX to perform all the necessary cleanup and composition of each set of images while the next exposure is being made. He has shown that this approach will allow well-located very large ground-based telescopes to observe extended objects such as galaxies to even fainter limiting magnitudes than the Hubble Space Telescope, and of course to cover much larger fields.

CCDs offer another way to achieve very deep penetration and to cover large areas of the sky, through the technique of drift scanning. A telescope tracking at less than the sidereal rate or even fixed on the meridian images a drifting field of the sky on a CCD, whose rows are clocked along at the same rate as the sky passes over the chip, therefore integrating each element of the sky down its full column of the chip. Readout is only off the last row. Strips of right ascension many hours long can be covered in a single night. Although the actual exposure for each small area is small on a given night, the results of an arbitrarily large number of nights can be added to give extremely deep penetration.

J. McGraw and colleagues at Arizona have built such a system fed by a 1.8-m telescope at Kitt Peak fixed on the meridian. A pair of RCA 320 x 512 CCDs observe the sky as it passes. In the course of a year the summed images will record all objects down to about 24th magnitude in the strip of right ascension being surveyed, as well as pick up a great many important transient events. Even this relatively simple system puts out around 400 Mbytes of data per clear night -- comparable to the expected daily output of the Hubble Space Telescope. To record this staggering flow, McGraw uses dual dedicated instrument-control minicomputers to put the data on magnetic tapes for transportation to Tucson. There they are processed by a dedicated Data General MV10000 computer and transferred for permanent storage to 12-inch Optimem WORM digital/optical disks having a gigabyte of capacity on each side (these now cost only $200 each).

To be useful in one of its most important functions -- namely the timely discovery of transient events or objects such as supernovae, novae, eruptions of cataclysmic variables, unusually interesting asteroids, etc, -- this avalanche of data must be at least partially reduced in something close to real time. Accordingly McGraw and A. Bernat are pushing the analysis concept of eliminating images as soon as possible, converting the information into a database format. Their approach is to regard any definable characteristic which can be extracted from the data as an element for a database. To extract each of these elements from the data is the work of different macros which have been "taught" to do their specific thing by the astronomer initially working with the system in an interactive mode. An an example, one macro isolates each discrete image and generates its coordinates, intensity, ellipticity, "radius of gyration", position angle on the sky, and the peak pixel brightness and position. All of this information then is fed to other macros which test for identification of the object. The system soon becomes quite artificially intelligent, able to classify most objects with high reliability. Yet another macro takes the anomalies, calls them to the attention of the astronomer, but also tries to form them into meaningful groups for analysis. This activity runs as a co-processor in the background -- a watchdog for serendipity. Other kinds of macros can do such chores as analyzing micro-shifts of rows or columns of pixels to study telescope vibration. The system is extremely flexible in application to other tasks set by the astronomers.

4. Speckle Interferometry

In lieu of the mechanical complexity of adaptive optics, a method exists at least in principle to extract the full diffraction limit from the images produced by any telescope. This concept of speckle interferometry has been known for nearly 20 years (called attention to by J. Texereau and first developed by A. Labeyrie) but only recently has begun to be utilized significantly. A brief outline of the cause of speckle formation in images can perhaps help to illuminate both the potential and the limitations of this approach.

A highly magnified, extremely short-exposure (typically about 0.01 seconds) star image formed at the focus of a large telescope is always seen to be composed of a large number of bright spots or speckles. Another exposure a few hundredths of a second later will show essentially the same overall image blur pattern, but a different set of speckles.

Where do the speckles come from, and why do they all fall within the envelope of the overall stellar-image seeing disk? An easy way to visualize the answer lies in the Fried parameter ro discussed above under the topic of Adaptive Optics. Recall that r_o is effectively the size of an elemental segment of plane (undistorted in phase) wavefront from a star which manages to get through the atmosphere to the telescope. At each successive moment, the primary mirror can thus be thought of as a transient set of small telescopes, each of aperture approximately r_o, lying along the curved surface of the primary. Each of these small telescopes sends its elemental segment of the full wavefront into the image plane as an Airy disk of diameter appropriate to its "aperture" of r_o. The shape of the primary mirror insures that all these elemental Airy disks overlap into a single stellar image. Consequently it is the size of the elemental Airy disks which produces the size of the stellar seeing disk formed under the ambient atmospheric conditions. The better the seeing, the larger the r_o and the smaller the elemental Airy disks which the primary mirror superimposes from all over its surface to create the fully illuminated seeing disk. However, these different plane segments of the initial full wavefront from the star have been shifted in phase from each other by passing through atmospheric blobs of different density. As a result the elemental Airy disks from each r_o segment of the full wavefront arrive out of phase and interfere with each other, creating within the seeing disk a very complex distribution of momentary maxima of light intensity -- the speckle pattern.

It may seem paradoxical that such a confusion of interference patterns can give useful information. The secret lies in the fact that over sufficiently small solid angles (the "isoplanatic patch") all incoming wavefronts have experienced essentially the same atmospheric r_o phase shifts. Consider a double star with sufficiently close members. At each instant <u>a virtually identical speckle interference pattern</u> is produced for each of these two point sources. Each pair of speckles in the pattern has the identical orientation and angular separation presented by the pair of double stars in the sky. The resolution displayed by each pair of speckles is proportional to the aperture of the telescope, since it is the interference between the most widely separated r_o elements across the full width of the primary mirror that determines the effective resolution of the spots within the speckle pattern.

In more complex cases than close double stars, the information contained in the speckle field can still be extracted, though usually only in part, through a two-dimensional Fourier transform of the pattern. Surface brightness distributions offer particularly challenging problems to speckle interferometry. Reconstruction of the image requires knowledge of phase of the transform. A number of techniques have been developed to solve this problem to varying degrees; these are summarized in an outline review by Chelli (1984).

The principal practical difficulties in speckle interferometry lie in the necessity of taking hundreds or thousands of relatively large and extremely short-exposure grey-scale images, recording and storing this enormous amount of data, doing the necessary transforms on each frame, and finally integrating and interpreting the result. A very high signal to noise ratio is required to extract meaningful information from any complex source pattern. The difficulties rise steeply with degradation of seeing, and the accompanying multiplication and overlap of speckles in the pattern. A partially hard-wired FFT system can be an important adjunct in making a speckle interferometer program feasible. But the effort is worth while. Successful speckle interferometry with the very large telescopes of the next several decades will lead to resolutions about ten times better than that of the Hubble Space Telescope, at least in small-field regions of the sky, and especially those having relatively simple spatial structure.

5. Interferometry

Interferometry began in optical astronomy with Michelson some 60 years ago, but until recently has been largely the province of radio astronomers. The big problem with large-scale utilization of so powerful a tool in optical astronomy has been that stable interference patterns require effectively coherent light sources, whereas the earth's turbulent atmosphere destroys much of this coherence over relatively small baselines. In addition, visual frequencies are so high as to have precluded as yet any effective use of heterodyning or of very long baseline techniques using ultra-accurate clocks as phase references.

Nevertheless phase-coherent (as opposed to amplitude) interferometry offers spatial resolutions on simple sources now approaching the milliarcsec range, later down to much finer resolution. The highest resolutions (microarcseconds) will have to be achieved in the vacuum of space, but new techniques on the ground have much promise as well. Because of the rapid improvement in atmospheric phase transmission toward longer wavelengths, groundbased interferometry with relatively long baselines of tens to hundreds of meters will certainly begin and may remain as the province of infrared astronomers. Pioneering projects along these lines include C. Townes' (1984) heterodyne interferometer system using a movable pair of 1.5-meter mirrors working at 10 microns, and the extensive long baseline Michelson interferometry experiments of A. Labeyrie at CERGA culminating also in a pair of 1.5-meter telescopes (Labeyrie et al. 1984).

The most extensive future plans along these lines are undoubtedly those of ESA in considering interferometry between the four 8-meter telescopes to be installed at La Silla over the coming decade (ESO 1985). This system should eventually produce, at wavelengths longer than perhaps 3 microns, images comparable to those of the VLA at resolutions down to about 1 milliarcsecond at the shortest useful wavelengths. Techniques of adaptive optics can clean up the wavefronts and allow work on sources perhaps as faint as $Mv = 19$. This level of resolution will begin to open up such problems as astrometric measurements on galaxies, nuclear structures in galaxies and quasars, and superluminal motions in quasars. Infrared resolutions of around 10 milliarc seconds will give clear pictures of such phenomena as accretion disks, bipolar flows, and cocoons around stars. The computational techniques required will be similar to those already developed by radio astronomers especially with the VLA.

V. SOME FUTURE PROBLEMS AND POSSIBILITIES

1. Historical Records

Photographic astronomy has the great advantage of compact, readily accessible, essentially permanent and easily understood storage of data. Great centers such as the Harvard collection, with its more than half a million plates, preserve a hundred-year record of the sky, much of it photographed many times over in the course of neary every year. These collections remain invaluable for their historical evidence of variability of stars, quasars, and other sources. However, they suffer from their analog character -- it is often tedious and difficult to get solid quantitative information from the plates.

Now that fast and accurate plate scanning machines are becoming available, it is not out of the question to visualize the digitizing of most of these old materials. To rough order of magnitude, there are probably two million historical photographs and spectra worth reducing and recording in this way. It is probably a high estimate to assume several megabytes of information on each. Accordingly the total storage problem would be only several terabytes -- a library of roughly a thousand large optical disks, each one containing in data-base form the complete century-long record of about 50 square degrees of sky. Some clever software would be required to locate, reduce and read out the complete record of any desired object, but given a well-ordered world, I would not be surprised to see such a project undertaken within the next twenty-five years.

2. Near-future Rates of Collection of Data

The 50 or so major optical telescopes which will be around in the 1990's are each capable of gulping and regurgitating partly digested data at the rate of several times 10^8 bytes/hr if equipped with large arrays of CCDs (most other uses would require orders of magnitude less in the way of data rates). Even reducing this data rate by an order of magnitude for the amount of time any single telescope might actually spend in such work, and by another order of magnitude for the number of telescopes which might actually be pressed into such service, still indicates a need for 10^{11} to 10^{12} bytes/yr of storage if one were to try to record it all. Still more seriously, merely an integrated 1° strip of sky recorded at arcscond resolution by a single large meridian-transit CCD also requires about 10^{12} bytes/yr.

At present, there is not in general use any good analog to the old photographic plate collection, which was considered to belong to the observatory, with observers expected to deposit their materials in the general collection when through with their analysis. With few exceptions, most of the modern electronic observers take their data, reduce some of it, and sooner or later erase the tapes or disks for some other use. Even the stored tapes and magnetic disks degenerate or simply become unreadable as the machines and codes change or are forgotten. Here is one of the most fertile areas for constructive invention and innovation in the use of computers in astronomy of the future. It used to be said of the Chicago stockyards that they made good use of every part of the pig but the squeal. In a sense this is the prospect of near-future digital astronomy with modern detectors. We are creeping up on the ability to record -- and utilize -- many of the photons coming to us much of the time from most of the sky. The astonishing pace of advance of computer and digital storage technology, and appropriately intelligent software, suggests that some approximation to this may be an attainable goal in the quite imaginable future.

Alternatively, the view is sometimes expressed that it has become cheaper and easier to repeat an observation than to archive and retrieve it. Here we see the outlook of those who approach astronomy mainly as a series of sharply pointed questions to be asked of the universe, if possible to be answered by similarly sharply directed observations. Much astronomical progress is indeed made in this way. But I believe the loss would be great if we try to do without the great synoptic collections, which allow truly massive and accurate statistical studies, the detection of the rare maverick cases which often open up entirely unexpected and fruitful new lines of research, and the record of variability and motion which play so large a part in our ultimate understanding of the universe.

3. Some Guesses for the Future

Toward the middle of the 21st Century most of the frontiers of observational astronomy will probably have moved permanently to space, particularly to geosynchronous orbit and to the lunar surface. Extraordinarily deep, massive and complete surveys will be made from these extraterrestrial observatories. However, many of the large ground-based telescopes will still be in very active use, in part because of their great economy of operation compared with similar facilities in space, and their convenience of access for use, testing of new instrumentation and training of students.

But fifty years from now, if not even much sooner, the mode of operation of groundbased telescopes will have become singularly like those in space. Telescopes and instruments will be so fully automated that, after instrument checkout and except for student training, astronomers will seldom be at their observatories. Likewise telescope assistants will be largely replaced by ultra-reliable robotic operators. As for the handling of data, expert systems will take care of nearly all routine chores such as classifying spectra, determining accurate proper motions, calculating fourier transforms, etc. Artificial intelligence of a high order will go far toward making a reality of the old joke among graduate students of wanting a telescope and computer system which will not only take and analyze the data, but also write the paper and transmit the results to ApJ in real time (for other computers to read and interpret?)! The prospect of all this coming to pass is a bit sobering. Without the occasional experience of spending long cold nights under the sky, personally grappling with telescopes, instruments, clouds, and recalcitrant raw data, will many of our grandchildren find the search for knowledge among the stars as challenging and exciting as it has been for us?

ACKNOWLEDGMENTS

It is a pleasure to record interesting and helpful contacts on these subjects with, among others, W. Cochran, D. De Young, P. Kelton, J. McGraw, J. Mead, R. Nather, F. Ray, E. Robinson, R. Tull, and A. Tyson.

REFERENCES

Chelli, A. 1984, "Infrared speckle methods", Proc. IAU Colloq. 79 (Garching, April 1984), 309.
Cochran, W. D., Smith, H. J. and Smith, W. H. 1982, "Ultra-high precision radial velocity spectrometer", SPIE Vol. 331 (Instrumentation in Astronomy IV), 315.
Crecine, J. P. 1986, "The next generation of personal computers", Science 231, 935.
ESO 1985, "Aperture synthesis (spatial interferometry) with the Very Large Telescope", Interim Report by ESO/VLT Working Group on Interferometry, VLT Report #42, (October 1985).
Fried, D. L. 1966, J. Opt. Soc. Am. 56, 1372.
Hardy, J. H. 1982, "Active optics -- don't build a telescope without it", SPIE Vol. 382 (Advanced Technology Optical Telescopes), 252.

Hauck, B. and Sedmak, G. 1985, "Data handling in astronomy and astrophysics" (Miramare, Trieste, July 1984), Mem. della Societa Astronomica Italiana 56, No. 2-3.

Kelton, P. 1984, IEEE 1984 Proc. of Real-time Systems Symposium, 83.

Labeyrie, A. et al. 1984, "Progress of the large interferometer at CERGA", Proc. IAU Colloq. 79 (Garching, April 1984), 267.

Mast, T. S. and Nelson, J. 1982, "Figure control for a fully segmented telescope mirror", Appl. Opt. 21, 2631.

Nather, R. E. 1973, Vistas in Astronomy 15, 91.

Noethe, L. et al. 1986, "ESO active optics system, verification as a 1-m diameter test mirror", SPIE Proc. 629 (Tucson, March 1986).

Ray, F. B. and Chung, Y. T. 1985, "Surface analysis of an actively controlled telescope primary mirror under static loads", Appl. Opt. 24, 564.

Ray, F. B. and Chang, J. H. 1986, "NASTRAN analysis for a sequence of cellular primary mirrors of the 8-meter class", SPIE Proc. 628 (Tucson, March 1986).

Roddier, F. 1981, "The effects of atmospheric turbulence in optical astronomy," Progress in Optics XIX (ed. E. Wolf), 283.

Townes, C. 1984, "Spatial interferometry in the mid-infrared region", J. Astrophys. Astr. 5, 111.

Tull, R. G., Choisser, J. P., and Snow, E. H. 1975, "Self-scanned Digicon: a digital image tube for astronomical spectroscopy", Appl. Opt. 14, 1182.

Vogt, S. S., Tull, R. G., and Kelton, P. 1978, "Self-scanned photodiode array: high performance operation in high dispersion astronomical spectroscopy", Appl. Opt. 17, 574.

Wilson, R. N., Franza, F., Noethe, L. and Tarenghi, M. 1984, "The ESO off-line telescope testing technique illustrated with results for the MPIA 2.2-m telescope II", IAU Colloq. 79 (Garching, April 1984), 119.

Woolf, N. 1984, "Adaptive Optics", IAU Colloq. 79 (Garching, April 1984), 221.

DATA ANALYSIS FOR THE ROSAT MISSION

H.U. Zimmermann, R. Gruber, G. Hasinger, J. Paul,
J. Schmitt, and W. Voges

Max-Planck-Institut fuer extraterrestrische Physik

Giessenbachstrasse 1, D-8046 Garching

SUMMARY

The German astronomical X-ray satellite ROSAT will perform the first all-sky survey with an imaging telescope. The large number of expected new source detections poses severe requirements on a fast and high quality data evaluation. In the context of the general mission goals the main data evaluation methods are described. Special emphasis is given to the organizational data handling structures applied to the Standard Data Processing.

INTRODUCTION

ROSAT is the abbreviation for Roentgen-Satellite and stands for an astronomical satellite built and operated by the German government with major contributions from the UK and US.

The scientific payload of the ROSAT spacecraft consists of a large X-ray telescope (XRT, 6 - 100 Angstrom) and a smaller XUV-telescope (XUVT, 70 - 700 Angstrom) which are looking parallel. The primary objective of the mission is to perform the first all-sky survey with an imaging X-ray telescope leading to an improvement in sensitivity by several orders of magnitude compared with previous surveys. The spectral band covered in the survey mode by both telescopes ranges from 6 to 200 Angstrom and can be divided into 6 "colours". A large number of new X-ray sources ($\sim 10^5$) is expected to be discovered and located with an accuracy of 1 arcmin or better, depending on source strength. The sources detected will represent almost all astronomical objects, from nearby normal stars to distant quasars.

After completion of the sky survey, which will take half a year, the instruments will be used for detailed investigations of selected sources with respect to spatial structures, spectra and time variability. In this pointing mode, which will be open for guest observers, ROSAT is expected to provide substantial improvements over the imaging instruments of the Einstein Observatory.

ROSAT will be launched by the Space Shuttle into a circular orbit of 56 degrees inclination and about 480 km altitude. The launch date has been strongly affected by the shuttle problems and is presently not expected before 1989.

THE GERMAN ROSAT SCIENTIFIC DATA CENTER (RSDC)

The following scheme of Control and Data Centers is under installation in order to ensure that the scientific objectives can be achieved and data and results be distributed to the observer community fast enough for reactions to the ongoing program.

The German Space Operation Center (GSOC at Oberpfaffenhofen) - in close collaboration with the RSDC at the MPI fuer extraterrestrische Physik at Garching (MPE) - will handle all items relevant to mission operation. At the RSDC, the Standard Data Processing of all XRT data will take place. Similar Data Centers for handling the UK and US data as well as the German part of the XUV-data are being installed.

Guest Observers (for the Pointed Observation Phase only) will directly communicate with their national data centers. German guest observers shall receive the output of the Standard Analysis procedure performed on their data usually about 3 weeks after the observation. Besides general information on the observation performed, the protocol will provide positions and intensities of all detected X-ray sources. For stronger sources results of spectral fits and time variability analysis will be given in addition. Perhaps quite a number of observers will already be satisfied with the information contained in this analysis protocol.

On request the observer will receive a ROSAT Observation Tape (ROT), which serves as input to further studies. If so, the observer may then take advantage of the ROSAT Image Processing System (RIPS) software package, which will be developed for extended X-ray data analysis purposes within our institute.

The collection of application programs contained in RIPS will be embedded and work within the IRAF and/or MIDAS Image Processing environments and be specialized to ROSAT instrumentation relevant analysis tasks.

It is part of our concept that the Image Processing workstations installed at the nodes of the planned Deutsches Astronetz (German Astronet) are fully compatible to the Image Processing stations we are going to provide for our internal needs and for guest observers coming to the institute. At the RSDC a LAN (local area network) of the main VAX 8600 computer (only partly used for ROSAT) and 6 Microvax II stations is planned to be installed for the mission.

The compatibility with the Deutsches Astronetz should then enable us to distribute software packages via the net to prevent that too many observers have to come to the institute to further analyze their data.

MAIN DATA HANDLING TASKS

During the 6 months survey phase ROSAT will continuously scan the sky in bands with a width of 2 degrees field of view for the XRT. These stripes are centered on great circles through the ecliptic poles. On the ecliptic equator a source will be seen by the XRT during about 32 consecutive orbits. At the poles sources will be seen during every orbit.

A major task of the survey data processing system is to overlap the individual scan stripes in a coherent way in order to produce "operational" sky maps. Their format will be 360 x 2 degrees for the XRT.

Subsequent tasks are source search, preliminary identifications, background mapping etc. All these tasks have to be carried out at least at the speed of data acquisition in order to avoid data backlog and to facilitate the use of the survey data for guiding the pointed observations.

One of the final output formats of the sky survey will be compatible with that of the optical sky surveys with Schmidt telescopes (Mount Palomar, ESO, SERC-UKSTU). This means that (for the compatibility with the Mount Palomar survey) computer printed sky maps of 6.6 x 6.6 degrees and 6 degrees period will be produced. It is anticipated that each of these about 1200 ROSAT survey plates will contain about 100 X-ray sources.

In addition to the ROSAT survey plates, there will be a source catalogue of the brighter X-ray sources. This catalogue will contain source location, error box size, intensity and spectral information, time variability and also possible identifications.

During the pointing phase of the mission, guest observations will be made. A major task of the pointed data processing is to perform a Standard Analysis of the data providing the guest observer with relevant information on the image and the detected sources to facilitate further analysis of the data.

The size of the tasks to be performed may be partly characterized by the amount of data to be handled. Primary and processed data to be stored in the ROSAT data archives will be close to 0.5 GBytes/day or about 500 GBytes during a 3 years mission lifetime for the XRT alone. Up to 1000 X-ray sources have to be handled every day during the Survey Phase to keep up with the incoming data stream.

Only an almost fully automated analysis scheme can fulfil those requirements. The next chapters will give a more detailed decription of our Standard Data Processing.

STANDARD DATA PROCESSING

All incoming XRT data from both the Survey and the Pointed Observation Phase will undergo a Standard Data Processing, similar to what has been done for the pointed observations in the EINSTEIN and EXOSAT projects. Experience from both satellite projects has been used to optimize the ROSAT system.

The Standard Processing software comprises a Data Normalization package and an Image and Source Analysis Package.

Data Normalization performs all the data installation, sorting, calibration and screening procedures necessary before image and source analysis can start.

The Image and Source Analysis package contains routines to search for sources and determine background properties, calculate source positions (including extent and error boxes) as well as source intensities in different energy bands. Possible counterparts from other wavelength regions will be added. For stronger sources spectral and timing analysis will be performed.

Before turning to a description of the data processing itself, a few comments on the analysis methods applied should be given.

Image Construction

The onboard X-ray detectors record the position and the energy of each detected X-ray event. Together with timing information these data are written to an onboard tape recorder and later dumped to ground. On ground, this time ordered series of detector events is then combined with the pointing information (attitude solution) and sorted into raster images. Typically data from one sky field are collected in several observation intervals and have to be merged to enhance the signal to noise ratio.

The appearance of a typical X-ray image is strongly affected by the background seen in the detectors, which itself varies heavily (about a factor of 7) over the orbit. The instrument response function also varies strongly with the off-axis angle, which might lead to problems of source confusion at large off-axis angles on longer sky exposures.

Detection Mechanism

We selected the sliding window technique to identify X-ray sources. The detected photons are sorted into an image with a pixel size optimized for highest sensitivity. A square window is slid across the image and the number of counts inside the window is compared to the prediction of the local background. If the signal exceeds a threshold value, the pixel is flagged as a possible source. The background prediction can either be taken from the pixels surrounding the detection cell or from a rather complicated model of the contribution from the different background components. We plan to use both methods.

The detection mechanism has to be performed in different energy bands in order to be sensitive for sources with different spectra. In addition, in order to be able to detect extended structures and to account for the variable point spread function (PSF) across the field of view (in the pointed mode), a series of detection processes with increasing bin size have to be run.

Estimation of Source Parameters

Source properties like position, error box, flux and significance are computed in a further step: the individual events from the regions of interest flagged by the detection process are handed over to a maximum-likelihood algorithm. For each individual photon, the PSF valid for the actual off-axis angle at the time of arrival is applied to calculate a likelihood function which is then maximized with respect to the source parameters position and flux. A confidence level for each parameter is also supplied, which can be used to reject sources with insufficient significance.

First tests show that this method is superior to other techniques of parameter estimation (center of mass or fitting procedures), because it takes into account all available information for each individual photon and can still be performed in reasonable amounts of computer time.

Simulations

In order to test and to calibrate the automatic analysis procedures, we carry out extensive simulations. In particular, a complete telemetry stream of four days of ROSAT survey mission has been simulated: the ROSAT telescope was moved across an artificial X-ray sky containing known sources and population of sources simulated according to observed log N/log S distributions (Maccacaro et al. 1984). For each source, arrival times were calculated according to a Poissonian distribution and individual photons were mapped into the focal plane with a ray-tracing program. Different sorts of background with their appropriate time dependence and spatial structure were added.

This dataset is extremely useful to check the data flow in the automatic analysis routines, to test and calibrate various source detection mechanisms, to derive the rate of spurious source detections, to obtain survey errorboxes as a function of source intensity and to determine sensitivity thresholds and log N/log S efficiency curves.

STEERING SYSTEM

The emphasis and manpower we put into the development of the Standard Data Processing package has been mainly influenced by the following considerations:

1. The ROSAT survey will enhance the number of known X-ray objects from a few thousand to something like a hundred thousand. Only an almost fully automated and extremely well designed and tested Software environment will allow data analysis to keep in step with the incoming data stream from the very begin of the mission. This will not rule out later reprocessing of data but should provide already a reasonably correct evaluation, urgently needed to guide the pointed observation phase.
2. The amount of data to be handled in the first processing as well as the computer resources needed during the survey data evaluation are so large that a special scheme is necessary to ensure homogeneous and quality standards exactly known over the whole sky.

As stated before, the daily tape we get from GSOC contains a series of time ordered events rather than an image. It covers a time interval of one ROSAT day or about 16 orbits and contains all housekeeping, aspect and calibration data in addition to the science data. The volume of incoming data is 250 Mbit/day.

About 40 tasks have been defined that perform the extraction of sources and the determination of their standard properties. Our goal is to analyze the data of one day per day. It is evident that such an analysis has to be an automatic one.

Clearly an automatic system is more than having a collection of application programs, that have to be run just one after the other. In the following section some principles are shown that guided our design of what we call the Steering System.

Goals

To achieve a smooth processing is one important goal of the automatic system, but in our opinion it is worth paying at least the same attention to achieve a high flexibility in case of problems.

In case of an error, the system should allow for easy recovery minimizing the necessary reprocessing.

It might be necessary in the early mission phases and also in the first phase of pointing mode to redo parts of the analysis with e. g. changed parameters or changed programs. The system should be able to keep track of such repetitions and should provide some flexibility in deciding which input files should be chosen out of various that may have been produced.

In addition, it should provide a complete logging and should be user friendly.

Segmentation

The application programs constitute something like the atomic level of the steering system. Nevertheless, to achieve flexibility we need still another kind of segmentation above the level of these programs. A natural way to divide the processing into several steps - characterized by different KEYs - is suggested by inspection of the structure of our analysis.

We decided to have processing steps as follows:
One step contains all tasks concerning the time ordered events like e. g. making detector corrections, calibration and applying amplitude to energy conversion (TEL).
Another step is formed by the tasks that process the data of one observation interval. This is essentially map processing, i. e. production of exposure map, background map and related tasks (OBI).
The next step contains all tasks that merge different observation intervals to get an image and extract sources (SEQ).
The last step determines properties of selected sources such as time variability or spectral properties (SOU).

Each of these processing steps splits up into a set of application programs.

Each data set used for processing of one of these steps is characterized by a "key number" (KEYNO) and the data of one target pointing are characterized by an OBI-number and the accumulated (merged) data of one observation are characterized by a SEQ-number.

Besides facilitating our job by achieving more flexibility, this division of the processing into several steps also poses some problems, which will be discussed in the following.

Configuration Management

In practice the processing will not be run in such a way that all the processing steps are carried out sequentially. It may easily be that a reprocessing of some data proves to be necessary with changed parameters or changed programs or simply because of errors. Thus, different versions of files will be produced and the system has to keep track of file versions and the set of parameters related to these file versions.

Each processing run thus must be characterized by a processing version number (PVRS) that stands for a set of parameters and program versions that have been used in the run.

Since the output of different versions PVRS may be used for input by the next step, an additional version number (NVRS) has to be attached to the key number KEYNO of the processing, characterizing the data input from former steps. This includes the case that some observation intervals have to be skipped from merging with other intervals to form an observation. This additional version is independent from the one that characterizes the set of parameter and program versions which the key number KEYNO actually will be processed with.

The existence of different kinds of versions requires checks for consistency that have to be done in advance in an automatic way. Thus, a configuration management is needed to handle the sets of versions and their interconnections in the actual processing. Also the file version handling is not controlled by the VAX automatic system but by the steering system.

One main task of the steering system thus – besides logging and error handling – is to provide a configuration management that allows being flexible on the one hand and that guarantees consistency on the other hand.

Logging

An important part of our steering system is a commercial relational database management system (INGRES). Within INGRES, not the data themselves are stored but only all steering information, e. g. file names, program names, their respective versions and – as a main point – the relations between these versions.

Apart from allowing to steer the processing, this information provides a complete logging of all data processing.

Each output file will reside within a directory which will be characterized by the key number and the run version of the processing. Since each repetition of a processing step automatically will get another processing version, the output files of the new processing version will reside within another directory. No chaotic deletion of information produced within a directory or within database is allowed. The concept of storing different versions in a clearly separated way allows us to clean the database and the directories in a controlled way from time to time.

Communication between Steering and Application Programs

In our design the application programs are standalone programs. The communication to the steering system will be done exclusively by means of a parameter file (PAR.PAR, s. fig. 1). This parameter file contains the full names and locations of the input and output files, all parameters needed by the application programs (or information where the parameters may be found) and all relevant steering information. It will be common to all application programs of one processing step and will be prepared by the steering system in advance. It resides in the directory which will automatically be established for the processing version and which will be declared as a current directory for the application programs. No direct connection between the application programs and the database management system exists.

Process Flow and Operator Interface

For each processing step a list within the database exists that represents the status of processing by various flags.

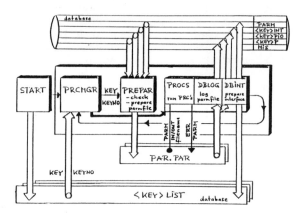

Fig. 1: ROSAT automatic processing system

By inspection of this list a "process manager" (PRCMGR, s. fig. 1) initiates the preparation (PREPAR) of the parameter files for all key numbers that are assigned to be ready within the list. It then submits a job for each key number. Within this job a selfstanding program (PROCS) successively executes the application programs of this step.

After successful execution of these programs, all information from the content of the parameter file will be logged within the database (DBLOG). Depending on the outcome of the processing, the list that constitutes the interface to the next step (DBINT) will be updated. The process manager will be called again and if some key number of the next step turns out to be ready, its processing may be started automatically by the process manager.

The observation intervals to be processed are determined in an automatic way from the actual tape and only after comparison with the mission time line they are brought into the steering system. On the other hand, after completion of an observation interval, the steering system checks whether all observation intervals expected to be part of an observation are ready. Only then the steering system updates the relevant list with the key number of this observation and starts its processing.

It is these lists that constitute the interface between operator and steering package. They are formatted on screen to allow an easy handling by the operator. For simplicity we use the screen form system provided by INGRES.

The operator has an easy possibility to set up a new run of the processing with only a subset of programs to be run, the output files of the missing programs being taken automatically from older processing versions. This allows the operator e. g. in case of an error to start again with the erroneous program - corrected in a suitable way - thereby using the results of the programs that already have been run. By various checks it is guaranteed that in using files already produced in an older processing version no inconsistencies are produced and that the logging stays complete.

S/W STANDARDS

We have realized the Steering system as a set of FORTRAN77 routines mixed up with DCL command files at a few points. It is well known that a project of this order needs well defined standards to facilitate development and maintenance.

To represent the overall flow of application programs we use bubble charts and minidescriptions of the application programs. We strictly apply the rules of structured programming and use Nassi-Shneiderman diagrams for graphical representation of the programs and subroutines. To guarantee that the diagrams are in accordance with the source code even after changes, we use a program package which automatically draws Nassi-Shneiderman diagrams from source code. This package has been developed at MPE.

We use FORTRAN77 with some restrictions for easy portability to other machines and with only a few well defined exceptions for better structured programming (DO WHILE...). No GOTOs are allowed. Standardized header masks allow our system manager to use utilities to check what system services a program is using, which subroutines it is calling etc. We developed error handling rules that allow error trace back as well as other tracing facilities that allow to follow data flow at run time. All I/O handling routines are highly standardized.

ARCHIVING

Finally a few remarks on the data archiving should be made. Concerning the hardware, we plan to use optical disks.

As to the software, we will use some kind of archiving integrated within our steering system. The reason is that the amount of data that we expect on one hand and the needs of quick reaction on user requests on the other hand require a rather high flexibility.

Since all I/O files of the key number of one processing step are stored within one directory, with each processing version having its own subdirectory therein, we will archive directories rather than single files. So we will store the archived files within the same directory with only the disk name being changed. This change of disk name has to be noted within the database tables used by the Steering system and only few additions should be necessary to allow to use "archived" files in the same way as use "living" files.

PREPARING ANALYSIS OF HUBBLE SPACE TELESCOPE DATA IN EUROPE

H.-M. Adorf[1], D. Baade[1], and K. Banse[2]

[1] Space Telescope - European Coordinating Facility
[2] European Southern Observatory
Karl-Schwarzschild-Str. 2
D-8046 Garching b. München, F.R. Germany

Abstract: Since no image processing system will ever be complete, the best system is the one which is the easiest to supplement. We therefore discuss various solutions to the typical problems encountered in porting application programs between systems: (a) standard interfaces to data bases and peripheral devices, (b) interface emulation, (c) physical replacement of calling sequences, and (d) restriction to subroutines only. The homogeneity of the importing system is seen as an important requirement for the success of any of these methods. Corresponding provisions within ESO's image processing system MIDAS are described.

INTRODUCTION

One of the tasks of the Space Telescope - European Coordinating Facility (ST-ECF) is to ensure that European HST observers have available data reduction capabilities which live up to the expectations put into HST data. In the following we consider two major aspects to this: (i) The provision of suitable algorithms and (ii) the definition/identification, development, and optimization of an environment in which those algorithms attain maximum efficiency.

THE SYSTEM: MIDAS

The Space Telescope Science Institute (ST ScI) is developing the Science Data Analysis System (SDAS, Albrecht 1983, 1985a,b; Shames, 1986), which at launch time will provide a minimum set of application programs for the reduction of HST data. SDAS is executed within the Interactive Reduction and Analysis Facility (IRAF) developed at Kitt Peak (Tody, 1986). As this solution did not appear suitable for Europe, the ST-ECF has adopted ESO's general astronomical data analysis system MIDAS - suitably modified to ensure compatibility with SDAS - to serve as the host system for the analysis of HST data in Europe. This means that the ST-ECF will primarily support the acquisition of already existing as well as the creation of new software with the aim of incorporating it into MIDAS, which should be viewed as "a general exportable system provided by ESO" for the European astronomical community. The general architecture of MIDAS is described elsewhere (Banse et al., 1983; Crane et al., 1985;

"The MIDAS Users Guide, 1986"); here we would like to highlight some efforts to make MIDAS especially suited for the analysis of HST data.

The reduction of data from new instruments will always pose special problems - in case of the HST the most prominent ones are the nature of the point spread function (PSF), especially its dependence on wavelength, and the different sampling of the PSF by the various instruments - for any data analysis system and will almost certainly trigger the development of new software. A data analysis system can aim to be fairly complete only with respect to instrument independent tasks, e.g. data display, simple statistics, etc., whereas it should provide an **open** and **flexible** environment for the inclusion of new applications, which is certainly the case with MIDAS. It is equally important that a system is **simple** so that even the inexperienced occasional FORTRAN programmer may easily incorporate his own dedicated application into the system and take advantage of all the general system "services". It is one of the strong points of MIDAS that is does not distinguish between a function provided by the system and an application program added by the user. Thus the expertise of both astronomers and software engineers can be integrated into one single system, to the advantage of the general user. As a particularly successful example we mention the image detection and classification suite of programs (INVENTORY etc., see "The MIDAS Users Guide, 1986", Chap. 11) written by A. Kruszewski while he was visiting ESO.

The success of a general data analysis system depends critically not only on the variety of its algorithms but at least as much on the "homogeneity" of its data structures. MIDAS supports a balanced assembly of essentially three generic data structures: Bulk Data Frames, Table Files and Keywords. Data in either format may be converted into any other. MIDAS table files (Grosbøl & Ponz, 1985) have proven to be extremely useful for the storage and further processing of results from image analysis. Recently included multivariate statistical functions (see "The MIDAS Users Guide, 1986", Chap. 13) further enhance the power of the MIDAS table system, which may now be viewed as a "statistical and scientific data base management system" (SSDBMS, see e.g. Shoshani & Wong, 1985). The statistical capabilities will help to extend image analysis as close as possible to the noise limit in assessing the significance of suspected conclusions. This is especially important for HST data where repeating the observations with a better suited telescope and/or instrument will usually be impossible.

In order to facilitate the planning of HST observations, the ST-ECF has developed detailed MIDAS based software models for the HST and several of its instruments (Rosa, 1985). They are relevant in the context of this paper insofar as they produce realistic images from **accurately known** input data; thus they allow testing of the reliability of the software and to answer the question about how much the original information may be recovered by various reduction strategies. This knowledge will also be important in assessing the significance of results derived from future real HST data.

SOFTWARE EXCHANGE

Much of the software used for the analysis of ground-based observations will also be useful for reducing HST data. Cost reasons, furthermore, preclude the development and testing (!) of completely new algorithms for each and every task. Thus the only realistic way to achieve a satisfactory degree of completeness is to include "foreign" programs into the given "domestic" environment. A common instrument and a common data base will additionally stimulate the quest for exchange of software and data.

While data may be satisfactorily transported using the widely accepted FITS standard (Wells & Greisen, 1979; Harten et al., 1985), the exchange of software between SDAS and MIDAS has only recently been achieved by a set of so called "Standard Interfaces" (see "ST-ECF 0-11, Standard Interfaces"), which are based upon the similarities displayed by the systems' global architecture (Fig.): Firstly, communication with external resources, i.e. terminal, graphics display, etc. and most importantly the generic data base files, is channelled through interfaces, which effectively isolate the application program from the external resources (principle of "information hiding"). Secondly, the algorithms which actually "do the work" are encapsulated in subroutines; information to and from these modules is passed only via the calling sequence (principle of "modularity").

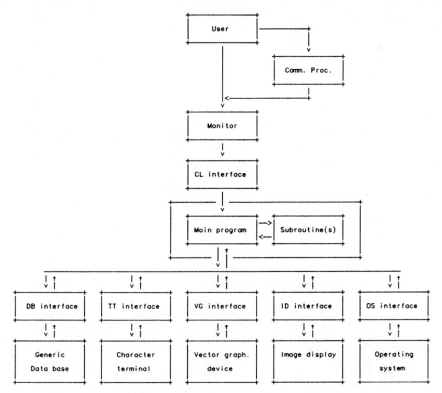

Fig. The system architecture of a modern data analysis system (e.g. IRAF/ SDAS or MIDAS). The user communicates with the monitor interactively or via command procedures. The monitor, which in principle may access any system resources, normally interacts with the database and the character terminal, and then invokes the application main program. The main program receives its instructions through an interface to the Command Language (CL), possibly interacts with system resources, and invokes one or more subroutines encapsulating the scientific algorithm.

Accordingly, an application program may then be ported from one system to the other as follows: The code of the main program and the subroutine(s) is left **untouched**. It is just **relinked** with a different subroutine library containing the local (SDAS or MIDAS) implementations of the standardized low-level interfaces. High-level interfaces (not displayed in the Fig.) may be assembled from the low-level ones and freely used within any applications program. As long as these high-level interfaces are distributed along with the calling program, portability is maintained.

The current set of Standard Interfaces, agreed between and obtainable from ESO/ST-ECF and STScI, provides calling sequences for communication between the application's main program and the **monitor** (CL Interface), a character **terminal** (TT Interface), and the generic **data base** files (DB Interface). Negotiations are going on between STScI, ESO/ST-ECF and the Rutherford Appleton Laboratory STARLINK node on calling sequences for communication between the application program and an **image display** (ID Interface). A draft agreement was presented (Terrett & Shames, 1986) at the Third ST-ECF Data Analysis Workshop.

Regarding a standardized interface for communication with a **graphic device**, the situation is more involved. MIDAS, on the one hand, has been using various graphic libraries in parallel, but recently work has started to convert to the ASTRONET Graphics Library (AGL, Fini, 1985) which provides hardware independent high-level graphics functions to the calling program. As AGL has evolved in an astronomical environment, it should be well tuned towards the special requirements of astronomically oriented application programming. It will continue to be developed further according to the needs of the European astronomical community. Graphics in SDAS, on the other hand, are currently built on a library of high-level graphics functions provided by the NCAR's SCD Graphics System (McArthur & Henderson, 1981). In the future both MIDAS and SDAS will use, or at least be compatible with, the Graphical Kernel System (GKS) the international device independent standard for graphics, which will make both data analysis systems usable with any graphics device, for which a GKS driver exists. AGL will achieve compatibility with GKS on the metafile level, whereas a version of NCAR's SCD already exists which internally calls GKS primitives.

The export of the MIDAS table system into SDAS illustrates that software exchange on a larger scale has already taken place. Similarly, we expect to import major SDAS packages into MIDAS, e.g. P. Stetson's DAOPHOT package for photometry on two-dimensional images.

The use of standardized interfaces requires mutual agreement on an subsequent use of predefined calling sequences, which is not always feasible. Fortunately the method of achieving portability via "standardizing interfaces" described above may be generalized to porting software via "emulating interfaces". If for every call to a subrouting of an interface library X there exists a fixed context-independent sequence of calls to subroutines from an interface library Y, then it is possible to "emulate" the interfaces from set X by subroutines which internally consist of nothing but calls to interfaces from set Y. Once a complete and error-free emulation X'(Y) has been written any application program originally linked with library X may be ported by relinking it with library X'(Y). Again, the code of the application program is left **untouched**. No testing of the algorithms is required after porting. Provided a complete and error-free emulation is used, the method ensures that the application program will yield the same results in either system.

Software which has been ported by emulating a subroutine library may suffer from inefficiency, because an additional and in principle superfluous layer of modules has been introduced. If for performance reasons an emulation must be avoided but all other conditions for an emulation are fulfilled, an application may be ported by "physically replacing" its old calling sequences by appropriate new ones. Whereas for a singular port the conversion can be done manually using an editor, it must be done in an automated way using a "macro preprocessor"* in case of repeated ports. Such a tool, which ideally is driven by nothing but a table of translation rules, i.e. a "dictionary", is very much required and will for instance allow an efficient import by "physical replacement" of much of the STARLINK software collection into MIDAS.

If none of the methods mentioned above is applicable, a final resort to import scientifically important algorithms, embedded in appropriate subroutines, consists of only "porting the subroutines" and reconstructing a driving main program (using a template), which handles the interactions between algorithm, user, data base, etc. To provide a general source of potentially useful code, qualified subroutines are also collected in, and available from, the "ST-ECF Software Library" (Murtagh, 1985).

Up to now practically no consideration has been devoted to the question of how to port high-level command procedures, which increasingly gain importance, as more and more image processing tasks are being solved on the command procedure level. This development currently threatens to undermine previous efforts to maintain software compatibility between MIDAS and SDAS. It will have a particularly strong impact on the user because it usually is the command procedures which contain most of the astronomical expertise. One possible solution would be to find a way of automatically translating command procedures**, another one to build or acquire a multi-lingual monitor (see e.g. Teuber, 1985).

We have only covered the issue of portability between astronomical sites of which at least the target site is capable and willing to run either MIDAS or SDAS. The problem of software exchange between sites which run neither system lies outside the scope of this work. However, the exploitation of the concept of Standard Interfaces and the usage of the other software porting methods indicated above should make the MIDAS-SDAS twin a large, if not the largest, pool of data analysis software available to astronomers.

---

* At the ST-ECF the MACRO utility from the Software Tools User's Group (available via DECUS), is used to semi-automatically convert software as described. Recent experiments show that the ISTMP macro processor from the NAG TOOLPACK/1 set (Cowell et al., 1985) seems to have solved all but one of the problems we had found in using the MACRO utility and which had prevented us from fully automating the conversion task.

** The state-of-the-art solution would be to use a lexical analyzer/parser generator, for instance LEX/YACC (Lesk, 1978; Johnson, 1975), to build a translator from a language grammar table and a set of translation rules. A grammar table for IRAF already exists, because it internally uses LEX/YACC.

REFERENCES

Albrecht, R., 1983, Data Analysis for the Space Telescope, in: "Technical Digest", Americal Astron. Soc. & Opt. Soc. America, ThA1-1.

Albrecht, R., 1985a, Science Data Analysis for the Space Telescope, Mem. Soc. Astron. Italiana, **56**:339.

Albrecht, R., 1985b, Space Telescope Science Data Analysis, in: "Data Analysis in Astronomy", V. Di Gesù, L. Scarsi, P. Crane, J.H. Friedman, S. Levialdi, eds., Plenum Press, New York and London, p. 191.

Banse, K., Crane, Ph., Ounnas, Ch., Ponz, D., 1983, MIDAS - ESO's Interactive Image Processing System Based on VAX/VMS, in: "Proc. Digital Equipment Computer Users Society", Zürich, p. 87.

Cowell, W.R., Hague, S.J., Iles, R.M.J., 1985, "TOOLPACK/1, Introductory Guide", Numerical Algorithms Group, Ltd., NAG Central Office, Mayfield House, 256 Banbury Road, Oxford OX2 7DE, UK.

Crane, Ph., Banse, K., Grosbøl, P., Ounnas, Ch., Ponz, D., 1985, MIDAS, in: "Data Analysis in Astronomy", V. Di Gesù, L. Scarsi, P. Crane, J.H. Friedman, S. Levialdi, eds., Plenum Press, New York and London, p. 183.

Fini, L., 1985, "ASTRONET Graphic Library. Reference Manual, (Vers. 2.0)" ASTRONET Documentation Facility, Osservatorio Astronomico di Trieste, Trieste, Italy.

Grosbøl, P., Ponz, D., 1985, The MIDAS Table File System, Mem. Soc. Astron. Italiana, **56**:429.

Harten, R.H., Grosbøl, P., Tritton, K.P., Greisen, E.W., Wells, D.C., 1985, Generalized FITS extensions with application to Tables, Mem. Soc. Astron. Italiana, **56**:437.

Johnson, S.C., 1975, Yacc - yet another compiler compiler, CSTR 32, Bell Laboratories, Murray Hill NJ.

Lesk, M.E., 1975, Lex - a lexical analyzer generator, CSTR 39, Bell Laboratories, Murray Hill NJ.

McArthur, G.R., Henderson, L.R., 1981, An Introduction to the SCD Graphics System, NCAR Technical Note - 161 + IA, Scientific Computing Division, National Center for Atmospheric Research, Boulder, Colorado.

Murtagh, F., 1985, The Software Library, ST-ECF Newsletter, **4**:10.

Rosa, M., 1985, Using the HST Software Model, ST-ECF Newsletter, **4**:7.

Shames, P.M.B., 1986, Data Analysis for the Space Telescope Mission, this volume.

Shoshani, A., Wong, H.K.T., 1985, Statistical and Scientific Database Issues, IEEE Transact. Softw. Engin., **SE-11**:1040.

"ST-ECF 0-11, Standard Interfaces", Space Telescope - European Coordinating Facility, Garching b. München, FRG.

Terrett, D., Shames, P.M.B., 1986, "An Image Display Interface for Astronomy, Draft Version 0.4", Space Telescope Science Institute, Baltimore MD.

Teuber, D., 1985, An Astronomical Data Analyzing Monitor, in: "Data Analysis in Astronomy", V. Di Gesù, L. Scarsi, P. Crane, J.H. Friedman, S. Levialdi, eds., Plenum Press, New York and London, p. 235.

"The MIDAS Users Guide, 1986, ESO Operational Manual No. 1", Image Processing Group, European Southern Observatory, Garching b. München, FRG.

Tody, D., 1986, to be published in SPIE Proc., **628**.

Wells, D.C., Greisen, E.W., 1979, FITS: A Flexible Image Transport System, in: "Image Processing in Astronomy", G. Sedmak, M. Capaccioli, R.J. Allen eds., Osservatorio Astronomico di Trieste, p. 445.

COMPASS: THE COMPTEL PROCESSING AND ANALYSIS SOFTWARE

SOFTWARE SYSTEM

R. Diehl[1], G. Simpson[3], and T. Casilli[5],
V. Schoenfelder[1], G. Lichti[1], H. Steinle[1].
B. Swanenburg[2], H. Aarts[2], A. Deerenberg[2], W. Hermsen[2],
K. Bennett[4], C. Winkler[4], M. Snelling[4*],
J. Lockwood[3], D. Morris[3], and J. Ryan[3]

[1]MPI fuer extraterrestrische Physik, Garching, West Germany
[2]Laboratory for Space Research, Leiden, Holland
[3]Space Science Center, University of New Hampshire, Durham, USA
[4]Space Science Department of ESA, Noordwijk, the Netherlands
[5]CRI A/S, Copenhagen, Denmark
*Present Address: British Aerospace, England

ABSTRACT

We describe the features of the COMPASS, the system for the processing and analysis of data from COMPTEL a gamma-ray telescope which will be put into orbit onboard the NASA Gamma-Ray Observatory. The main features of the COMPASS are: high host system independence, data integrity maintenance facilities, a menu-controlled user shell, and data access via data abstractions.

1. INTRODUCTION

COMPASS is the software system which will support the analysis of the data from COMPTEL, one of the four astronomical experiments which will make up the NASA Gamma Ray Observatory (launch date 1988/1990). COMPTEL is a Compton telescope, the first such instrument in space. It is the joint product of the Max-Planck-Institut fuer extraterrestrische Physik, the Laboratory for Space Research in Leiden, the Space Science Department of the European Space Research and Technology Centre, and the University of New Hampshire. COMPTEL will analyse the sky in a region of the electromagnetic spectrum which is virtually unexplored, that of gamma rays from 1 to 30 MeV. The instrument delivers a considerable body of information on each photon which it detects, and it will detect about 10^9 photons during its nominal two year lifetime. The COMPASS tasks of processing the COMPTEL information into a scientifically usable form and supporting the scientific analysis of the data will require considerable machine and software resources.

COMPASS will be required to handle a volume of data in the range of 20-50 gigabytes. The size of the software system is estimated to be 100.000 lines of executable Fortran. Due to the complex nature of the COMPTEL analysis, COMPASS will be continually evolving during the mission and after its completion. The design, implementation and operation of COMPASS will be shared among the four collaborating institutes. These facts make adaptability, maintainability, and portability of the system central requirements.

The scientific objectives of the mission define the COMPASS outputs, as shown in Figure 1. The specific goals we have defined, in addition to the qualitative concerns just mentioned, are:

1. Handle the volumes of data required to meet the scientific objectives within the machine resource constraints of the 4 institutes.
2. Guarantee the integrity of the data, especially against faults induced by the evolving software system.
3. Provide user interfaces which direct the attention of the scientific analysts to the scientific objectives, rather than to operating system details.

Real-time response is not a requirement of the COMPASS, except in the area of user interactions.

In this paper we describe some of the aspects of the COMPASS which may be of interest to other astronomical database developers.

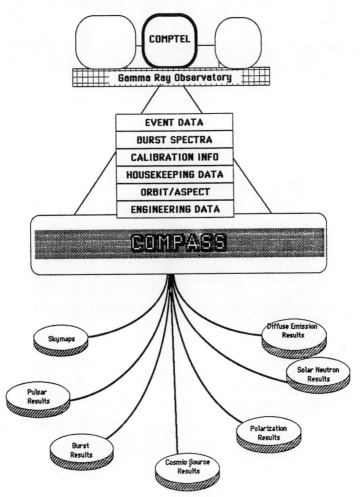

Figure 1: TRANSLATION of COMPTEL DATA to SCIENTIFIC RESULTS

2. SYSTEM CONCEPT

The system concept is seen in Figure 2, where the COMPASS is represented as a series of shells around the host computers. The inner shells isolate the application programs from the specific host and data storage system. The user shell protects the astrophysicists from the time-consuming detail associated with the specific host command language, and helps guarantee consistent treatment of the data. The shells also protect the data and program resources. The innermost shell is the operating system of the host computer. We have 3 different operating systems in the collaboration: IBM`s VM/CMS and MVS/TSO, and PRIME`s PRIMOS.

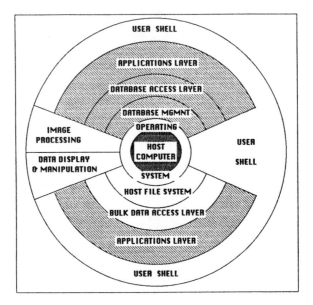

Figure 2:

COMPASS SYSTEM CONCEPT

The host file system is used to store the bulk of the COMPTEL data, to take advantage of the speed which can be achieved via direct use of the host operating system functions. The "Bulk Data Access Layer" around the host file system presents the applications software with a logical data access interface which is independent of the operating system in use. The Database Management System is used primarily for "dataset descriptors" (described below), and not for the bulk data. This is because the logical relations which need to be managed are among the parameters of the data files, rather than among the data items themselves.

The Applictions Layer contains the programs which perform the various steps of data processing and analysis. There are 16 scientific subsystems in this layer. The applications programs will be identical for all the machines.

There are two special zones of the diagram which do not go through the normal shells. The User Shell's access to the operating system (to initate jobs for the User) is a system-dependent task, which therefore must bypass the shells. The Image Processing system and the Data Display and Manipulation subsystem are shown accessing the local operating systems because of the need to maximize use of existing hardware and software. These systems read datafiles from the COMPASS and are exercised locally to manipulate the data for interactive studies.

3. SCIENTIFIC APPLICATION PROGRAM INTERFACES

The input/output services required by the application programs are grouped into the following categories:

- user communication (command and parameter input, messages)
- database management services (queries, updates)
- bulk data access
- data display

All user command and parameter inputs are handled by a user shell menu system. The menus will be functionally identical at all the sites, and will allow the user to quickly walk through a tree-structured command system to reach the desired data analysis facilities. At the end of such a walk is a menu on which the user identifies a specific high-level task or set of tasks to be executed. Within the User Shell, User Profiles control which of the COMPASS facilities are available to each user. Clearly, the undergraduate help must not have access to the facility for clearing the database, for example.

Controlling information for each application program is gathered by an application-specific user shell "setup" task, which collects user inputs using screen forms. The result of the setup task is a "Task Parameter Table", which contains all the controlling information that the application requires, including both user-defined parameters and system information which the user does not control (such as host file names). The Task Parameter Table stays with the output data, forming an important part of the data descriptor (see below). Once the complete Task Parameter table is assembled, the user shell initiates a batch process to execute the task, returning the user to the COMPASS menus for further commanding.

The bulk data access layer, in addition to providing optimized access to data storage, is used to enforce data abstraction. This is one way in which the issues of maintainability and adaptability mentioned in the introduction are addressed. Data abstraction means hiding the information about the physical representations of the data, in order that the software which is using the data not become dependent upon the particular form in which the data are stored. Rather, the application program calls data access functions which contain only the logical data items which are meaningful to the application, while the access routines take care of the links to the physical data. Each of the COMPTEL subsystem interfaces is implemented as a data abstraction, in which only a small set of functions are privy to the information about the physical layout of the data. All access to the data is via these functions, which are rigidly specified and maintained. (We specify our access functions following the specification language described in Litkov and Guttag: Specification and Design of Software, (MIT), 1985). New data access functions may be added to the library, and the organization of the data may be altered (for example to improve perfomance), but the applications programs are protected because the specifications of the access functions are frozen.

Database Management facilities are used to free the applications programs (and the users!) from file names: dataset references and queries are made based on the logical description of the data as available via the DBMS descriptors rather than by searching or separately indexing the physical data. The same holds true for software functions. Whenever an application program generates an output dataset or a new software object is added to the COMPASS a user shell function is called to create the appropriate descriptor and update the database accordingly.

For data display, the scientific applications programs have access to several libraries, ranging from GKS graphics primitives to high-level statistical distribution handling software from the CERN HBOOK library. In addition to these, a special workstation-based interactive Data Display and Manipulation subsystem exists. This subsystem will allow the user to examine data directly from the data base using interactive HBOOK facilities, or alternatively the user may choose to download files for local study using commercial multi-dimensional animation graphics and spreadsheets.

4. DATA and SOFTWARE MANAGEMENT

The COMPASS presently consists of a total of 19 subsystems, each of which includes a variety of subfunctions. During the COMPASS lifetime we expect to encounter: evolution of the basic subsystems (both because of software evolution and improved scientific methods), new subsystems (new scientific objectives), and the detection of various sorts of errors. Changes in the software driven by these factors will threaten the COMPASS data integrity, in two ways:

1) software may be in error, producing wrong results, which must therefore be erased from the data base, along with all subsequently derived data; and
2) different software modules, while individually correct, may be inconsistent with one another, causing incompatibility between the data which they generate.

The treatment of the data by the users can also threaten the data integrity. There is virtually an infinite number of different treatments which can be applied to process and analyse the data. Without some control which harmonises the treatments occurring in the different subsystems, and without some precise method of verifying the compatibility of the inputs to each user-initiated process, inconsistent results may be produced.

In addition to these factors, circumstances may arise (e.g.: on board the spacecraft) which can make certain processes invalid. This is not strictly a software matter, but it is a concern which the COMPASS must address.

COMPASS combats these problems using a relational database management system. A catalog of all the COMPASS files, including data, software, documentation, and control tables is embedded in the DBMS. COMPASS then controls the data flow and maintains the data integrity using the following specific tools:

Data Set Descriptors
Task Profiles
Task Parameter Tables
Selection Sets
Processing Definitions Tables

Commentaries
Menus & User Profiles.

Some of these have been described in sufficient detail above. Here we complete the discussion of the COMPASS by describing the rest of these features.

DATA SET DESCRIPTORS are the key entities used to facilitate data access, defend the data, and assure coherency among the user task and the various data selected for analysis. As illustrated in Figure 3, each COMPASS data file (there will be approximately one million) will be given a descriptor on which is recorded not only the scientific and technical parameters of the file, but also the software and data heritage. The software heritage item identifies the software which was used to generate the data. The data heritage item lists every data set which was used in the creation of the data. User requests for data satisfying scientific and/or technical criteria are thus facilitated, and the COMPASS system can track down all the datasets affected by any software or data fault.

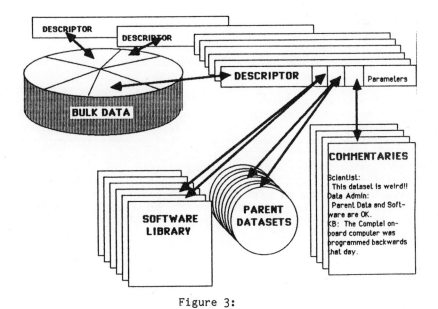

Figure 3:
DESCRIPTOR CONCEPT

The relational capability of the DBMS over the dataset descriptors will be used to interrogate the database on the basis of logical relationships. The user or application program will, for example, be able to find the set of all skymaps which address a specific region of the sky in a specific energy regime, before he embarks on a program to produce such maps himself.

Certain descriptors will also be used in a hierarchical mode. Much of the data may be organized in a variety of hierarchies: of time (orbits, days, observation periods, the whole mission), or energy (small intervals gathered into larger intervals), etc. Single high level descriptors will be used to point to many lower-level descriptors, thus allowing the collection to be considered as a single logical entity.

TASK PROFILES are the User Shell mechanism for connecting the selections which users make on their menus with the applications software modules and system command streams needed to carry out the requested action. There is a library of task profiles, each one referenced to a particular menu item. Selecting the menu item invokes the Task Profile Handler, which initiates the setup dialog for the task. The Task Profiles are kept up-to-date, so that obsolete and incompatible versions of the software are not invoked, and so that the output datasets can be tagged with their software history.

SELECTION SETS: the majority of the COMPASS data types are eventlike, i.e.: a series of records of identical logical content with varying values. One of the most common activities under the COMPASS will be selecting a subset of the data which has a particular range of parameters. For example, one may want to form the subset of all the gamma rays which have energies in a range which depends on the geomagnetic latitude of the instrument, to optimize the signal-to-noise ratio for a particular investigation. Often, two streams of information must be merged; if the selections which have been applied to these two streams are not compatible, the analysis will be nonsense. The selection set is a very general concept for handling this kind of problem. A selection set consists of 4 parts:

1. a unique name,
2. a Fortran function which processes a single "event" returning a logical value (accept/reject),
3. a set of parameters which the function uses, and
4. a documenting comment.

There may be a great variety of different selection functions, and each function may be used with a wide variety of its specific parameter values, but each combination is uniquely identified. Management of selection sets within the COMPASS is critical to ensuring the integrity of the scientific results. To achieve this, selection sets are identified and catalogued like the data, using descriptors. There will be an "official" library of selection sets, endorsed by the COMPTEL scientific community as appropriate for standard analysis, as well as a vaiety of development selection sets, the products of which do not enter into the regular flow of COMPASS information.

PROCESSING DEFINITIONS TABLES provide a degree of coherency among the products of the different subsystems by defining "standard" values of certain instrumental and data treatment parameters.

COMMENTARIES: from the scientist's point of view, all of the above functions should be taken for granted. He is much more concerned with communicating and documenting the conclusions which have been drawn regarding the scientific and technical content of the data. The commentaries are a facility for doing just that. Via the DBMS, one or more text files may be associated with specific data descriptors. (See Figure 3.)

5. CONCLUSION AND ACKNOWLEGEMENTS

The COMPASS will undergo its first "live" testing in early 1987, when it will be applied to the analysis of the instrument calibration data. At that time, we expect to have the User Shell and the first quarter of the applications subsystems in production. The goal for completion of the full system is early 1988; but the system will evolve after launch of the spacecraft.

The COMPASS is the brainchild of the COMPTEL Data Reduction Group, aided by Computer Resources International of Copenhagen, It is in large part a product of a user-developer community of astrophysicists with a broad range of scientific and computing backgrounds. The authors express their substantial appreciation to all those who have contributed to the ideas described here.

THE STRASBOURG ASTRONOMICAL DATA CENTRE (CDS)

AND THE SETTING UP OF A EUROPEAN ASTRONOMICAL DATA NETWORK

André Heck

Observatoire Astronomique
11 rue de l'Université
F-67000 Strasbourg
France

INTRODUCTION

Following what I heard during the previous interventions and in order to avoid any misunderstanding, I feel necessary to begin this communication by making a distinction between the different types of data an astronomer can deal with.

Observational data are those that essentially constitute the bulk of archive centres (as the IUE, EXOSAT, etc. data banks). Catalogues ("logs") of observations are their natural complements. Finally we could call astrophysical catalogues those gathering values of reduced or extracted parameters.

All these data are complementary, but it should be clear that maintaining the corresponding data banks brings up completely different problems. CDS is essentially concerned by the last two types of catalogues.

Users with account numbers on SIMBAD, the CDS astronomical data base, can now be found in twelve countries. They are located at places like the STARLINK node at Rutherford Appleton Laboratory, the European Space Agency IUE Observatory in Madrid and EXOSAT Observatory in Darmstadt, the European Southern Observatory and the Space Telescope European Coordinating Facility in Garching, the Space Telescope Science Institute in Baltimore, the NASA Goddard Space Flight Center in Greenbelt, etc.

SIMBAD contains currently more than 70,000 non-stellar objects (essentially galaxies), besides about 600,000 stars. SIMBAD is permanently updated and provides cross-identifications and many different kinds of astronomical data from about 400 interconnected catalogues, and bibliographical references from the 85 most important astronomical periodicals covering the period from 1950 to the present date.

Moreover, more than 500 individual catalogues (including the IUE log and IRAS observational material) are available on magnetic tape. Some of them are also available on microfiche. The CDS Special Publications offer specific catalogues compiled by CDS collaborators or tutorials on statistical methodologies.

Pioneering the field, CDS has set the standard for astronomical data centres in other countries, often helping them to establish themselves. Formal collaborations and permanent exchanges of data have been secured with several of them.

CDS is involved in several important space projects like HIPPARCOS and TYCHO, and is collaborating with other space centres like those mentioned above. CDS plays also a role in the various discussions that should lead in a near future to the setting up of a European astronomical network.

A LITTLE HISTORY

In 1972, the French National Institute of Astronomy and Geophysics (INAG, now the National Institute of the Sciences of the Universe - INSU) decided to create a "Centre de Données Stellaires" with the following aims:
- to compile the most important stellar data available in machine-readable form (positions, proper motions, magnitudes, spectra, parallaxes, colours, etc.),
- to improve them through critical evaluation and comparisons,
- to distribute the results to the astronomical community, and
- to carry out its own research.

The centre has been installed at Strasbourg Observatory and is headed by a Director (presently C. Jaschek) who is responsible to a Council composed of six French and six foreign astronomers.

Besides collecting astrometric, photometric, spectroscopic and other catalogues, the first important accomplishment of the CDS has been to construct an enormous dictionary of stellar synonyms called the "Catalogue of Stellar Identifications" (CSI). Some stars have more than thirty different names. This catalogue has been complemented by the "Bibliographical Star Index" (BSI) giving, for each star and from the major astronomical periodicals from 1950 onwards, the bibliographical references to the papers mentioning this star. On the average a star is cited in five publications, but some stars are quoted in more than five hundred papers.

Taking advantage of the fact that, through the CSI, any identification can give access to all connected catalogues, and thus to their data, a user-friendly conversational software system has been built around it, leading to the present dynamic and unique configuration of the SIMBAD (Set of Identifications, Measurements and Bibliography for Astronomical Data) base accessible from remote stations.

Subsequently, data on non-stellar objects were included, together with their bibliographical references (from 1983 onwards). Taking account of this, and in order to retain well-known abbreviations like CDS, CSI and BSI, the word "stellar" appearing in them has been replaced by "Strasbourg".

PRESENT STATUS

Thus CDS represents much more than a mere accumulation of catalogues. Presently, SIMBAD is probably the largest astronomical data base in the world. It contains about 700,000 objects (including more than 70,000 non-stellar objects - mostly galaxies) for which more than 2,000,000 identifications have been recorded. These figures will be quickly out-of-date with the planned inclusion of the Guide Star Catalogue of the Hubble Space Telescope (of the order of 30 million objects).

The table of synonyms and the connected catalogues can be accessed through any object designation (about 400 different types) or by object coordinates, equatorial, ecliptic (at any equinox) or galactic. In the latter mode, one may request to get all objects within a rectangle or a circle of given dimensions around a given position. Criteria can also be specified on parameters such as magnitude, existence of various types of data, etc. Such sampling allows SIMBAD to be a powerful tool for statistical studies. Moreover maps can be produced. Thus SIMBAD can be also a precious auxiliary for identifying fields and preparing ground or space observing runs or proposals.

The bibliographic index contains references to stars from 1950 to 1983, and to all objects outside the solar system from 1983 onwards. Presently there are more than 500,000 references to more than 150,000 objects. These are taken from the 85 most important astronomical periodical publications.

SIMBAD, the CDS data base, is accessible through data networks, including the French TELETEL public service. To the present date, there are about seventy users (including a few amateur astronomers) in twelve countries regularly interrogating SIMBAD. The figures are rapidly increasing. Apart from a simple terminal, the only requirement for accessing SIMBAD is obtaining an account number from CDS which will be used for invoicing. Costs are however not very important. Astronomers without access to a data network can mail their request to the Data Centre which will then return a printout with the corresponding data.

In the same way, copies of individual catalogues (from a list of more than 500) can be obtained on magnetic tapes. Some of them (about 50) are also available on microfiche.

SIMBAD is continuously growing and kept up-to-date, not only by the Strasbourg CDS staff, but also by many cooperating persons in other institutions. All the catalogue available at CDS have been produced by specialists, so that their high quality is guaranteed. Some catalogues, prepared at CDS itself and available as CDS Special Publications, are made in fields where the Strasbourg personnel has specific qualifications. Thus the "Catalogue of Stellar Groups" lists some 30,000 stars according to their spectral peculiarities. The CDS Special Publications series includes also tutorials on statistical methodologies and directories gathering all practical data on astronomical societies and institutions.

Collaboration with other institutes having specialization in specific fields is a natural consequence of the CDS policies. This is particularly the case for Bordeaux, Meudon and Paris (bibliography), Genève and Lausanne (photometry), Heidelberg (astrometry) and Marseille (radial velocities).

To encourage exchanges with other countries, formal agreements have been signed, in particular with NASA (USA), the Astronomical Council of the USSR Academy of Sciences and the Potsdam Zentralinstitut für Astrophysik (German Democratic Republic). CDS is also collaborating with Japan (Kanazawa Institute of Technology) and the United Kingdom (STARLINK). The goal of these agreements is to allow all astronomers in the world to have access to all existing catalogues.

OTHER CDS ACTIVITIES

On a much larger scale, CDS takes also an active part in space projects like HIPPARCOS and TYCHO which are heavily dependent on SIMBAD for the preparation of their respective input catalogues. The Space Telescope GSSS team is collaborating with the CDS for the inclusion of the stellar cross-identifications from SIMBAD in their Guide Star Catalogue. CDS acts also as the European disseminator of the IRAS observational material and has been requested by the European Space Agency (ESA) to homogenize the IUE log of observations.

The ESA IUE Ground Observatory in Madrid was actually the first foreign station connected to SIMBAD which was used operationally for checking target coordinates and as an open service to visiting astronomers. ESA EXOSAT Observatory was later connected, followed by other space centres like the Space Telescope Science Institute in Baltimore and NASA Goddard Space Flight Center in Greenbelt. The Space Telescope European Coordinating Facility in Garching is also connected to SIMBAD through its host, the European Southern Observatory.

Apart from smaller scale scientific meetings taking place twice per year, Strasbourg Observatory has organized several important colloquia centred on data collection, dissemination and analysis. The last one, in 1983,

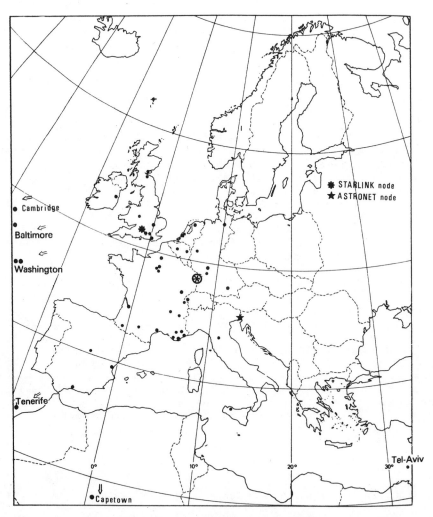

The map represents the European locations of CDS user stations.
Non-European stations are arrowed.

was co-sponsored by the European Space Agency and was devoted to the statistical methodology in astronomy. All these meetings were great successes and revealed the importance that the astronomical community attaches to this type of work.

The interest in CDS work is also shown by the growing number of astronomers visiting it, either to get to know the CDS or to set up a collaborative project. CDS stays in touch with its users and other interested persons by a six-monthly Bulletin distributed free of charge. Apart from keeping readers updated on CDS services and latest developments, it contains also general papers and news about other data centres' activities. The CDS Bulletin issues are also including now the IAU Commission 5 (Astronomical Data) Newsletter as well as the Newsletter of the Working Group for Modern Astronomical Methodology.

Recently CDS has initiated the implementation of a reduced astronomical data base for the general public accessible through the French public TELE-TEL service.

A EUROPEAN ASTRONOMICAL NETWORK?

CDS is quite naturally involved in the various discussions that should bring soon in operation a European astronomical data network. A specific colloquium has actually been organized by CDS in NOvember 1985 on the theme "Astronomical Data Networks". The various communications are gathered in CDS Bulletin n° 30. They give a review of the various networks in operation, being set up or in project.

A follow-up meeting is organized on 29 & 30 May 1986, again in Strasbourg, under the auspices of the European Science Foundation. A number of projects have indeed seriously evolved meanwhile and it is also necessary to keep an adequate pressure on the national telecommunication administrations in order they upgrade the international protocols for data exchange.

The European Space Agency is considering to set up a network called ESIS (European Space Information System) and consisting of a pilot experiment gathering four centres: the IUE data bank at VILSPA, the EXOSAT data bank presently at ESOC in Darmstadt (later at ESTEC in Noordwijk), the ST archive that will be located at the ST-ECF (at ESO, München) and finally the CDS. Gateways to SPAN and STARLINK are also foreseen.

The objectives of this network are directed towards an integration as deep as possible, so that an astronomer could get all fundamental data and the bibliography from the CDS SIMBAD, as well as pointers towards IUE images, EXOSAT data, etc.

It seems obvious that the astronomical community will follow the development of the communication era and that astronomers, even in small institutions equipped only with a terminal, will have access to very powerful means of data retrieval and, hopefully soon, of data processing.

AND ...

If you are interested in the CDS services or for any enquiry, do not hesitate to get in touch directly with us at the following address: C.D.S., Observatoire Astronomique, 11 rue de l'Université, F-67000 Strasbourg, France (Phone: +33-88.35.82.00, telex: 890506 starobs f, EARN FRCCSC21.U01117).

AN INDEXING ALGORITHM FOR LARGE MELTI-DIMENSIONAL ARRAYS

Pat Moore

National Radio Astronomy Observatory*
P.O. Box 0
Socorro, New Mexico 87801   USA

When using multi-dimensional arrays in high level languages a mapping has to be performed between the array indices and the address space of the machine. Except for some special purpose processors this address space is one-dimensional, so the mapping is from a multi-dimensional space to a one-dimensional space. There are many possible mappings that can be used, and all high level languages use one particular mapping that is easy to implement - the row/column mapping where either rows or columns are stored in consecutive memory locations.

This row/column mapping is efficient and convenient in many cases, but there are several algorithms in which more efficient mappings can be devised. An example is the gridding algorithm where randomly ordered quantities are convolved onto a regular cartesian grid. This operation requires a number of pixels to be accessed. These pixels are the adjacent pixels along all dimensions of the array. To achive this with row/column mapping requires that many rows or columns be available. If these data are not in central memory there is a substantial I/O overhead, much of it for pixels that are not needed by the algorithm. If it were possible to store pixels adjacent in ALL dimensions nearby in the computer address space, not just the row or column dimensions, it would be possible to greatly reduce this unnecessary I/O overhead.

To illustrate this point let us limit the discussion to power of two size axes. The conventional row/column method for mapping many dimensions onto one dimensional address space is to take one dimension at a time and pack each complete dimension index into the final address with one dimension occupying all the low significant bits of the address. The following illustrates how the address for a two dimensional array ARRAY(4,4) is conventionally formed:

---

*The National Radio Astronomy Observatory is operated Associated
 Universities Inc. under contract with the National Science
 Foundation.

```
              Dim 1 Bit 1    (LSB)
              Dim 1 Bit 2
              Dim 2 Bit 1
              Dim 2 Bit 2    (MSB)
```

This is the reason for the different status of the two diminsions -
only dimension 1 moves through address space in small increments. The
way to obtain the most equitable status for each dimension is to
interleave the bits from the various dimensions in building up an
address. In the previous example this would be done as follows:

```
              Dim 1 Bit 1    (LSB)
              Dim 2 Bit 1
              Dim 1 Bit 2
              Dim 2 Bit 2    (MSB)
```

This has the advantage of putting all the low significant bits from
all dimensions near the low significant end of the address. Hence for
any small increment in any dimension addresses will be reasonably near
to each other. This can considerably reduce the physical memory
requirements when stepping through an array in a non-optimal way.
Hence paging is minimized in a virtual memory environment and disk IO
can be reduced where block IO is being done on a large array. It is
clearly easy to generalize the interleaving to an arbitrary number of
dimensions each of arbitrary size subject to the power of two
constraint. Non-square arrays can be dealt with by treating the low
significant bits as above but packing the high significant ones. E.g.
the three dimensional array ARRAY(4,32,2) would use the following
mapping:

```
              Dim 1 Bit 1    (LSB)
              Dim 2 Bit 1
              Dim 3 Bit 1
              Dim 1 Bit 2
              Dim 2 Bit 2
              Dim 2 Bit 3
              Dim 2 Bit 4
              Dim 2 Bit 5    (MSB)
```

   This is only one particular example of how more efficient mapping
algorithms can be devised for particular cases. What we would ideally
like is to be able to tell a high level language what mapping to use,
and to provide suitable hardware to implement these mappings. For the
case of power of two size axes this is very easy to do, as only simple
bit manipulations are required. It does, however need changes in both
existing high level languages and computer hardware. We can still ask
the question though, what can be done given the limitations of
existing languages and hardware? For the general case we can do very
little. Substantial overheads soon build up doing all the necessary
bit manipulations. If we limit the question to a simple common example
though, we can make some progress.

   One of the most common addressing modes used is to step through
an array one pixel at a time. In particular consider stepping through
the second dimension of the array in the last example. We need to add

an increment to the address such that when a carry occurs it skips over bits not associated with dimension number two. We can simulate this by setting all bits not associated with dimension number two in our increment while ensuring that none of these bits are ever set in our dimension number two address. The following will illustrate the method:

Address	Increment	Mask	
0	0	0	not dimension 2
0	1	1	increment bit
0	1	0	not dimension 2
0	1	0	not dimension 2
0	0	1	bit to receive carry
0	0	1	bit to receive carry
0	0	1	bit to receive carry
0	0	1	bit to receive carry

If we calculate Address = (Address + Increment) the increment will be added into the address bit 2. We now have to clear all the bits that are not associated with dimension 2 so that subsequent increments will still have the required carry property i.e. Address = (Address .AND. Mask). We now have the following:

Address	Increment	Mask	
0	0	0	not dimension 2
1	1	1	increment bit
0	1	0	not dimension 2
0	1	0	not dimension 2
0	0	1	bit to receive carry
0	0	1	bit to receive carry
0	0	1	bit to receive carry
0	0	1	bit to receive carry

We have correctly incremented the address for dimension number two. If the operation is repeated i.e. Address = ((Address + Increment) .AND. Mask) we get the following:

Address	Increment	Mask	
0	0	0	not dimension 2
0	1	1	increment bit
0	1	0	not dimension 2
0	1	0	not dimension 2
1	0	1	bit to receive carry
0	0	1	bit to receive carry
0	0	1	bit to receive carry
0	0	1	bit to receive carry

Again the address has been correctly incremented. We can continue to do this and step through dimension number two of the array. Of course we need to combine the dimension number 2 address we have calculated with addresses for the other dimensions but this is easily achieved with a logical OR operation.

There are some important points to realize. Only the address variable is changing. The increment and mask variables are calculated once for the required step size. Also it is easy to generalize this to use any step size - positive or negative. We can calculate an increment and mask for a specific step along a particular dimension and use them to step as far as we like along that dimension. As an example the following is the FORTRAN code required to fill a two dimensional array with a constant:

```
      REAL ARRAY(*), CONST
      INTEGER I1, I2, DIM(2), ADDR(2), MASK(2), INC(2)

      ADDR(1) = 0
      ADDR(2) = 0
      DO 2 I2 = 1,2**DIM(2)

         DO 1 I1 = 1,2**DIM(1)
            ARRAY(1 + (ADDR(1) .OR. ADDR(2)) ) = CONST
            ADDR(1) = MASK(1) .AND. ( ADDR(1) + INC(1) )
 1       CONTINUE
         ADDR(2) = MASK(2) .AND. ( ADDR(2) + INC(2) )
 2    CONTINUE
```

It is assumed that MASK and INC have been previously set to appropriate values. The offset of one in the index of ARRAY is due to FORTRAN arrays conventionally starting at element number one whereas the indexing algorithm described here starts at zero. Note that in this case of filling the whole array it is not necessary to reinitialize ADDR(1) each time the inner loop begins. This is because it is possible to set up INC so that address calculations are always performed modulo the dimension size. Note also that two important assumptions heve been made here. First, INTEGER variables must be represented internally using two's complement convention, and second, the FORTRAN logical operators (.OR. and .AND.) must operate bitwise on INTEGER variables. Neither of these assumptions are part of the ANSI FORTRAN standard, although they are frequently the case.

Having devised this scheme for stepping through multi-dimensional arrays it is possible to generalize it still further. There is no necessity for the bits to be totally interleaved. It is possible to combine the bits from each dimension in an arbitrary way subject only to the constraint that the order of bits from each dimension in the final address is preserved. (It is not possible to implement a bit-reverse algorithm.) So, for example, a two dimensional array ARRAY(256,256) could be declared in patches each 8*8 in size with each patch stored in conventional row/column order. The following mapping will achieve this:

```
      Dim 1 Bit 1     (LSB)
      Dim 1 Bit 2
      Dim 1 Bit 3
      Dim 2 Bit 1
```

```
Dim 2 Bit 2
Dim 2 Bit 3
Dim 1 Bit 4
Dim 1 Bit 5
Dim 1 Bit 6
Dim 1 bit 7
Dim 1 Bit 8
Dim 2 Bit 4
Dim 2 Bit 5
Dim 2 Bit 6
Dim 2 Bit 7
Dim 2 Bit 8      (MSB)
```

In order to achieve this with conventional FORTRAN it is necessary to declare the array with 4 dimensions ARRAY(8,8,32,32) and to use 4 nested DO loops to step through the complete array. Using the scheme described here it is possible to store the array in precisely the same order but have it declared with only 2 dimensions and loops. In the extreme case it is possible to simulate the conventional FORTRAN storage order! This is in fact done below in the timing tests as it provides an accuarte comparison of the loop and indexing overheads.

The following show some benchmark tests run using FORTRAN on a VAX 11/780 running VMS. The tests involved filling a two dimensional real array of size 256 by 256 with a constant value and the times refer to repeating this operation 10 times over. Three types of array declaration are compared. First is conventional FORTRAN with the array declared ARRAY(256,256) using normal FORTRAN indexing. The other two cases both use the indexing technique described here. The second addresses memory in the same order as conventional FORTRAN so that paging overheads will be the same. The third uses fully interleaved bit indexing to show the reduced paging that can be achieved. In each case the two DO loops were nested in both possible ways - best and worst cases. All the times are CPU time in seconds.

	Best	Worst
Conventional FORTRAN	5.06	268.33
Simulated FORTRAN	7.87	271.47
Fully interleaved	8.30	8.50

In the best case the technique described here is inferior to conventional FORTRAN. This is hardly surprising as a considerable degree of optimization has been built into both the compiler and hardware. It should be realized, however, that the times expressed here for the best cases are dominated by array index calculation, loop overheads and memory access time. The test is as trivial as it could be made. In many cases loop timings are dominated instead by other more useful calculations so the difference will be less pronounced. This can easily be domonstrated. The interleaved access case is about

a factor of 1.6 slower than conventional FORTRAN in the above best
case. Adding just a single real multiply to the inner loop in each
method reduces this factor to about 1.4.

In the worst case the interleaved bit method shows a spectacular
advantage. The reason for the huge time increase in two of the worst
cases is operating system overheads involved in paging. The technique
of splitting an array into patches each stored in conventional
row/column order can also show a similar advantage in paging
overheads. The advantage of the method described here is that the
index increments are independent of the starting index. With the
patched access technique it is necessary for a programmer to be aware
of patch boundaries and explicitly step over them.

A further advantage of this algorithm is that all dimension
addresses are combined to form the total address in the same way.
These dimension addresses can be specified in any order. They only
need logical OR operations to combine them. This is very useful when
dealing with arrays which have been transposed or have different axis
definitions. For example it is easy to write a simple routine to
extract a one dimensional spectrum from a data array of arbitrary
shape – provided that one of the dimensions is frequency. All that is
needed is the INC and MASK parameters to step through the frequency
axis. The loop structure is identical regardless of which axis is the
frequency axis. This is not the case for other methods of array
indexing where the loops will generally look different depending on
which axis is the frequency axis.

A combination of interleaved bit array indexing and a good
virtual memory environment can provide a very convenient and efficient
way of accessing multi-dimensional arrays. There are a number of
applications where this technique may be of use. One example is that
it may remove the need to transpose large multi-dimensional arrays as
it is equally efficient to move along any axis. Another example is
algorithms that require access to a local area of an array such as the
gridding algorithm discussed earlier.

THE EUROPEAN SCIENTIFIC DATA ARCHIVE FOR THE HUBBLE SPACE TELESCOPE

G. Russo[1,2], A. Richmond[1], and R. Albrecht[1,3]

[1] Space Telescope European Coordinating Facility, European Southern Observatory, Garching bei München, FRG
[2] On leave from Dipartimento di Fisica, Università di Napoli, Italy
[3] Affiliated to Astrophysics Division, Space Science Department, European Space Agency

ABSTRACT

The implementation at the ESO/ESA ST-ECF in Munich of the Data Management Facility (DMF) for the scientific archive of the Hubble Space Telescope (to be put in orbit within 1987) is described. The DMF is using a relational database machine for the Catalogue and optical disks for the Archive; these systems are accessed through an External Interface (EIF) to put data in, and a User Interface (UIF) to retrieve data. The architecture of the UIF is described in detail.

INTRODUCTION

The Hubble Space Telescope (HST), which will be launched within 1987 by NASA by means of the Shuttle, will provide, through the five scientific instruments on board, an impressive mass of information which cannot be obtained from ground-based observatories.

Observing time on the HST is therefore an extremely valuable opportunity. In the effort to maximize the scientific return, NASA has established that, unless there are compelling scientific or operational reasons, no target will be re-observed by different researchers, but data will be made publically available from a science data archive. Since an oversubscription rate of about a factor of 10 is expected for the HST observing time, archival research will be in heavy demand.

Data to be archived consist of:

1. Science Verification Data, obtained in the early days after launch;

2. Calibration Data, periodically acquired by the HST staff;

3. Scientific Observations, in raw and calibrated form;

4. Engineering and other ancillary data.

The scientific observations will be carried out according to approved Guest Observer (GO) proposals, whose investigators will have proprietary right for a period of one year, after which the data are publically available. The size of the dataset for each of the five scientific instruments on-board (WF/PC, FOC, FOS, HRS, HSP) and their incoming rate, according to the first 6 months of the planned Guaranteed Time Observation (GTO) programs, is given in the table below.

To these figures, one must add the (non-proprietary) parallel mode observations which the WF/PC can carry out while any of the other instruments is in use, and which are of the order of 20 per day.

These figures lead to 300-500 Gbyte/year of data, a volume which obviously needs a dedicated, reliable and user-friendly system able to allow general Archival Researchers easy access to the list of observations (Catalog) and eventually to the data themselves (Archive). Moreover, these needs are to be satisfied shortly after the HST launch, because of the early release of the Calibration and Science Verification data.

A working agreement has been reached between the Space Telescope Science Institute (ST ScI) in Baltimore, USA, and the Space Telescope European Coordinating Facility (ST-ECF) in Garching bei München, FRG, to jointly develop a Data Management Facility (DMF) to support a basic retrieval capability immediately after the launch of the HST. As data becomes nonproprietary, the ST ScI, which has the overall responsibility for the HST, will transfer a copy of the archive data to the ST-ECF.

However, this does not mean that the two implementations will be identical, and this paper will in fact deal with the DMF implementation at ST-ECF only.

GENERAL

The DMF comprises two major systems: the Archive and the Catalogue; the former will store all the data, which are one- or two-dimensional images, while the latter will index the Archive, cataloguing sufficient information to permit a variety of intelligent queries. The Catalogue is implemented as a relational database on a Data Base Machine, while the Archive will use the newly available write-once, read many times (WORM) digital optical disks.

This kernel system is interfaced to the outside world by means of an External Interface (EIF) to put data into the system, and by a User Interface (UIF) to retrieve the data. Since the most important use of the DMF as implemented at ST-ECF will be archival research, particular emphasis has been devoted to the architecture of the UIF. The internal organization of the DMF is outlined in Figure 1.

Instrument	raw file size	calibrated file size	units	avg. no. per day
WF/PC	5.1	5.1	Mbyte	6
FOC	5	10	Mbyte	8
FOS	100	200	Kbyte	12
HRS	100	200	Kbyte	12
HSP	125	250	Mbyte	3

HARDWARE

The DMF system runs on a DEC VAX-11/785 computer under the VMS operating system, but will be made accessible to other computers, both local and remote, running either VMS or UNIX; the use of system independent software interfaces will help ensure long term compatibility.

The Catalogue uses a Britton-Lee IDM 500/1 database machine with associated local magnetic disks (eventually also a tape unit); the IDM is connected to the VAX through an IEEE 488 parallel interface.

The Archive uses optical disks for its primary storage requirements; the current unit is an Alcatel-Thomson GIGADISC GD1001, connected to the VAX through an Emulex UC13 interface. The unit accepts single or double sided preformatted media, with a capacity of 1 Gbytes per side; the format to be used for such devices has been devised jointly by ST-ECF and ST ScI.

Standard I/O devices on the VAX complete the system; in addition there is a connection to the German Packet Switching Network, Datex-P (X-25 protocol), and several modems (300 and 1200 baud) for dial-up remote connection. Figure 2 summarizes this hardware environment.

USER INTERFACE

There are surely as many opinions on the essential features of the World's Most Friendly User Friendly Interface, as there are Users. In designing a User Interface which at least comes reasonably close to minimising the "offence" function (defined as the sum of the products of complaints and loudness of voice), we hope to identify features which can be accepted by a **majority**; that is not necessary the same as what the majority would have chosen had everyone been asked to list their favourite features!

In particular, we tried to design an Interface that, while not copying VAX/VMS in detail, is certainly not "orthogonal" to it, and hopefully, may be found to be more "comfortable" after very little experimentation by the potential User.

STARCAT (an acronym for Space Telescope ARchive and CATalogue) is our implementation of the user interface to the DMF. As suggested by the name, it shall provide the following major facilities for General Observers and Archival Researchers:

1. Queries on the contents of the Catalogues.

2. Files retrieved by name from the Optical Disk Archive.

Currently (March 1986), only the first facility is provided; access to the optical disks will be possible in a few months.

The **goals and objectives** are, in approximate priority order, to **maximise**:

1. **convenience** of access to HST data.

2. **software quality**: reliability, maintainability, and usability.

3. **compatibility** with other systems, e.g. MIDAS, IRAF

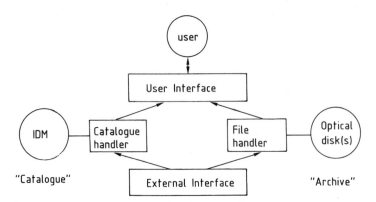

Figure 1: Internal Organization of the DMF

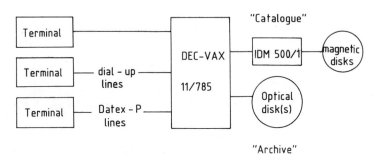

Figure 2: Hardware Environment of the DMF

4. **portability**, i.e. software / hardware independent.

Terminal Interface

Starcat provides full screen display (tty mode later) with the following command / parameter interfaces:

- Dialogue - question and answer; program guides user unless responses provided on command line;
- Cursor and Keypad - cursor position selects item, keypad identifies function (Catalogs querying);
- Command/qualifiers (to be implemented).

The dialogue mode of interaction is provided because this places less burden on the user to remember a multitude of "qualifiers" as required in the typical VMS environment (although a good HELP facility alleviates the problem somewhat, it is tedious to have to switch from the task in hand, to search the HELP displays). In the dialogue mode, it is up to the program to guide the user at every step **if necessary,** though the experienced user can by-pass the prompts by use of the "type-ahead" feature. However, to accommodate those users who are "at home" in VMS, a limited form of qualifiers will be provided.

The **Dialogue Mode** features:

- user **control by command**: a command specifies the required program function
- **dialogue interaction**: Starcat is run on a "question and answer" basis
- **type-ahead**: several questions can be answered with one input, suppressing prompts which are known in advance
- **line editing**: e.g. characters on the user input line may be deleted
- **"help facilities"**: program commands may be displayed with detailed descriptions
- **error reporting**: if the user enters invalid data or the program encounters any error, the user is informed with a clear error message, and the last prompt is repeated
- **command abbreviation**: only need to type enough first characters to distinguish the command from others; a special character, e.g. ESC may be provided to "complete" the command if un-ambiguous
- **recall of commands**: e.g. by numeric range in session buffer, or by abbreviation
- **file input**: questions may be answered from a file

Starcat normally presents a full-screen display. The introductory displays have the basic appearance shown in Fig. 3. **Text within angle brackets indicates a variable field.**

The first screen displayed on starting Starcat corresponds to the command **Help Introduction.**

```
+-------------------------------------------------------+
| Space Telescope Archive and Catalog (version id.)     |
| Command: < command name >          < date and time > |
|                                                       |
        < item name >   < brief description >

| <Prompt:> <user input>                                |
|           < help / error message >                    |
+-------------------------------------------------------+
```
             Figure 3:   Basic Screen Format

Commands

A command selects a particular function required of the program. For example, the command **Help** causes the program to provide more detailled help on all aspects of Starcat.

A command is selected by answering the prompt

            Command :

with its name, or an **abbreviation**; that is, only enough first characters of the command name are needed, to distinguish it from other commands in the same program.

The commands will be displayed by entering **Help commands**. Note that either, or both of the "words" entered may be abbreviated.

The **Catalogs** command, after a catalogue name has been specified, hands over control to a commercial package called OMNIBASE. Although this package provides some "hooks" which enables us to intercept some actions like user input and data retrieval, the basic philosophy of this package is clearly beyond our immediate control (but we may have reasons for trying to overcome this, e.g. to allow queries to be issued from command procedure files...).

                                        Help topic: *Introduction*

```
       +--------------------------------+
       |                                |
       |    Welcome to  S T A R C A T   |
       |                                |
       +--------------------------------+
```

If you are using this system for the first time, and / or you would like to be 'led through', then follow the arrows:
                    -> type HELP

Many thanks to those of you who have commented on this program. Please note that you can also help us by telling us what features are the most positive assets (so that we keep them...) REPORT has been modified to automatically select the destination and subject for your comments; you only need to enter your text, and press CTRL/Z to exit from REPORT.

The IUE Catalogue has been updated to: 31/10/85 for VILSPA;
                                        31/08/85 for GSFC.

A list of intended user functions is shown in Table 1; current functions may be shown on-line by the command "Help Commands".

Software Development

We have worn our Software Engineer's hard hats during this development (which is by no means finished), and adopted a (semi-)formal methodology comprising Requirements Definition, Architectural Design, Prototyping, Reviews, etc...

In order to facilitate the attainment of these goals, we have placed emphasis on the **software development methodology** employed for this project. The methodology is reflected by the choice of project standards, which were derived from the standards of ESA and ST ScI, extended to incorporate the modern **object-oriented** software design technique. Some of the major features of the methodology / standards are:

- visibility of the software development process, e.g. by reviews, early documentation, and prototyping;

Table 1: Commands List

Command	Description
Calc	Arithmetic calculator.
Catalogs	Access to Catalogs.
Exit	Exit from program.
Fetch	Issue archive retrievals from filename list, to files or tape. The filenames list is prepared either by use of Query or Modify.
Help	Like VAX help and menu. We may envisage different levels for novices and experts.
IQL	Straight - through IQL interface to Catalogue, saveable on file.
Log	Enable / disable session logging.
Report	Record gripes and praises for feedback to the system manager.
Setup	Change the 'setup' parameters. The program mode, e.g. novice expert, may be kept in an initialisation file (as with edtini.edt). This mode could be changed during the program run, e.g. for familiar / new commands.
Modify	Changes to filenames list. This may be some simple editor – like facility, allowing to change existing text, or to add new text, i.e. filename. Note that this also allows to specify filenames without having issued any Query.
Query	'Sugar coated' IQL-like statements which place filenames resulting from queries on the Catalogue, into a list (buffer) accessible from Modify and Fetch.
Read_File	Read further input from file. This will allow the user to store commonly used command sequences in a file. It might be refined by a form of 'macro' facility, e.g. when a special character is detected in the file, the following text, up to some terminator, is shown to the user, like a prompt, and his / her reply replaces that text.
Spawn	Commands may be passed directly from the user, to the host operating system, without leaving Starcat.
Stop_File	End read_ -- or write_file.
Tables	List / display catalogue relations and their meanings.

Help topic: *Help*

```
      You may get detailled HELP on the following topics:
       -> Commands
                Calc
                      Functions Variables
                Catalogs (Subitems are available)
                          ESOUPPS HALLEY  HR      IRAS    IUE   POSS    SAO
                          SCHMIDT SW      SWLIB UKS       VERON WILSON

                      Keypad
                      MIDAS
                          Format
                Exit Help IQL Report
                Setup
                      Tracing
          Introduction
                Overview
                Status
```

- emphasis on the early phases of the software life-cycle, i.e. requirements definition, prototyping, and design specification, to minimise costly changes and "debugging";
  One of the major questions which prototyping should help to resolve, is the exact structure of the user dialogue for queries; relational databases provide the capability to query the data in virtually every way imaginable, but the query language (IQL) is relatively low-level and complicated. We shall try to identify the most useful types of query, and then to provide a suitable astronomer-oriented language;

- a layered software structure to facilitate comprehensibility, software object sharing, and portability. The layering approach has the major benefit of very clearly separating independent aspects of the software, i.e. de-coupling them. Each layer acts as a **virtual machine** to the layers above it; hiding implementation details, while providing services in terms closer to the application requirements. The higher levels are application-oriented, the "inner" levels maintain the key data structures, and the lower levels manage communications with the host environment and provide basic utilities (e.g. string handling);

- rigorous code management. The software is written in C, but with the use of the C macro substitution facilities to make the code more readable. Every function is kept extremely short, typically being entirely displayable on a VDU screen, and "gotos" have not yet been found useful. Each file and each function begins with a standardised header. A preprocessor (written by François Ochsenbein) reads the headers and extracts information **from the code**, which is transferred to a database on the IDM.

CONCLUSIONS

The DMF presented here is being implemented at ST-ECF, and therefore some changes are still possible, in order to better satisfy general users requests.

PARALLEL PROCESSING

Chairman : V. Cantoni

CELLULAR MACHINES - THEORY AND PRACTICE

M. J. B. Duff

Department of Physics and Astronomy
University College London
Gower Street, London WC1E 6BT

INTRODUCTION

The idea of cellular automata or iterative arrays of processors is far from new and owes much to the pioneering work of John von Neumann[1] in association with others, such as Stanislaw Ulam[2] who saw, in the example of the human brain, an almost unbelievably powerful computing structure which owed its strength to a regular organisation of fundamentally simple computational units (neurons). However, it could be said, with the benefit of hindsight, that von Neumann's contribution was impeded by a technologically premature realisation that large arrays of computing elements would be subject to serious failure unless some self-repair mechanism could be built into them. He devoted the last years of his life, until 1957, studying the complexity requirements necessary to enable array elements to reproduce themselves, this being an effective way of achieving self-repair. In fact, much can be usefully done with cellular arrays before the need to incorporate self-repair is an important factor and it is to be regretted that the brilliant von Neumann did not apply his mind to the more practical development of his subject.

In relation to image processing, two papers by S. H. Unger[3,4] and others describing the frog's visual system by J. Y. Lettvin et al.[5] and M. B. Herscher and T. P. Kelley[6] showed how cellular arrays might be used to effect parallel processing on images. Unger's theoretical proposal, although not generally appreciated at the time it was published, pre-empted many of the ideas expressed in more practical systems devised and sometimes constructed in the following twenty-five years; the biological studies and hardware emulation of frog vision showed that simple image processing was indeed a practical possibility using cellular arrays.

Unfortunately, the technology available to implement these ideas in the 1960s was grossly inadequate and the earliest attempts were fraught with difficulties. B. H. McCormick's ILLIAC III[7] was never completed and arrays built by S. Levialdi[8] who was studying connectivity and propagation, and by the present author[9,10] in work initially directed towards the analysis of tracks of high-energy charged particles, were small and inflexible. An alternative approach, which has had many followers, was adopted by M. J. E. Golay[11] and K. Preston Jr.[12] who used small sub-arrays (typically with 7 or 9 elements) to emulate larger arrays, scanning the small processor array through the large image data array. Only in the

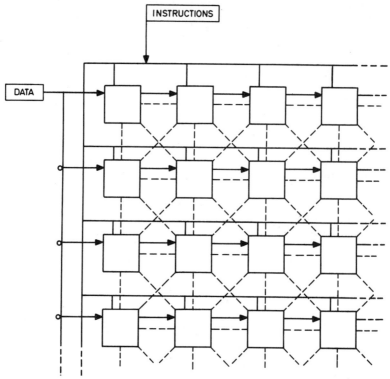

Fig. 1. Principal connections in a typical cellular array used for image processing.

last few years has it been possible to harness integrated circuit technology to the task of array construction, with the result that quite large processor arrays are becoming a cost-effective means of carrying out real-time image analysis.

CELLULAR ARRAYS

A cellular array consists of a regular array of simple processing elements (PEs). In the simplest systems, each PE is connected to its immediate neighbours (4 or 8 in square arrays, 6 in hexagonal arrays) and instructions are broadcast from a central controller through connections to each PE, so that each PE executes the same instruction at the same time (see Fig. 1). A small amount of memory is located with each PE, providing local data storage and I/O buffers. There may also be fast I/O channels linking the PEs with external data.

Array Size

Ideally, an N x N array will be used to process images with N x N pixels, but for large images this would imply excessively high numbers of PEs. Until very recently, 128 x 128 has been the largest PE array to be constructed. In order to process larger images, three alternative schemes have been tried:

1. An N x N pixel image is subdivided into blocks of size M x M
   (N being an integral multiple of M) and a two-dimensional
   scanning path used to process all parts of the image, block
   by block, by means of an M x M array of PEs.

2. An N x N pixel image is subdivided into blocks of size N x M
   (N being an integral multiple of M) and a one-dimensional
   scanning path used with an N x M array of PEs.

3. An N x N pixel image is subdivided into blocks of size M x M
   (N being an integral multiple of M) and a PE array of size
   N/M x N/M used to process the data, associating the top
   left-hand corner M x M block of data with the top left-hand
   corner PE, and so on.

In the first two schemes, the inter-PE connectivity is important and edge stores are built around the PE array so as to present the array with the appropriate boundary conditions in going from block to block. Thus the user will see the scanned array as being logically equivalent to a full N x N array of PEs with all the inter-PE connections maintained. In the third scheme, it is less straightforward to treat the processing in this way since neighbourhood array data is sometimes located in the same PE and sometimes in a neighbouring PE. The setting up of neighbourhood relationships is therefore a more complex, although still possible, process. Every scheme, including the fundamental N x N PE array for an N x N pixel image, has both advantages and disadvantages so, where choice is possible, it will obviously be sensible to select the scheme which is optimum within any imposed constraints (cost, performance, task specification and so on).

Processor Complexity

Unless a very small array is to be scanned over an image (such as the earliest systems in which only one PE was used in association with a 3 x 3 memory array, thus allowing neighbourhood data to be presented rapidly to the PE), it will be necessary to work with simple PEs in order to keep costs down. In practice, most arrays have used single-bit PEs which perform single-bit arithmetic or logical functions between arrays of single-bit data (bit-planes). Grey-level operations are carried out bit-serially. Thus in CLIP4[13], the PE includes two Boolean processors which can be configured to form a single-bit full adder. Starting with this simple specification, which should enable all image operations to be programmed into the system, it is obviously possible to increase PE complexity in many ways in order to obtain superior performance. In particular, the single-bit ALU can be replaced by, say, an 8-bit ALU; operations such as multiplication and division can be speeded up by incorporating a shift register; internal data paths can be upgraded to allow parallel transfers of multibit data between memory, registers and processor.

In general, the overall performance of the system is likely to be in direct proportion to the performance of the individual PE which is, in a sense, a predictable and therefore unexciting result. It may well be that a more useful direction for seeking improved performance is in increasing processor interconnectivity.

Processor Interconnectivity

Since local neighbourhood operations on images are important, it is natural to structure an array so as to facilitate them. In a local neighbourhood operation, an array of new pixel values is computed in which each value is some function of the original values in the immediate locality of each pixel, the computations being carried out in parallel across the whole array (if the PE array is large enough). A suitable local neighbourhood in a square array is the 3 x 3 sub-array centred on each pixel so it is convenient to connect each PE to its immediate eight neighbours. With relatively small reduction in performance, the interconnections can be limited to four neighbours, usually N, S, E and W of each pixel. In either case, it is necessary to include switches in every interconnecting path so that, under program control, the required data transfers can be selected.

If connections between PEs are confined to local neighbourhoods, then, in an N x N array, of the order N steps (O(N) steps) are needed to transfer data from one side of the array to the other. This can be very time-consuming and various alternative connection schemes have been proposed to reduce the distances involved. For example, pyramid arrays[14,15], in which layers of processors are stacked one above the other, reducing the number of PEs by a factor 4 between successive layers so as to form a pyramid, reduce the distance to O(2 log N) steps, although it is not always possible to take full advantage of this reduction since all paths converge through the apical PE. Another suggestion is to use a hypercube topology. A discussion of many such schemes has been prepared by L. Uhr[16]. Whatever scheme is eventually adopted, it is clear that the improvement in performance will be strongly task-dependent and often difficult to estimate. Benchmarking image processor performance is not an easy exercise[17] and no general analytical methods for handling array performance calculations have yet been established.

Array Peripherals

In an image processing system, the cellular array will be imbedded in a structure which will be required to supply it with data and instructions and to accept its outputs. For simple image operations (edge finding, for example), it will often be the case that the processing time in the array is negligible compared with the time needed for image acquisition (from sensor to array) and display. It is also sometimes necessary to reconfigure data before presenting it to the array, especially when the image is larger than the array. A typical method of introducing an 8-bit pixel image into an array is to enter the data a bit-plane at a time, column by column, shifting the single-bit data in parallel through each row of the array. This involves N steps in an N x N array and may therefore be unacceptably slow. In principle, a parallel output camera, with one wire for each pixel, could connect directly to each PE, but this is usually ruled out by the lack of suitable cameras. Nevertheless, it is important to realise that the performance of a system can be severely degraded if the array support is inadequate and that a considerable amount of engineering design must be undertaken if real-time image analysis is to be achieved.

PROGRAMMING CELLULAR ARRAYS

It would be misleading to suggest that low-level programs for cellular machines are constructed in the same way as those for serial machines although the differences are less marked than is sometimes imagined. For example, in CLIP4, the sequence:

```
LDA 1
LDB 2
SET A+B
PST 3
```

takes a bit-plane from memory address 1, ORs it with a bit-plane from address 2 and puts the resulting bit-plane (PST - process and store) in address 3. The principal difference between this short program and a similar one written for a serial processor is the complete absence of any mechanism for scanning through the image array; the process is performed in parallel at every pair of pixels in the two bit-planes. In the following sequence:

```
LDA 1
LDB 2
SET P.A,[1-8B]P.A
PST 3
```

every pixel in the bit-plane in address 1, which is in the connected set containing a marked pixel in address 2, is extracted and stored in address 3. The details of the coding are unimportant for this discussion but it will be noted that the implied propagation process in this program removes the need for explicit exploration of paths of connection between the pixels in the image.

In the examples above, low-level instructions are able to carry out what would elsewhere be regarded as high-level operations and, particularly in the second example, exploit the close relationships between the PE array structure and the structure of both the data and the algorithm. As a general rule, efficiency can only be achieved in cellular machines if the need to preserve these relationships is observed. Since, for a given machine, the structure of the array will presumably be fixed (but see comments on reconfigurability later), the only available variables in the equation are the algorithm and the data structures.

Algorithm Structure

It is in fact feasible, although unwise, to implement any algorithm on a suitably general-purpose cellular array. Thus a 'raster scanner' can be programmed into the array which will then visit every pixel in turn, mimicking the role of a conventional serial machine. Adoption of this policy would almost inevitably result in processing speeds which are significantly inferior to those obtainable using conventional computers. Instead, the challenge for the low-level programmer is to devise algorithms which obtain the required results by routes which make good use of the natural structures of the array and of the data. In typical arrays, this will imply exploiting propagation whenever possible (propagation is the process in which a PE conditionally transmits data between its neighbours, without intermediate memory references) and employing any algorithms which make effective use of the array's parallel processing capability with one PE per pixel. However, since PEs are there whether they are in use or not (much as is memory in serial computers), it is not important to avoid algorithms which do not use all the PEs for all the time. What *is* important is that there shall always be a PE available when and where it is needed.

In practice, any difficulties in programming cellular machines lie in the invention or selection of efficient algorithms; once this has been done, the coding will usually be reasonably simple.

Data Structure

As has already been suggested, much of the success of two-dimensional cellular arrays used as image processors has been due to the close match between the N x N array of processors and the N x N array of pixels to be processed. It has been found by bitter experience that reducing the number of PEs per pixel by a factor S reduces performance by a factor kS, where k is always larger than 1. Some processes suffer worse than others, propagation being particularly badly affected. Overall, k might be expected to be in the range 5-10, depending on details of the control and memory structure.

The real weakness of cellular arrays emerges when, after early processing, the image data is reduced so that it no longer forms a two-dimensional array. At this point, so-called intermediate-level processing starts and there is still much debate as to how computers should be structured so as to perform well in this region of processing.[18] One possible solution, which also has possible merits even in some of the earlier processing steps, is to incorporate reconfigurability into the array. The PE interconnections are, typically, taken through an interconnection network which then allows different connections to be established under program control.[19] However, these methods are normally used with fairly small numbers of PEs which barely merit the description 'cellular arrays'.

FUTURE DEVELOPMENTS

In some ways, it might be thought that there is not much further to go in the design of cellular arrays. The future would seem to consist largely in building larger and larger arrays of increasingly complex PEs, as new technology permits. It is not unreasonable to conclude that interest in the subject is therefore nearly exhausted. On the other hand, there are possible ways in which development might take place:

Independent Control of PEs

There is no reason why all the PEs should be controlled in 'lock-step'. Each could execute individual programs in response to current local data (permitting different parts of the image to be processed in accordance with local conditions) or PEs could be redirected to 'help' in regions of high complexity.

Multidimensional Systems

Some multidimensional systems are already emerging (pyramids, hypercubes and so on). Layers of PEs configured for high performance in the later steps of processing could be grafted on to the efficient low-level two-dimensional arrays.

Analogue Systems

It is not clear why all image computers should be digital. The future may be in fuzzy processing using multivalued logic and analogue circuitry.

Biotechnology

There is still an enormous difference in scale between the largest arrays built so far and the mammalian brain. Could it be that increase in scale will result in a qualitative, rather than quantitative, improvement in performance? If so, will this imply the need for the self-repair mechanisms envisaged by von Neumann?

CONCLUSIONS

Cellular arrays in their many forms are making an outstanding contribution to image analysis and would still seem to have a bright future. As they increase in complexity, so will the problems experienced in programming them increase. It already is possible to devise arrays which are, to all intents and purposes, almost unprogrammable. The big challenge now is to advance our understanding of the complex relationships between computer structure, algorithms and data so that programming will appear as a natural process. If this is achieved, then the future for cellular arrays will be very bright indeed.

REFERENCES

1. J. von Neumann, Theory of automata: construction, reproduction, homogeneity, Part II of "The Theory of Self-Reproducing Automata", A. W. Burks, ed., University of Illinois Press, Urbana (1966).
2. A. W. Burks, ed., "Essays on Cellular Automata", University of Illinois Press, Urbana (1968).
3. S. H. Unger, A computer oriented towards spatial problems, Proc. IRE 46:1744 (1958).
4. S. H. Unger, Pattern detection and recognition, Proc. IRE 47:1737 (1959).
5. J. Y. Lettvin, H. R. Maturana, W. S. McCulloch and W. H. Pitts, What the frog's eye tells the frog's brain, Proc. IRE 47:1940 (1959).
6. M. B. Herscher and T. P. Kelley, Functional electronic model of the frog retina, IEEE Trans. MIL-7:98 (1963).
7. B. H. McCormick, The Illinois pattern recognition computer - ILLIAC III, IEEE Trans. EC-12:791 (1963).
8. S. Levialdi, CLOPAN: a closed-pattern analyser, Proc. IEE 115:879 (1968).
9. M. J. B. Duff, B. M. Jones and L. J. Townsend, Parallel processing pattern recognition system UCPR1, Nucl. Instr. and Meth. 52:284 (1967).
10. M. J. B. Duff and D. M. Watson, Automatic design of pattern recognition networks, in: "Proc. Electro-Optics '71 International Conference, Brighton, England", Industrial and Scientific Conference Management Inc., Chicago, Ill., p. 369 (1971).
11. M. J. E. Golay, Hexagonal parallel pattern transformations, IEEE Trans. C-18:733 (1969).
12. K. Preston Jr., The CELLSCAN system - a leucocyte pattern analyzer, in: "Proc. Western Joint Computer Conf.", p. 173 (1961).
13. M. J. B. Duff, CLIP4: a large scale integrated circuit array parallel processor, in: "Proc. 3rd Internat. Joint Conf. on Pattern Recognition", Coronado, Ca., USA, p. 728 (1976).
14. S. L. Tanimoto, Programming techniques for hierarchical parallel image processors, in: "Multicomputers and Image Processing", K. Preston, Jr. and L. Uhr, eds., Academic Press, New York, p. 421 (1982).

15. V. Cantoni, M. Ferretti, S. Levialdi and F. Maloberti, A pyramid project using integrated technology, in: "Integrated Technology for Parallel Image Processing", S. Levialdi, ed., Academic Press, London, p. 121 (1985).
16. L. Uhr, Multiple-image and multimodal augmented pyramid computers, in: "Intermediate-Level Image Processing", M. J. B. Duff, ed., Academic Press, London, p. 129 (1986).
17. M. J. B. Duff, How not to benchmark image processors, in: "Evaluation of Multicomputers for Image Processing", L. Uhr, K. Preston, Jr., S. Levialdi and M. J. B. Duff, eds., Academic Press, New York (to be published).
18. M. J. B. Duff, ed., "Intermediate-Level Image Processing", Academic Press, London (1986).
19. J. T. Kuehn and H. J. Siegel, Multifunction processing with PASM, in: "Intermediate-Level Image Processing", M. J. B. Duff, ed., Academic Press, London, p. 207 (1986).

THE PAPIA PROJECT

    Stefano Levialdi

    Department of Mathematics
    University of Rome
    Ple. Aldo Moro 2
    00185 Rome, Italy.

ASTROPHYSICAL COMPUTATION

    Since the advent of electronic computing man has had the possibility of exploring theories and testing ideas with a new particular tool: the computer. This tool allowed to add a new methodology to the existing ones: namely theory and experiment[1]. Computation lies between abstract thinking and concrete realization; within Physics, the laws that may be discovered through laboratory work or hypothesis formation with paper and pencil can be tried on a man-made model and then, if those laws do not exhibit the expected results, they can be changed. All this may be done by a program (on a computer) and this kind of work is made everyday easier today by new tools (software and hardware) which raise the level of the machine so as to improve the human-machine communication in efficiency and effectiveness.

    In no field of science the analogy with a model-of-the-Universe is closer than in Astrophysics, where one may formulate such a model and then introduce it in a program to test the different theories (and corresponding laws) on specific acquired information from all the available sensors (satellites, telescopes, etc.).

    In Astrophysics research both the computations and the data that must be processed are of a very high dimension. Typically, sets of coupled partial differential equations with considerable nonlinearity must be solved giving rise to multiple time and space scales relevant to the solution[2]. Another aspect of the required massive computation refers to the graphical part which involves stellar pictures, two-dimensional projections of three-dimensional objects (typically with overlapping problems), visual wave-lengths of electromagnetic radiation and of non visual radiation like infrared, ultraviolet, X-ray and gamma-ray. In the past, only problems that allowed the introduction of symmetry (to simplify the computation) and therefore would use the plane to represent three-dimensional objects, or even better, one single dimension (like a star studied along its radius) had been treated. Recently, the new, lower cost of conventional machines (and the drastic reduction in memory cost) as well as the bright future offered by new and different computer architectures, improved the quantity and quality of the work that may be done by astrophysicists.

The activity in this area, as in most research areas, is not a production work where the same program is run many times on different data but, instead, the program is constantly being modified and improved. Typically, the computer is used in batch-mode and therefore no advantage has been gained from the new interactive facilities offered through remote terminals.

As an example of the present computational needs of astrophysicists we may quote the analysis of a spherical star requiring about 200 spatial grid points, 25 nuclei with 10 additional variables at each point (about 5 iterations per time step) and 10000 time steps to follow the star through its hydrostatic phases: a CDC 7600 or a Cray-1 computer are typical machines on which this kind of work is being done today. But if two- dimensional problems are considered, then grids of 200 x 200 points are required and therefore we have one order of magnitude more than in the previous case and must use some kind of pipeline computer or a CDC Cyber 205 or an array processor hosted by a large machine. What is needed is not only more computer power (for number-crunching) but better software tools for improving portability of programs, readability, and friendliness to suggest and try new ideas more than just to run programs in shorter times. We also need good visual aids that help to see what the program is doing (program animation), to modify parts of the program, to check partial results and, in a near future to write new correct programs with less effort[3].

PYRAMIDAL COMPUTATION

Computation systems may be seen as the physical implementation of a computational model. As the recent literature has amply documented[4,5], an interconnected number of processors (of varying complexity and coupling degree) is the starting point of new parallel computing systems, whereas their interconnection pattern is still open to debate mainly because of the proven weight of the interprocessor communication times within a complex computation task. This last issue has generated a wide variety of interlinking schemes, of configurations (fixed and variable) adequate to classes of problems (with fixed data structure and variable algorithms) as shown by a number of research projects.

Within the area of pictorial information systems and due to the known and widely appreciated flexibility of existing mammalian retinas, the so-called multiresolution approach[6] has been introduced some years ago based on the following considerations[7]:

a) the hierarchical nature of the visual pathway which has a reduction of 100:1 between retinal vision and optical nerve leading to the cortex,
b) the logarithmic reduction of communication time between any two

elements (processors) in a pyramid structure, since the longest distance n (for n^2 processors) will be reduced to log n in an exponentially tapered pyramid of planes (each one being smaller than its underlying plane by a constant factor),

c) the pyramid structure is particularly well suited to tree (or sets of trees) data structures and therefore this architecture may naturally embed tree oriented algorithms,

d) the recent progress in automating VLSI custom circuitry by means of new, interactive, friendly programs, has simplified the design and production of chips containing large numbers of simple processors or small numbers of elaborate processors.

A wide number of groups in the world have decided to build, after some preliminary simulation experiments, multicomputer systems based on a pyramidal structure. In fact, very shortly at Maratea (May, 1986) a NATO Advanced Research Workshop on Pyramidal Systems for Image Processing and Computer Vision will take place and about forty different projects will be reported and discussed from both architectural and algorithmic points of view.

Many years have elapsed since the first cellular computers were suggested, designed and built (see for instance[8]) and perhaps a shorter number of years has provided know-how in the construction of pipeline systems[9] which are advantageous whenever large sequences of images must be processed by means of local operations (of reduced neighborhood) and the input/output times become a relevant part of the total processing time. Basically, a pipeline computing system may feed and process nearly simultaneously (in slightly overlapped times) but the processing weight must be equally distributed among processors in order that no significant delay (on any one processing stage) takes place. Conversely, image data must be of the same size of the array processor to fully exploit its capabilities and local operations must be the core of the processing task in order to make the best use of this particular architecture.

The pyramid architecture may be seen as an interesting combination of an array processor (at the base) and a pipeline of gradually smaller arrays (each one, a section of the pyramid, on top of the other) until a single processor (at the vertex) is reached. A small number of different variations of this theme may be considered: overlapping arrays where each processor is connected to a number of lower processors that overlap and non-overlapping arrays which are partitioned in such a way that each group of processors (for instance four) are connected to exactly one processor on the upper plane until one plane but the last is made of four processors directly connected to the vertex, apex of the pyramid on the last plane.

It is for the above reasons and after a preliminary study of possible multiprocessor architectures that, in Italy, we have decided to design and build a pyramidal multicomputer (named **PAPIA** after the old name of Pavia town and also standing for **P**yramid **A**rchitecture for **P**arallel **I**mage **A**nalysis[10]). This

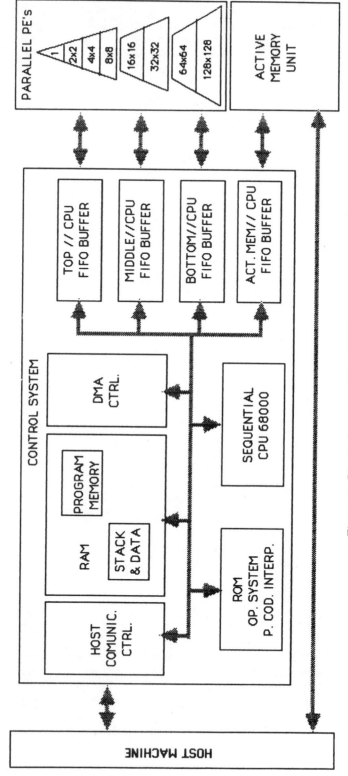

Figure 1 Block diagram of the PAPIA system

effort is supported by a National Research Program which funds Universities and other State Research Organizations (mainly CNR, Italian National Council of Research) yearly for the study and development of multiprocessor systems for image and signal analysis. Our National group is presently made of three units: one in Pavia, one in Rome and another in Palermo having two groups, a first one at the Institute of Cosmic Physics of the Italian National Council for Research, (supporting this Workshop) and a second one at the Department of Mathematics of the University, Faculty of Sciences. Recently, another National Organization (ENEA, responsible for developing new technologies for energy supply, both traditional and non-conventional ones) has decided to support the PAPIA project increasing the funding level (by a factor of 100) and is aiming at the production of an engineered version of this system for a potential application as the eye module of an automatic manipulator operating in a radioactive environment.

Presently, about twelve persons are working full-time on this project, ranging from the architecture layout, design and testing to the simulation and invention of new pyramidal algorithms for image analysis, perception simulation, pictorial data base indexing, etc. Two international cooperation projects are underway on these last two subjects.

PAPIA HARDWARE

The PAPIA system is made of a pyramid of processors which are driven by a control system that allows the simultaneous execution of each instruction on all processors in SIMD mode whilst image data are fed through an active memory that is directly connected to the base of the pyramid. A sequential host computer views the above components as a part of its peripherals together with an input device (a telecamera), an external data base, an output device (display) and a conventional terminal for the system user. See figure 1 which shows all the above components of the system.

The PAPIA system, as we will shortly see in more detail, may also operate in a multi-SIMD mode since different couples of pyramid planes can execute different instructions (in multi-SIMD mode) since a different controller has been provided for each couple of planes.

The basic chip (the PAPIA chip) contains five processors: four of them are connected to a central one (on the chip surface) which represents the "father" or top processor, in this way the unit may be recursively used to build the pyramid by interconnecting a number of chips which depends on the chosen size of the base of the pyramid. In the first version of PAPIA, a base of 32x32 processors is foreseen, therefore the pyramid will have 5 planes (excluding the apex).

The design of the chip started in 1984 and was based on 4μ-Si gate NMOS technology ocupying an area of 5 x 5 mm^2. Figure 2 shows the chip

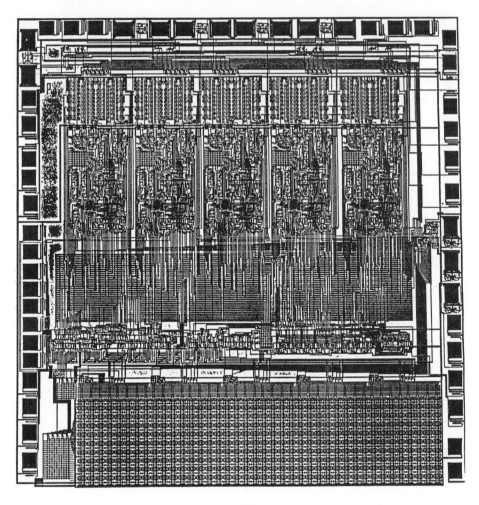

Figure 2 PAPIA chip layout

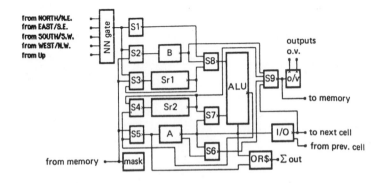

Figure 3 Block diagram of the PAPIA processor

layout containing the five processors (father and four children), the RAM memory of 256 bits (approximately) for each processor, which can be directly accessed only by the corresponding processor, and the 48 pins of the dual-on-line package that was chosen for this chip implementation. Due to the small number of pins, a gating technique was required for allowing the vertical (or horizontal) interprocessor connections, the first ones from children to father (or viceversa) and the second ones using a 4-connectivity pattern from each processor to its four brothers (or viceversa). Moreover a gating technique allows simultaneous access to a subset of neighbours (implementing some boolean functions).

Figure 3 illustrates a block diagram of the processor unit showing the input/output gating block, the different registers (one bit and shift registers), the ALU, the global OR and the masking register. The top left gating block allows the input from five different processors (from the father, from the four brothers) in Vertical mode or, in Horizontal mode, from four different processors (NE, SE, SW and NW neighbours). The two variable shift registers are used for bit serial arithmetic allowing local computations without requiring to store intermediate results. Boolean and comparison facilities have also been included.

The masking registers may enable/disable a full plane of processors or a single processor (by providing a specific bit plane pattern).

The control of an iterative process (for termination) may be performed by a global EX-OR function on all the states of the processors belonging to a single plane before and after the ith operation: if no changes occur, then the process should stop since it is useless.

Finally, two clocks are available in the chip: a computing one and a communicating one (the second being faster, by a factor two, than the first one). Since the two clocks overlap, loading/unloading of images and computing may be performed concurrently. The images are loaded in columnwise fashion, from the base, via the active memory which, in turn, receives image data from the host computer either directly during acquisition or, indirectly from an external memory when the images have been previously digitized.

In a future version of PAPIA the base will contain 128 x 128 processors and therefore the system may be viewed as made of three sections (each one able to execute a different instruction in truly multi-SIMD mode). The top section (see figure 4 for a schematic view of the full pyramid) containing the vertex, a 2 x 2 plane, a 4 x 4 and an 8 x 8 plane, the middle section with a 16 x 16 plane and a 32 x 32 plane and, finally, the bottom section with 64 x 64 and 128 x 128 processor planes.

The full PAPIA system therefore contains a Global Control and a Scalar Processing unit (GCSP) which interfaces with the host computer, hosts the user program, the scalar data and runs the programs executing the scalar code sections; the Pyramid Control Unit (PCU) which performs all the hardware functions required to feed each pyramid section with instructions and control signals; the Multiprocessor Pyramid Unit (MPU) which works as a collection of arithmetic logical units for square arrays of different sizes of data and, lastly, the Active Memory Unit (AMU) devoted to the acquisition and loading/unloading of images also enabling the processing of images with size greater than the one of the pyramid base.

The next chip to be designed and built will contain 21 processors, therefore embedding a three layer pyramid having a higher integration level than the actual one which was deliberately kept at a conservative value for reasons of simplicity and reliability since it was our first prototype.

PAPIA SOFTWARE

The software environment, a fundamental issue for obtaining comfortable programming and interaction, may be seen as made of an operating system, particular utilities aimed at the special peripherals previously mentioned and a high level language to express the pyramid algorithms that will inevitably be intertwinned with conventional, sequential, computation.

The kernel of an existing multitask real-time system (the Philips PSOS™) was chosen because of the availability of primitives for dynamic processes, for memory management, for timing and interrupt service. An external layer of this kernel is supposed to contain a set of routines for the PAPIA multiprocessor applications that become the user interface to his programs allowing him to perform system startup, machine configuration, program loading, serial and task control, flag and message exchange, error logging, etc.

The basic machine instructions for operating the pyramid of processors (12 bit long) may be subdivided into instructions that have a direct access to the memory and those that have no relation to memory which, in turn, may be further subdivided into register instructions, arithmetical and comparison instructions and local operation instructions[12].

The memory instructions are used for loading into registers and storing into memory (they have the leftmost bit set to 1). The remaining instructions (they have the leftmost bit set to O) allow to set the logical registers, to set the mask registers, to set the length of the shift registers, to perform information transfer between registers. Arithmetical, logical and comparison operations may be activated as well as a selection of the registers by other instructions; still other instructions allow shifting, clearing and setting specified registers. Finally, local operations may be performed by instructions which code the connection orientation (vertical or horizontal), the logical operator (AND, OR, NOT, EXOR), etc.

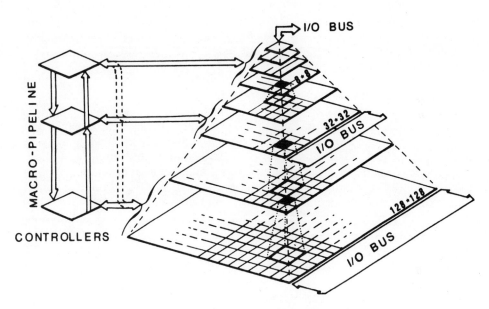

Figure 4 Schematic view of the full pyramid

At the machine level, a PMACRO has been defined for the parallel code and its assembler translator has been implemented. A 68000 C compiler, a preprocessor and a library of 68000 assembly language routines are used to build complete tasks including serial and parallel code as well as system control instructions. PMACRO is a low level language designed to allow PAPIA users to program the machine pyramid control units, the active memory unit and the global control unit.

Turning now to higher level languages, obviously needed to make the PAPIA system practical and convenient, a number of possibilities has been analyzed: a C extension which is called PCP (Parallel C for PAPIA) and a problem oriented language which is termed POL. Both languages will be implemented in Pascal and will produce an object code to be run on the system.

The PCP language will contain parallel instructions and special data types, particularly matched to the pyramid architecture. There will be limitations, due to the hardware configuration, in the maximum number of variables that may coexist at a certain time, in the maximum working accuracy, etc. The reasons for choosing to extend the C language are the following:

i) portability, ii) availability and iii) structured, top-down programming that closely reflects the hierarchical nature of the architecture.

The POL language belongs to the procedural family of languages and will code the most used image transforms that are generally part of any working system: point and local operations, geometrical transforms, statistical computations, storage of a pictorial data base and retrieval of images. These particular program segments will be incorporated into the PCP coded program so that, finally, it may be compiled and each part of the program will run on the appropriate hardware component (controller, interpreter, sequential host computer, active memory and pyramid section). The program will be written with the help of a standard editor (on the host computer) that will allow its final storage on a corresponding file.

HIGH LEVEL VISION WITH PAPIA

The typical applications of multiprocessor systems in computer vision have generally been in the area of low level vision where the tasks were the detection of objects, the counting of objects, the extraction of topological features, etc. Recently[13], a proposal was made to mimic the perceptual abilities of the human visual system by means of the simultaneous computation and detection of the grey level histogram modalities (number of histogram modes) at the different levels of the pyramid. In the human eye this capability accounts for the preattentive vision which is described by a school of perception researchers by a set of laws generally known under the name of Gestalt Laws. These are: the laws of similarity, proximity, good continuation and closure. The operation of these laws can be simulated by means of two-dimensional grouping algorithms which perform in a way which is similar to the one of the human eye.

As an example, for the law of similarity, if an image contains two types of local patterns (even in the presence of noise) the eye is able to quickly visually merge all "similar" patterns into a single group therefore perceiving two, separate, groups.

In the case of the proximity law when looking at an image containing some dots which are close to each other and other dots which are far apart from the previous ones and between themselves, we perceive close dots as if belonging to a single object and sparse dots as if they represented another, different, object.

It is the working hypothesis of this approach (PPA: pyramidal perceptive algorithms for high level vision) that the detection of bimodality in an image histogram accounts for the visual perception of the "groups" and that if the pyramid may also reveal the bimodality (without requiring the histogram evaluation) we may simulate the perceptual behaviour at a very fast rate.

The pyramid computation is performed as follows[14]: each cell at a level i will receive the inputs from its "children" cells at level i-1 and will find a partition (into two subsets) of the input values such that the variance of each subset about its mean is minimized; if the means of the subsets are μ and β, the sizes of the subsets r and s and the standard deviations $\emptyset$ and π, the cell will compute the "Fisher distance" $d = |\mu - \beta| / \sqrt{\emptyset^2 + \pi^2}$ between the two subsets. If this distance is high (above a certain threshold) then the population is bimodal (the means are many standard deviations apart) otherwise the cell will signal "not bimodal". This process is repeated on the level above (i+1) and, again bimodality (if it exists) will be detected and "passed on" to the level above until the vertex of the pyramid is reached. At some levels there will be cells that have detected bimodality (on the basis of the previous partition of subsets) and others (having a Fisher distance under the given threshold) which will signal no bimodality. The basic idea is that the brain computes such bimodality and considers it relevant if a cell on a high level of the pyramid has a bimodal population.

In some cases there may be more than two modes (trimodality, etc) but the computation required would be much more complex and this model accounts for the gross estimation of variances of subsets (where the means are distinctly separated) assuming the brain (in a fast succession of operations) computes the population sizes in order to perform "perceptive grouping".

The detection of compact regions may also be done by computing, recursively, the means and variances by cells on different levels of the pyramid and comparing the ratios of variances over included areas so that if, at a given level, the ratios are all higher than the one corresponding at the level below in this level we have detected an homogeneous region (its variance along the corresponding area gives a lower ratio). At the base of the pyramid noise may influence the computation of the variances but as the process climbs up the local values average and the global estimation of variances becomes more reliable. The nodes which have a variance-to-area ratio smaller than the one of the nodes in the level below are called root nodes and indicate the presence of compact regions. A top-down process (of a tree growing process) will detect all the leaves (at the base of the pyramid) corresponding to the root nodes which are the pixels of the compact regions.

This approach may help to bridge the gap between low level vision algorithms on cellular arrays and high level vision algorithms on pyramid machines: the essential feature of being able to have the information both from local regions and from global regions plus the fact that many computations may be performed concurrently make the pyramid machine a very adequate tool to embed new algorithms for image processing and understanding regardless of the fact that our own visual system may behave differently.

CONCLUSIONS

This work has briefly presented the motivation of the PAPIA project and its guidelines as well as a recent international research project which aims at using pyramidal algorithms to perform high level vision tasks, specifically those concerning the perceptual level as described by Gestaltists. The complexity of assembling all the required knowledge to design, build, test and use a pyramid multiprocessor system is remarkable and the level of organization required is, perhaps, above our present possibilities: yet we feel that this project will proceed, perhaps with the help of a new funding agency (like the Italian ENEA), so that all the work done until now by all the contributing groups (in Pavia, Rome and Palermo) may evolve into a real working system.

REFERENCES

1. S. Wolfram, *Introduction* to Com. ACM, Special Section on Computing in Theoretical Physics, 28, 353,(1985).
2. W. D. Arnett, *Computational Astrophysics*, in Com. ACM, 28, 354,(1985).
3. S. L. Tanimoto, *Designing Iconic Programming Systems: Representation and Learnability*, TR-85-07-05, Dep. of Computer Science, University of Washington, Seattle.
4. *Computer*, Special Issue on Highly Parallel Computing, 15, 1:1982.
5. P. E. Danielsson, S. Levialdi, *Computer Architectures for Pictorial Information Systems*, IEEE Computer, 53, (1981).
6. A. Rosenfeld, *Some Useful Properties of Pyramids*, in: "Multiresolution Image Processing and Analysis", A. Rosenfeld, edit., Springer-Verlag, Berlin, 12:2 (1984).
7. S. L. Tanimoto, *Sorting, Histogramming, and Other Statistical*

Operations on a Pyramid Machine:, in 6. , 12:136.
8. M. J. B. Duff, D. M. Watson, T. J. Fountain, G. K. Shaw: *A cellular logic array for image processing* , Patt. Recog. **5**:229 (1973).
9. S. R. Sternberg: *Pipeline Architectures for Image Processing*, in: "Multicomputers and Image Processing", K. Preston, Jr., L. Uhr, edits., Academic Press, New York, 291(1982).
10. V. Cantoni, S. Levialdi, *PAPIA: a case history*, in: "Pyramid Multi-Computers", L. Uhr, edit., Academic Press, New York, to appear, (1987).
11. V. Cantoni, L. Carrioli, O. Catalano, L. Cinque, V. Di Gesù, M. Ferretti, G. Gerardi, S. Levialdi, R. Lombardi, A. Machì, R. Stefanelli, *The PAPIA mage Analysis System*, SPIE International Technical Symposium on Image Processing, Cannes, (1985).
12. A. Canobbio, *Definizione dell'Architettura, Progetto Elettrico e Funzionale del Chip di PAPIA*, Thesis in Electronic Engineering, Department of Informatics and System Sciences, University of Pavia, (1984).
13. A. Rosenfeld, *Pyramid Algorithms for Perceptual Organization* , NATO ARW on Pyramid Computing Systems, Maratea, (1986).
14. A. Gross, *M.S.Thesis*, in preparation, University of Maryland, (1986).

LANGUAGES FOR PARALLEL PROCESSORS

Anthony P. Reeves
School of Electrical Engineering
Cornell University
Ithaca, New York 14853

INTRODUCTION

The effective programming of parallel computers is much more complex then the programming of conventional serial computers. There are two fundamental models of highly parallel computer architectures: single instruction stream-multiple data stream (SIMD) in which a single program control unit is used to control a set of slave processing elements and multiple instruction stream-multiple data stream (MIMD) in which a set of interconnected independent processors cooperate on a single task. The high level programming language constructs appropriate for each model are discussed.

There is a large amount of similarity between most conventional serial computers when organizing programs. Consequently, algorithms formulated for one system are easily ported to other systems and a common high level language can be used. Advantages of high level languages, compared to a machine specific assembly language, include productivity (faster more reliable program development, plus common knowledge by many programmers), use of established subroutine libraries, and portability. Common high level languages have evolved for these systems over the last 30 years, starting with Fortran; more recent additions include Pascal and Ada. In the search for higher computing speeds many different types of parallel processors have been developed. These systems have very different architectures and algorithms developed for one system may be very inefficient for another. It is very difficult to design an effective high level language suitable for a range of parallel processor architectures. Furthermore, the efficient porting of algorithms from one system to another is not guaranteed. For SIMD systems an array language based on a matrix algebra notation is usually used. More recently developed MIMD systems offer much more flexibility; however, high level software is at a very early stage of development, especially in the area of multiple dynamic tasks.

In order to take advantage of highly parallel computers (i.e., systems with in the order of 100 or more processors) new programming language concepts must be used. Programmers must consider programs in terms of parallel operations, especially for SIMD systems. For MIMD systems, a number of different conceptual frameworks are possible. In the next section an outline of different parallel computer architecture types is presented and then, in the following two sections, programming languages for highly parallel SIMD and MIMD schemes are discussed in turn.

PARALLEL PROCESSORS

When considering parallel programming languages the underlying parallel computer architecture should also be considered. There are four main types of parallel architecture: (a) fast vector, (b) highly parallel processor array, (c) tightly coupled MIMD system, and (d) loosely coupled MIMD system.

Fast Vector Computers

Fast vector machines are usually based on a conventional serial machine with an interleaved memory and a fast arithmetic pipeline processor. The goal is to speed up vector operations which frequently occur in scientific programs. Programming of such systems is done either with a vectorizing Fortran or, in less sophisticated systems, with a set of library subroutines. For most systems an "extended" version of Fortran with some vector statements is available to permit programmers to directly use the fast processor. Such extensions are usually highly non-portable between different computers. The great advantage of such systems is that old programs may be run faster than before and no parallel programming expertise is needed. A major factor in the speedup thus obtained may often be due to the high speed technology used to construct the processor rather than the parallel processing architecture. The disadvantage with this approach is that the degree of parallelism is limited and the implementation cost is usually very high when compared to other parallel techniques. A speedup in the range 4 to 10 over the serial computer without the vector processor may be contemplated. Examples of such systems include supercomputers such as Cray and Cyber 205, add-on array processors such as the FPS 120, and array processor cards for microprocessor systems.

Highly Parallel Processor Arrays

The concept of these systems is to use a large number of ALU's to simultaneously process a number of data elements; i.e., the SIMD mode of operation. Usually, these systems consists of a set of Processing Elements (PE's) each of which contains an ALU and some local memory. There is a single "host" program control unit which broadcasts the same instruction to all PE's. The advantage of such systems is that very large numbers of PE's (many thousands) may be efficiently used for suitable applications. The main disadvantage is that users must very carefully map their problems onto the system. Conventional serial languages such as Fortran are out of the question. Usually an array based high level language is used. Frequently, this language is very specific (non-portable) for the particular processor for which it was designed and may have built in constants relating to the hardware. For example, it is typical to be able to declare arrays which have the same number of elements as there are PE's. Since there are a very large number of PE's the interconnections between them are limited to a small number of permutations. A major programming problem is to effectively use the primitive permutations; many of the parallel languages include these permutations as primitives. Examples of these systems include Illiac IV, STARAN, DAP and the MPP.

Tightly Coupled MIMD Systems

In tightly coupled MIMD systems a number of independent processors share a common data memory. This provides the most convenient programming environment for MIMD processing but is difficult to implement. The problem is to give fast access for a large number of processors to a single shared memory environment. This is difficult even when processors are accessing different memory modules and may be impossible when they all make requests to the same module. Usually only a small number (4 to 16) processors can efficiently share a common memory. An exception to this is the NYU Ultracomputer design and the RP3 implementation which use a combination of cache memories and special hardware in the memory switches to reduce shared memory conflicts. For some of these systems a standard Fortran compiler with automatic vectorization is available. More usually, an extended language is available with process synchronization and communication primitives. Concurrent languages such as Ada and C with Unix system calls have these features.

Loosely Coupled MIMD Systems

The final system type is a distributed computer system although, in this context, the term loosely coupled MIMD system is to be preferred. In general, the main concern with distributed systems is to effectively share a set of resources with a number of users performing different tasks. For large scale scientific problems we are concerned with a single user effectively using all the resources to achieve a minimum throughput time for a single task. This distinction means that much of the work done in the area of distributed systems is not directly applicable to our problems. For this system type a set of independent processors work on a common task; however, they do not share a common memory but communicate through message bassed communication channels. The potential for very highly parallel systems exist but the main problem is to minimize the interprocessor communication since this may dominate the computation time. Highly parallel loosely coupled systems have only recently emerged for large scale scientific problems; most are based on VLSI microprocessor nodes with a hypercube interconnection scheme. These systems require a special programming language for effective utilization. Most systems currently use a Fortran or C with extended interprocess communication facilities. These extensions are similar to the message passing primitives and task control features found in distributed processing systems. Unfortunately these languages are not well matched to scientific applications and much work is needed in this area.

SIMD PROCESSOR ARRAY SYSTEMS

SIMD systems perform the same operation on a number of data elements at the same time. For most scientific applications it is convenient to consider the data items to be stored in an array and the high level language is similar to a matrix algebra with the arrays as matrix variables. In general, the language extensions define operations on aggregate data structures.

Operations on aggregate data structures

Most serial high level languages define primitive data types such as integer, real, boolean and aggregate data types such as array and record; In general, operations can on applied directly to the primitive data types. A simple technique to increase the parallelism in a language is to permit operations to be also applied to the aggregate data structures.

There are two types of aggregate data structures: static and dynamic. The static data structures include arrays and records. Arrays of primitive data type elements are the simplest to specify operation semantics. In general, two array operands must be conformable, i.e. have the same shape or one operand may be a scalar which is then extended to an array of the shape of the array operand. The result is an array of element to element operations. This extension is the most useful for scientific applications. Records may be treated in a similar manner; however, conformability rules are more complex since records may contain different variant parts. Furthermore, since a record may contain different data types, it is possible for an operation to be valid for some record elements but invalid for others. Static nested structures of arrays and records introduce no new problems.

Dynamic data structures are created with pointers and usually involve a dynamic memory allocation procedure e.g. new in Pascal. The size of a dynamic structure is not known at compile time since pointers may be joined to other pointers when the program is running. It is more usual for dynamic structures to consist of a set of records which have pointer fields which link them to other records. Note that dynamically allocated data structures involving arrays or records but which do not involve pointers are not a problem in this context since their size is known at compile time.

The semantics for operations between list structures are not, in general, simple extensions of the scalar operators since the operands usually have different shapes. List based languages such as Lisp have operators which are defined for these structures.

One of the first matrix algebra programming languages was APL. It was not designed for parallel processing but as an extended formal notation for expressing programs. It contains a large number of very flexible multi-dimensional matrix operators. One of the implementation problems with this language, especially for highly parallel systems is the dynamic array range feature which means that array shapes (rank and range) are not known at compile time. There have been several efficient implementations of APL on computers with limited amounts of parallelism.

Parallel Pascal

For highly parallel SIMD systems it is important that the programming language permits a direct mapping between the algorithm and the architecture. Parallel Pascal [1] was designed as a high level language for a class of SIMD systems. It was originally targeted to the Massively Parallel Processor (MPP) which is a mesh connected processor with 16384 PE's organized in a 128 x 128 mesh. Parallel Pascal does not restrict the rank or range of arrays; although these must be defined at compile time as in standard Pascal. The version of Parallel Pascal for the MPP does restrict the last two dimensions of parallel arrays to be 128 x 128 to conform with the hardware and a library of subprograms is used to simplify the processing of other sized arrays.

Parallel Pascal is an extended version of the conventional serial Pascal programming language which includes a convenient syntax for specifying array operations. It is upward compatible with standard Pascal and involves only a small number of carefully chosen new features.

There are three fundamental classes of operations on array data which are frequently implemented as primitives on array processors but which are available in conventional programming languages; these are: data reduction, data mapping or permutation, and data broadcast. Furthermore, the concepts of array data selection and conditional execution have to be extended for the parallel case. These operations have been included as primitives in Parallel Pascal. In the following, the main features of an SIMD language are outlined using Parallel Pascal syntax.

Parallel Statements

The prime feature of all SIMD languages is that parallel statements are permitted; for example consider three arrays which are declared as follows:

   var a, b, c: array [1..10] of integer;

the following statement

   a := b + c + 1;

is equivalent to

   for i := 1 to 10 do
      a[i] := b[i] + c[i] + 1;

The arrays in an expression must have the same shape (i.e., the same rank and ranges); any scalars in a parallel expression are extended to have the same shape as the arrays.

Data Mapping

Array data mapping and permutation functions are a new feature necessary for languages for highly parallel SIMD architectures. On a conventional computer data mappings are usually specified by index expressions in a *for* loop. Highly parallel SIMD systems usually implement a limited number of data permutations directly by hardware. Frequently, a key to efficient algorithm implementation is to express it in terms of these permutations. In Parallel Pascal four data mapping operations are available as primitive standard functions; however, for some Parallel Processors it may be necessary to specify more primitive functions for efficiency. The standard Parallel Pascal functions for data permutation and distribution are given in table 1.

Table 1: Data Mapping and Distribution Functions

Syntax	Meaning
shift(array, S1, S2, ..., Sn)	end-off shift data within array
rotate(array, S1, S2, ..., Sn)	circularly rotate data within array
transpose(array, D1, D2)	transpose two dimensions of array
expand(array, dim, range)	expand array along specified dimension

The *shift* and *rotate* functions are found in most high level parallel languages and are efficiently implemented by most parallel systems. The shift function shifts data by the amount specified for each dimension and shifts zeros (null elements) in at the edges of the array. Elements shifted out of the array are discarded. The rotate function is similar to the shift function except that data shifted out of the array is inserted at the opposite edge so that no data is lost. The arguments of these functions specify the array to be shifted and the amount to shift in each dimension.

For example, for the arrays declared by:

```
var
  a, b: array [1..5,0..9] of integer;
  c, d: array [0..9] of integer;
```

the statement

```
a := shift(b, 0, 3);
```

is functionally equivalent to

```
for i := 1 to 5 do
  begin
    for j := 0 to 6 do
      a[i,j] := b[i,j+3];
    for j := 7 to 9 do
      a[i,j] := 0;
  end;
```

and the statement

```
c := rotate(d, 3);
```

is functionally equivalent to

```
for i := 0 to 9 do
  c[i] := d[(i + 3) mod 10];
```

The shift and rotate functions are the keystone primitive functions for many highly parallel computers; in a number of parallel languages these data mappings are expressed as parallel index expressions rather than a function call. The function call approach was used in Parallel Pascal so that other primitive mapping functions could easily be added to the language without involving any syntax modifications. For example, a shuffle permutation might be a useful primitive function for an architecture which involves an Omega like network or a hypercube interconnection scheme. It is much simpler to specify an operation such as the fast Fourier transform (FFT) by directly using such primitives. Quick program porting to other Parallel Pascal implementations can be achieved by writing library functions which implement the shuffle permutation by using the shift and rotate primitives. The efficiency of such a ported algorithm may not be very high and some reprogramming of the algorithm may be desirable.

Given the wide diversity of parallel processor architectures it is perhaps unreasonable to expect that a single algorithm specification can be efficiently compiled for all systems. The scheme proposed above for Parallel Pascal permits simple porting of all algorithms to different architectures for verification and provides a consistent notation for reprogramming an algorithm to tune it for a particular architecture.

While *transpose* is not a simple function to implement with many parallel architectures, a significant number of matrix algorithms involve this function; therefore, it has been made available as a primitive function in Parallel Pascal. The parameters to transpose specify the the array to by manipulated and the dimensions which are to be interchanged.

Data Distribution

Data broadcast is achieved implicitly in Parallel Pascal by using scalars in a parallel expression. There is also an array rank increasing primitive function in Parallel Pascal for data distribution called *expand*. This function increases the rank of an array by one by repeating the contents of the array along a new dimension. The arguments to this function are the array to be expanded, The number of the new (expanded) dimension and the range of this dimension.

This function is used to maintain a higher degree of parallelism in a parallel statement; which may result in a clearer expression of the operation and a more direct parallel implementation. In a conventional serial environment data distribution functions would simply waste space without achieving any change in performance.

For example, the following statement adds a vector to all rows of a matrix

a := b + expand(c, 1, 1..5);

The above statement is functionally equivalent to the following

for i := 1 to 5 do
    a[i,] := b[i,] + c;

The important difference between the above two forms is that the expand computation has parallelism of 50 while the parallelism of each cycle of the loop is only 10. For parallel systems with 50 or more PE's it is much simpler to efficiently implement the statement with the expand function; however if only 10 PE's are available then the second form is more convenient. In general, it is much simpler for a compiler to match an explicitly highly parallel statement to a machine with less parallelism than vice versa.

Data Reduction

Array reduction operations are achieved with a set of standard functions in Parallel Pascal which are listed in table 2. The first argument of a reduction function specifies the array to be reduced and the following arguments specify which dimensions are to be reduced.

For example, given the array declarations

    var
        a: **array**[1..10,1..5] **of** integer;
        b: **array**[1..5] **of** integer;
        c: integer;

the following are example Parallel Pascal statements

    b := sum(a, 2);   (* sum the rows of a *)
    c := sum(a, 1, 2); (* sum all elements of the array a *)
    c := max(b, 1);   (* find the maximum value of b *)

Each dimension parameter of a reduction function implies that there will be one less dimension in the result array; a scalar is considered to be an array without any dimensions in this context.

Table 2: Reduction Functions

Syntax	Meaning
sum(array, D1, D2, ..., Dn)	reduce array with arithmetic sum
prod(array, D1, D2, ..., Dn)	reduce array with arithmetic product
all(array, D1, D2, ..., Dn)	reduce array with Boolean AND
any(array, D1, D2, ..., Dn)	reduce array with Boolean OR
max(array, D1, D2, ..., Dn)	reduce array with arithmetic maximum
min(array, D1, D2, ..., Dn)	reduce array with arithmetic minimum

Array reduction functions are found in many matrix algorithms and most parallel array languages. Usually, the ideal goal of using n PE's is to achieve an $O(n)$ speedup when compared to a single serial processor. However, if n data elements are to be reduced then the best computation time for the reduction is $O(\log_2 n)$ and for some architectures the data routing time is $O(n)$. Therefore, the speedup is, at best, $O(n/\log_2 n)$. The situation is better for reducing k.n data elements, where $k > 1$, since all PE's locally reduce k data points with parallelism n before reducing the partial result. The speedup in this case is proportional to $(k.n)/(k + \log_2 n)$.

Data Selection

Selection of a portion of an array by selecting either a single index value or all index values for each dimension is frequently used in many parallel algorithms; e.g., to select the ith row of a matrix which is a vector. Specification of a single index value is the usual indexing method in standard Pascal. In Parallel Pascal all index values can be specified by eliding the index value for that dimension. For example, given the arrays declared by:

var a,b: array [1..5,1..10] of integer;

in Parallel Pascal the statement

a[,1] := b[,4];

assigns the fourth column of b to the first column of a and the following statement means assign the second row of b to the first row of a.

a[1,] := b[2,];

Vector high level languages, especially those designed for fast vector computers, often permit the specification of more complex data selection. This option reflects the greater flexibility of the vector processors to more efficiently process sparse vector data. For example, some vector processors directly implement scatter and gather operations. Consider the following array declarations:

var x: array [1..n] of integer;
    y: array [1..m] of integer;
    indexx: array [1..m] of 1..n;
    indexy: array [1..n] of 1..m;

the scatter operation is specified by:

for i := 1 to n do
   y[indexy[i]] := x[i];

and a gather operation is specified by

for i := 1 to m do
   y[i] := x[indexx[i]];

The syntax for the specification of the parallel versions of these operations is simply:

   y[indexy] := x;
and
   y := x[indexx];

The above are very powerful selection operations, especially for sparse vector operations in which, for example, n << m. When n = m, they permit the dynamic specification of an arbitrary data mapping between x and y. Such flexibility may be realizable with a reasonable cost on a vector computer with a low degree of parallelism; however, there is no efficient way to implement such operations on highly parallel processor arrays such as the MPP. Furthermore, note that if we have n data elements distributed between n PE's then the cost of performing an operation on a small subset of data elements is the same as performing the operation on all data elements (and masking the operation as described in the next section). The extension of vector indexing to higher dimensioned arrays may be defined in a number of different ways.

Conditional Execution

An important feature of any parallel programming language is the ability to have an operation operate on a subset of the elements of an array. In a serial language each array element is processed by a specific sequence of statements and there are a variety of program control structures for the repeated or selective execution of statments. In a parallel language the whole array is processed by a single statement; therefore, an extended program control structure is needed. Our experience with developing Parallel Pascal [1] indicated that a single parallel control structure, an extended **if** construct, was adequate for programming parallel algorithms. The conditional execution statement was renamed **where** due to semantic differences with the standard **if** statement.

The syntax of the Parallel Pascal **where** statement is as follows:

> **where** array-expression **do**
> statement
> **otherwise**
> statement

where array-expression is a Boolean valued array expression and statement is a Parallel Pascal statement. The **otherwise** and the second controlled statement may be omitted.

The execution of a **where** structure is defined as follows. First, the controlling expression is evaluated to obtain a Boolean array (mask array). Next, the first controlled statement is evaluated. Array assignments are masked according to the boolean control array. If there is an otherwise statement it is then evaluated; in this case array assignments are masked with the inverse of the control array.

For example, given the array declarations:

var a, b, c:**array** [1..10] **of** integer;

the following statement

> **where** a < b **do**
> c := b
> **otherwise**
> c := a;

is functionally equivalent to

```
for i := 1 to 10 do
  if a[i] < b[i] then
    c[i] := b[i]
  else
    c[i] := a[i];
```

The main semantic difference between the **where-do-otherwise** structure and the **if-then-else** structure is that with the former both controlled statements are evaluated, independent of the value of the control expression, while with the latter only one of the two controlled statements is evaluated.

Most parallel languages have at least a **where** statement or an equivalent; in some languages all serial control structures are extended to accept array control parameters. Where statements are usually defined to conform to one of two sets of semantic rules: *conditional evaluation* or *conditional assignment*. For the conditional evaluation case, only computations where are necessary to contribute to results are performed. This form is favored for those fast vector computers which can efficiently implement scatter and gather operations. The evaluation of complex controlled expressions becomes very difficult with this form since the shape of the partial results in an array expression may not be the same shape as the final results; this complicates the determination of which partial results should be computed. Further complications of this form are introduced when subprograms are called from within a controlled parallel statement.

For the conditional assignment case, only the visible assignments in the controlled statement are affected by the controlling array; all expressions, both parallel and serial, are evaluated in the conventional manner. The semantics of this operation are much cleaner and simpler to implement; however, there are two potential problems with this scheme: First, processing is done on all array elements; therefore, the processing done on the non-selected elements is wasted. This is, in general, not a problem with highly parallel processor arrays since the execution time is the same with or without the extra computation (also true for Cray 1 which does not directly implement scatter or gather operations). Furthermore, a clever compiler may be able to avoid some unneeded work on a system which implements scatter and gather. Second, arithmetic overflow may occur in processing some of the results which are not needed.

For example, the following statement will successfully execute if any element of a is zero and conditional evaluation rules are obeyed but will cause a divide by zero error if conditional assignment is used since the reciprocal of a is computed for all elements.

```
where a <> 0 do r := 1/a;
```

In Parallel Pascal the conditional assignment scheme is used since it is more suitable for highly parallel systems.

MIMD SYSTEMS

Loosely coupled MIMD systems consist of a set of independent processors (nodes) interconnected by a message based switching network. For systems designed for parallel scientific applications there is usually a host node which performs the main system functions such as I/O and task initiation, and a set of identical processing nodes. Typically, each node is programmed with a conventional serial language which has some extended message passing and task management system calls.

For an example we consider the programming language for the Intel iPSC system [2], which based on the Caltech hypercube [3]. The serial language used is either C or Fortran. An operating system on each node manages the routing and queueing of messages between the various processes.

The parameters used in message communication are listed in Table 3. A process is identified by the *node* on which it resides and a process identifier *pid* which may be chosen by the user. Messages have a *type* identifier associated with them which is chosen by the user. A process may open a number of channels, each of which is associated with a channel identifier ci.

The main process communication functions are given in Table 4. Messages may be sent and received either asynchronously using *send* and *recv* or synchronously using *sendw* and *recvw*. In the asynchronous cases, the function returns before the action has been completed; hence the program may continue immediately. Operation completion can be tested with the *status* function; this function must return true before the channel can be used for further messages. In the synchronous case the function blocks until the transaction (as far as this process is concerned) has been completed. The *probe* is useful in this context to check if data is available before executing a blocking receive. All messages are sent with a user defined type identifier. A receive function will only accept a message which has the same type as its type field. For effective programming of a MIMD system it is convenient to have both synchronous and asynchronous mechanisms.

Table 3: iPSC Parameter Names

Name	Meaning
ci	channel identifier
pid	process identifier
node	node identifier
type	type identifier
buf	data buffer
len	length of message
cnt	length of received message
result	a boolean test result

The main process control functions for the iPSC are given in Table 5. Currently, these functions can only be executed by host processes; however, it is possible for a node process to request execution of these functions by sending a message to the host. The *load* function creates a new process by loading the executable *file* in its local memory onto the specified node and starting it. The *lkill* function terminates a process and the *lwait* function waits for a process to terminate and will return in the parameters cnode, cpid, and ccode the node, pid and completion code of the terminated process respectively.

Table 4: iPSC System Communication Functions

Syntax	Meaning
ci = copen(pid)	open a channel
send(ci, type, buf, len, node, pid)	send a message
sendw(ci, type, buf, len, node, pid)	blocking send
recv(ci, type, buf, len, cnt, node, pid)	receive a message
recvw(ci, type, buf, len, cnt, node, pid)	blocking receive
result = status(ci)	test channel free
result = probe(ci, type)	test message available

Table 5: iPSC Host Process Control Functions

Syntax	Meaning
load(file, node, pid)	create a new process
lkill(node, pid)	terminate a process
lwait(node, pid, cnode, cpid, ccode)	wait for process termination

Two basic strategies which have been used for programming these systems are outlined below and a third strategy is proposed. First, the basic primitives as outlined above can be directly used; see [3] for an example. Other programming tools can be added to the primitives listed above; for example, a function to automatically select the next node to which a process will be allocated. In general, however, it is very difficult to program complex scientific programs at this level. Furthermore, advanced MIMD concepts, such as dynamic load balancing, are very difficult to introduce to such programs. Second, it is possible to simulate an SIMD system. To do this arrays are uniformly distributed to all nodes; each node runs a single process. The distributed nature of the environment, including all message passing, is hidden from the user by a set of system functions such as the *shift* function in Parallel Pascal. In fact, Parallel Pascal could be efficiently implemented on these systems. For an example of this approach see [4] in which a set of matrix operations have been developed for hypercube systems. This scheme is very effective if the problem can easily be specified with the usual SIMD constraints; the disadvantage is that the additional flexibility of the distributed system cannot easily be used.

A third approach is the concept of a task oriented operating system. With this approach the user programs a problem by means of a sequence of task oriented system primitives. The user only specifies the computation process and the format of the input and output data. The system performs all subtask decomposition, allocation and scheduling to best fit with the available resources. While this is the most difficult approach to implement it offers the greatest promise for the future. There are a number of system capabilities which are very difficult to program at the user level; for example, dynamic load balancing and reconfiguration for fault tolerance. Furthermore, there are problems, especially those which manipulate dynamic list structures, for which the SIMD model of computation is not appropriate. Another consideration is that all nodes may not be equal, special function units which are highly efficient for perhaps a single operation may be attached to some of the nodes. A clever operating system would direct operations of this type to the special nodes. In summary, the effective general programming of these systems should be managed by a task oriented operating system which has an understanding of both the available resources and the algorithms relating to the task to be computed; it should also be able to dynamically adapt to maintain optimal resource utilization.

CONCLUSION

The organization of a high level programming language for parallel systems depends upon a large extent to the type of parallel architecture. Fast vector systems and most tightly coupled MIMD systems do not offer very high degrees of parallelism and, consequently, standard programming languages with a few extensions can be effectively used. The major gains for parallel systems will be made with the highly parallel computer architectures; this means SIMD arrays for problems which can be effectively mapped to these structures and loosely coupled MIMD systems for other problems. For SIMD arrays various language schemes exist which, in general, are based on a matrix algebra syntax. These languages are very well matched to the computer architecture and offer a reasonable degree of portability between systems. The situation for loosely coupled MIMD systems is much less developed. The currently available communication primitives are at much too low a level for the scientific user and algorithms need to be developed to take advantage of the high degree of flexibility in these systems. Two strategies to improve this situation have been considered: an SIMD high level language and a task oriented operating system.

REFERENCES

1. A. P. Reeves, "Parallel Pascal: An extended Pascal for Parallel computers," *Journal of Parallel and Distributed Computing* **1** pp. 64-80 (1984).
2. Intel Corporation, *iPSC System Overview*, Order Number:175278-002, 1986.
3. C. L. Seitz, "The Cosmic Cube," *Communications of the ACM* **28**(1)(1985).
4. O. A. McBryan and E. F. Van de Velde, "Hypercube Algorithms and Implementations," *2nd SIAM Conference on Parallel Computing*, (1985).

THE MASSIVELY PARALLEL PROCESSOR:

A HIGHLY PARALLEL SCIENTIFIC COMPUTER

Anthony P. Reeves

School of Electrical Engineering
Cornell University
Ithaca, New York 14853

INTRODUCTION

The Massively Parallel Processor (MPP) [1, 2] is a highly parallel scientific computer which was originally intended for image processing and analysis applications but it is also suitable for a large range of other scientific applications. Currently the highest degree of parallelism is achieved with the SIMD type of parallel computer architecture. With this scheme a single program sequence unit broadcasts a sequence of instructions to a large number of slave Processing Elements (PE's). All PE's perform the same function at the same time but on different data elements; in this way a whole data structure such as a matrix can be manipulated with a single instruction. The alternative highly parallel organization, the MIMD type, is to have an instruction unit with every PE. This scheme is much more flexible but also much more complex and expensive.

Computers based on the SIMD scheme are usually very effective for applications which have a good match to the constraints of the architecture. Furthermore, they are usually also extensible in that it is possible to increase the performance for larger data structures by simply increasing the number of PE's.

In order to utilize the features of a SIMD system, as with all computer designs, it is important for the programmer to have some knowledge of the underlying architecture; for example, it is important to know that some matrix operations have the same cost as a scalar operation. For these systems a special programming environment is usually used and, in general, serial programs designed for conventional serial computers must be reformulated for highly parallel architectures. The MPP is programmed in a high level language called Parallel Pascal [3]. Therefore, the main advantage of SIMD systems is a much more cost effective method for doing scientific computing than conventional computer or supercomputer systems. The major disadvantage is that the user must become familiar with a new kind of programming environment.

A major consideration in the design of a highly parallel computer architecture is the processor interconnection network. Processors must communicate with a speed which does not impede data processing; however, a general processor interconnection network is usually prohibitively expensive when a large number of processors are involved. A major design task is to design a restricted network which is adequate for the anticipated tasks for the system. The mesh interconnection scheme, used on the MPP, is simple to implement and is well suited to a large number of image processing algorithms.

The first section of this chapter describes the architecture of the MPP. Then the convenient data structers which can be manipulated by the MPP are outlined and the high level programming environment is discussed. The performance of the processor interconnection network for important data manipulations is considered in detail since this indicates which algorithms can be efficiently implemented on the MPP. Finally, some current applications of the MPP are outlined.

The Impact of Technology

The implementation of highly parallel processors is made possible with VLSI technology. The MPP was designed with the technology available in 1977. The custom processor chip has 8000 transistors on an area of 235 x 131 mils and contains 8 bit-serial PE's. It is mounted in a 52 pin flat pack and requires 200 mW at 10 MHz; 5 micrometer design rules were used. The SIMD mesh architecture can directly take advantage of the ongoing major advances in VLSI technology. A number of more advanced chips have been developed since the MPP design. For example, the GAPP chip [4] developed with the technology of 1982, has 72 bit-serial PE's, each having 128 bits of local memory. This chip requires 500 mW at 10 MHz. ITT [5] is predicting that by 1987 they will be able to make 16 16-bit PE's (with four spare) on a single chip with each PE having 1k words of local memory. This chip would use 1.25 micrometer design rules and would involve 600,000 transistors on an area of 450 x 600 mils. For the more distant future, advantage can be taken of wafer scale integration as soon as it becomes economically available. Techniques for dealing with the fault tolerance needed with such a technology have already been considered [6].

THE MPP ARCHITECTURE

The Massively Parallel Processor consists of 16384 bit-serial Processing Elements (PE's) connected in 128 x 128 mesh [1]. That is each PE is connected to its 4 adjacent neighbors in a planar matrix. The two dimensional grid is one of the simplest interconnection topologies to implement, since the PE's themselves are set out in a planar grid fashion and all interconnections are between adjacent components. Furthermore, this topology is ideal for two dimensional filtering operations which are common to low level image processing such as small window convolution.

The PE's are bit-serial, i.e. the data paths are all one bit wide. This organization offers the maximum flexibility, at the expense of the highest degree of parallelism, with the minimum number of control lines. For example, as an alternative to the MPP consider 2048 8-bit wide PE's (on the MPP one chip contains 8 1-bit PE's). The 8-bit version would have a less rich set of instructions restricted to predefined byte operations while the bit-serial processors can process any data format. The advantage gained with the 8-bit system is that full processor utilization is achieved with arrays of 2048 elements while arrays of 16384 elements are required for full utilization of the MPP. The MPP PE is well matched to low level image processing tasks which often involve very large data arrays of short integers which may be from 1 to 16 bits.

The effectiveness of the MPP architecture for various interprocessor data manipulations is considered. The MPP offers a simple basic model for analysis since it involves just mesh interconnections and bit-serial PE's. The minimal architecture of the MPP is of particular interest to study, since any architecture modifications to improve performance would result in a more complex PE or a more dense interconnection strategy.

The MPP Processing Element

The MPP processing element is shown in Fig. 1. All data paths are one bit wide and there are 8 PE's on a single CMOS chip with the local memory on external memory chips. Except for the shift register, the design is essentially a minimal architecture of this type. The single bit full adder is used for arithmetic operations and the Boolean processor, which implements all 16 possible two input logical functions, is used for all other

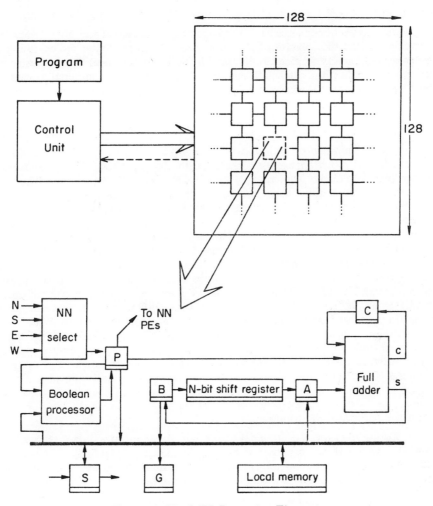

Figure 1. The MPP Processing Element

operations. The NN select unit is the interface to the interprocessor network and is used to select a value from one of the four adjacent PE's in the mesh.

The S register is used for I/O. A bitplane is slid into the S registers independent of the PE processing operation and it is then loaded into the local memory by cycle stealing one cycle. The G register is used in masked operations. When masking is enabled only PE's in which the G register is set perform any operations; the remainder are idle. The masked operation is a very common control feature in SIMD designs. Not shown in Fig. 1. is an OR bus output from the PE. All these outputs are connected (ORed) together so that the control unit can determine if any bits are set in a bitplane in a single instruction. On the MPP the local memory has 1024 words (bits) and is implemented with bipolar chips which have a 35 ns access time.

The main novel feature of the MPP PE architecture is the reconfigurable shift register. It may be configured under program control to have a length from 2 to 30 bits. Improved performance is achieved by keeping operands circulating in the shift register which greatly reduces the number of local memory accesses and instructions. It speeds up integer multiplication by a factor of two and also has an important effect on floating-point performance.

Array Edge Connections

The interprocessor connections at the edge of the processor array may either be connected to zero or to the opposite edge of the array. With the latter option rotation permutations can be easily implemented. This is particularly useful for processing arrays which are larger than the dimensions of the PE array. A third option is to connect the opposite horizontal edges displaced by one bit position. With this option the array is connected in a spiral by the horizontal connections and can be treated like a one-dimensional vector of 16,384 elements.

The MPP Control Unit

A number of processors are used to control the MPP processor Array; their organization is shown in Fig. 2. The concept is to always provide the array with data and instructions on every clock cycle. The host computer is a VAX 11/780; this is the most convenient level for the user to interact since it provides a conventional environment with direct connection to terminals and other standard peripherals. The user usually controls the MPP by developing a complete subroutine which is down loaded from the VAX to the main control unit (MCU) where it is executed. The MCU is a high speed 16-bit minicomputer which has direct access to the microprogrammed array control unit (ACU). It communicates to the ACU by means of macro instructions of the form "add array A to array B". The ACU contains runtime microcode to implement such operations without missing any clock cycles. A first in-first out (FIFO) buffer is used to connect the MCU to the ACU so that the next macro operation generation in the MCU can be overlapped with the execution in the ACU. A separate I/O control unit (IOCU) is used to control input and output operations to the processor array. It controls the swapping of bitplanes between the processor array and the staging memory independent of the array processing activity. Processing is only halted for one cycle in order to load or store a bitplane.

I/O and the MPP Staging Memory

The staging memory is a large data store which is used as a data interface between peripheral devices and the processor array; it provides two main functions. First, it performs efficient data format conversion between the data element stream which is most commonly used for storing array data to the bitplane format used by the MPP. Second,

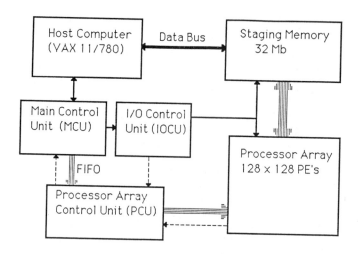

Figure 2. MPP System Organization

it provides space to store large (image) data structures which are too large for the processor array local memory.

The I/O bandwidth of the processor array is 128 bits every 100 ns; i.e., 160 Mbytes/second. When fully configured with 64Mb of memory the staging memory can sustain the MPP I/O rate. Currently the MPP is configured with 32 Mb and can sustain half the optimal I/O rate. In addition to data reformating, hardware in the staging memory permits the access of 128 x 128 blocks of data from arbitrary locations in a large (image) data structure. This feature is particularly useful for the spooling scheme which is outlined in the following section.

MPP DATA STRUCTURES

The 128 x 128 array is the equivalent to the natural *word* size on a conventional computer since elementary MPP instructions manipulate 128 x 128 bitplanes. Stacks of bitplanes representing integers and 32-bit real numbers are also basic instructions in the MCU which are supported by runtime subroutines in the microprogrammed control unit.

The fundamental data structures for the MPP are shown in Fig. 3.a. The long vector format is supported by the hardware spiral edge connections. However, for long data shifts the vertical interprocessor connections can also be used. The 128 x 128 matrix is the most natural data structure for the MPP. Higher dimensional data structures may also be implemented. If the last two dimensions of the data structure are 128 x 128 then higher dimensions are simply processed serially. If the last two dimensions are less than 128 x 128 then it may be possible to pack more than two dimensions into a single bitplane. For example, a 16 x 8 x 8 x 4 x 4 data structure can be efficiently packed into a 128 x 128 array. Convenient data manipulation routines for this data structure can be efficiently developed at the Parallel Pascal level.

Frequently, the data to be processed by a parallel processor will be in the format of arrays which exceed the fixed range of parallelism of the hardware. Therefore, it is necessary to have special algorithms that will deal with large arrays by breaking them down into blocks manageable by the hardware, without loosing track of the relationships between different blocks.

There are two main schemes for storing large arrays on processor arrays: the *blocked* scheme and the *crinkled* scheme; these are illustrated in Fig. 3.b. Consider that a M x M array is distributed on an N x N processor array where K = M / N is an integer. In the blocked scheme a large array is considered as a set of K x K blocks of size N x N each of which is distributed on the processor. Therefore, elements which ar allocated to a single PE are some multiple of N apart on the large array. In the crinkled scheme each PE contains a K x K matrix of adjacent elements of the large array. Therefore, each parallel processor array contains a sampled version of the large array. For conventional array operations which involve large array shift and rotate operations both blocked and crinkled schemes can be implemented with only a very small amount of overhead. The crinkled scheme is slightly more efficient when shift distances are very small and the blocked scheme has a slight advantage when the shift distance is of the order of N.

The third type of data structure which can be manipulated on the MPP is the huge array which is much too large to fit into the 2 Mb MPP local storage. This scheme, the *spooled* organization, involves the staging memory and is illustrated in Fig 3.c. In the spooled scheme the data is stored in the staging memory and is processed one block at a time by the processor array. The I/O operations to the staging memory are overlapped with data processing so that if the computation applied to each block is large enough then the cost of spooling will be negligible. However, if only a few operations are applied to each block the the I/O time will dominate. For near neighbor operations one possibility is to perform a sequence of operations on each block without regard to other blocks. The boundary elements of the result array will not be valid. This is circumvented by reading overlapping blocks and only writing the valid portions of the result blocks back to memory.

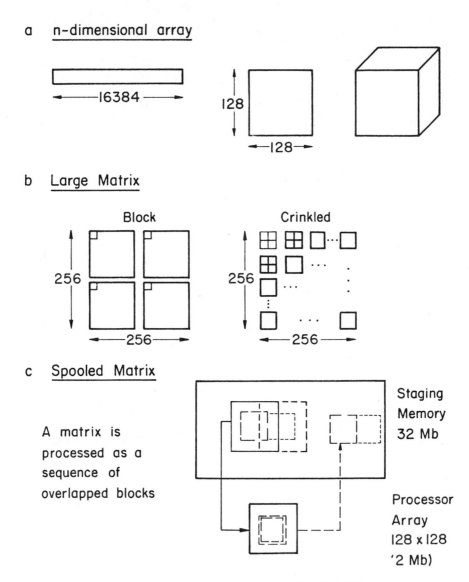

Figure 3. Data Structures for the MPP

THE MPP PROGRAMMING ENVIRONMENT

There are three fundamental classes of operations on array data which are frequently implemented as primitives on array computers but which are not available in conventional programming languages, these are: data reduction, data permutation and data broadcast. These operations have been included as primitives in the high level language for the MPP called Parallel Pascal. Mechanisms for the selection of subarrays and for selective operations on a subset of elements are also important language features.

High level programming languages for mesh connected SIMD computers usually have operations similar to matrix algebra primitives since entire arrays are manipulated with each machine instruction. A description of Parallel Pascal features is given elsewhere in this proceedings [7]. A brief synopsis of important features of this language follows.

Parallel Pascal is an extended version of the Pascal programming language which is designed for the convenient and efficient programming of parallel computers. It is the first high level programming language to be implemented on the MPP. Parallel Pascal was designed with the MPP as the initial target architecture; however, it is also suitable for a large range of other parallel processors. A more detailed discussion of the language design is given in [3].

In Parallel Pascal all conventional expressions are extended to array data types. In a parallel expression all operations must have conformable array arguments. A scalar is considered to be conformable to any type compatible array and is conceptually converted to a conformable array with all elements having the scalar value.

In many highly parallel computers including the MPP there are at least two different primary memory systems; one in the host and one in the processor array. Parallel Pascal provides the reserved word *parallel* to allow programmers to specify the memory in which an array should reside.

Reduction Functions

Array reduction operations are achieved with a set of standard functions in Parallel Pascal. The first argument of a reduction function specifies the array to be reduced and the following arguments specify which dimensions are to be reduced. The numeric reduction functions maximum, minimum, sum and product and the Boolean reduction functions any and all are implemented.

Permutation Functions

One of the most important features of a parallel programming language is the facility to specify parallel array data permutations. In Parallel Pascal three such operations are available as primitive standard functions: *shift, rotate* and *transpose*.

The shift and rotate primitives are found in many parallel hardware architectures and also, in many algorithms. The shift function shifts data by the amount specified for each dimension and shifts zeros (null elements) in at the edges of the array. Elements shifted out of the array are discarded. The rotate function is similar to the shift function except that data shifted out of the array is inserted at the opposite edge so that no data is lost. The first argument to the shift and rotate functions is the array to be shifted; then there is an ordered set of parameters, each one specifies the amount of shift in its corresponding dimension.

While transpose is not a simple function to implement with many parallel architectures, a significant number of matrix algorithms involve this function; therefore, it has been made available as a primitive function in Parallel Pascal. The first parameter to transpose is the array to be transposed and the following two parameters specify which dimensions are to be interchanged. If only one dimension is specified then the array is flipped about that dimension.

Distribution Functions

The distribution of scalars to arrays is done implicitly in parallel expressions. To distribute an array to a larger number of dimensions the *expand* standard function is available. This function increases the rank of an array by one by repeating the contents of the array along a new dimension. The first parameter of expand specifies the array to be expanded, the second parameter specifies the number of the new dimension and the last parameter specifies the range of the new dimension.

This function is used to maintain a higher degree of parallelism in a parallel statement which may result in a clearer expression of the operation and a more direct parallel implementation. In a conventional serial environment such a function would simply waste space. For example, to distribute a N-element vector A over all rows of a N x N matrix, the expression is "expand(A,1,1..N)"; as an alternative, to distribute the vector over the columns, the second argument to expand should be changed to 2.

Sub-Array Selection

Selection of a portion of an array by selecting either a single index value or all index values for each dimension is frequently used in many parallel algorithms; e.g., to select the ith row of a matrix which is a vector. In Parallel Pascal all index values can be specified by eliding the index value for that dimension.

Conditional Execution

An important feature of any parallel programming language is the ability to have an operation operate on a subset of the elements of an array. In Parallel Pascal a *where - do - otherwise* programming construct is available which is similar to the conventional *if - then - else* statement except that the control expression results in a Boolean array rather than a Boolean scalar. All parallel statements enclosed by the where statement must have results which are the same size as the controlling array. Only result elements which correspond to true elements in the controlling array will be modified. Unlike the if statement, both clauses of the where statement are always executed.

System Support

In addition to the MPP Parallel Pascal compiler there is a Parallel Pascal translator and a library preprocessor to aide high level program development. The translator translates a Parallel Pascal program into a standard Pascal form. In this way, conventional serial computers can be used to develop and test Parallel Pascal programs if they have a standard Pascal compiler.

Standard Pascal has no library facility; all subprograms i.e., procedures and functions, must be present in the source program. A library preprocessor was developed to allow the use of libraries without violating the rules of standard Pascal.

MPP Compiler Restrictions

The Parallel Pascal compiler for the MPP currently has several restrictions. The most important of these is that the range of the last two dimensions of a parallel array are constrained to be 128; i.e., to exactly fit the parallel array size of the MPP. It is possible that language support could have been provided to mask the hardware details of the MPP array size from the programmer; however, this would be very difficult to do and efficient code generation for arbitrary sized arrays could not be guaranteed. Matrices which are smaller than 128 x 128 can usually be fit into a 128 x 128 array by the programmer. Frequently, arrays which are larger than 128 x 128 are required and these are usually fit into arrays which have a conceptual size which is a multiple of 128 x 128. For example, a large matrix of dimensions (m * 128) x (n * 128) is specified by a four dimensional array which has the dimensions m x n x 128 x 128. There are two fundamental methods for packing the large matrix data into this four dimensional array (see Fig. 3.b), this packing may be directly achieved by the staging memory in both cases.

Host programs for the MPP can be run either on the main control unit (MCU) or on the VAX; in the latter case the MCU simply relays commands from the VAX to the PE array. The advantages of running on the VAX is a good programming environment, floating point arithmetic support and large memory (or virtual memory). The advantage of running on the MCU is more direct control of the MPP array.

Compiler directives are used to specify if the generated code should run on the MCU or the VAX. With the current implementation of the code generator, only complete procedures can be assigned to the MCU and only programs on the MCU can manipulate parallel arrays. Therefore, the programmer must isolate sections of code which deal with the PE array in procedures which are directed to the MCU.

MPP PERFORMANCE EVALUATION

The peak arithmetic performance of the MPP is in the order of 400 million floating point operations per second (MFLOPS) for 32 bit data and 3000 million operations per second (MOPS) for 8-bit integer data. In order to sustain this performance the data matrices to be processed must be as large as the processor array or larger and the amount of time transferring data between processors should be relatively small compared to the time spent on arithmetic computations. For image processing the former constraint is rarely a problem; however, the latter constraint requires careful study.

In order to analyze the effectiveness of the interconnection network for different manipulations it is necessary to characterize the processing speed of the PE and the speed of the interconnection network. On the MPP both of these are data dependent; we have considered three representative cases: single-bit *Boolean* data, 8-bit *integer* data and 32-bit floating-point (*real*) data. For each of these data types we have estimated a *typical* time for an elemental operation. These estimates are of a reasonable order for this minimal PE architecture but are not very precise. For example, the instruction cycle time for a memory access and operation on the MPP is 100 ns. An elemental boolean operation may be considered to take 100 ns.; however, it may be argued that an operation should involve two operands and have all variables in memory in which case three memory accesses (instructions) would require 300ns. For our analysis a two instruction (200 ns) model was used to represent Boolean instruction times. For the real and integer data a convenient number midway between the times for addition and multiplication was used; this was 5 μs. for an integer operation and 40 μs. for a real operation. It should be remembered that elemental operations also include many other functions such as transcendental functions since these can be computed in times comparable to a multiplication on a bit-serial architecture. By adding a large amount of additional hardware to each PE it is possible to increase the speed of multiplication by 10 times or more [8].

For each of the data manipulations considered, times for the three different data types was computed. The performance of the MPP for each manipulation is indicated by the ratio of the data transfer time to an elemental PE operation on the same data type; this will be called the *transfer ratio*. One way to look at this ratio is the number of elemental data operations which must be performed between data transfers for the data transfers not to be the dominant cost for the algorithm. On the MPP data may be shifted between adjacent PE's in one instruction time (100 ns.) concurrently with a PE processing instruction.

Shift and Rotate Operations

The only permutation function which is directly implemented by the MPP is the near neighbor rotate (or shift). The direction of the rotation may be in any of the four cardinal directions. The rotation utilizes the toroidal end around edge connections of the mesh. The *shift* function is similar except that the mesh is not toroidally connected and zeroes are shifted into elements at the edge of the array; therefore, the shift function is not a permutation function in the strict sense. The concept of the rotate and shift functions extends to n dimensions; on the MPP the last two dimensions of the array correspond to the parallel hardware dimensions and are executed in parallel, higher dimension operations are implemented in serial. The cost of the rotate function is dependent on the distance rotated. It also depends on the size of the data elements to be permuted.

The transfer ratios for the shift operation are given in Table 1. Ratios are given for shift distances of 1 and 64 elements; 64 is the largest shift which will normally be required in a single dimension on a 128 x 128 matrix since a shift of 65 can be obtained with a rotate of -63 and a mask operation. The worst case figures for a two dimensional shift is 64 in each direction; i.e., twice the figures given in Table 1.

For single element shifts the interconnection network is more than adequate for all data types. For maximum distance shifts the ratio of 33 for Boolean data could cause problems for some algorithms but the situation is much better for real data.

Table 1: The Cost for Shift and Rotate Operations

Shift distance	Operation cost in μs.			Transfer Ratio		
	Boolean	integer	real	Boolean	integer	real
1	0.2	1.6	6.4	1.0	0.32	0.16
64	6.5	51	210	33	10	5.2

Important Data Manipulations

A simple algorithm to perform any arbitrary data mapping on the MPP is as follows. Start with the address of where the data is to come from in each PE. For each PE compute the distance that the data must be moved to reach that PE. Using the spiral interconnections, rotate the data 16384 times. After each rotation compare the distance in each PE with the distance moved and if they match then store the data for that PE. The transfer ratios for this algorithm are 82000, 10000 and 4200 for Boolean integer and real data types respectively. Obviously, this is much too slow for most practical applications. Fortunately, for most applications only a small number of regular data mappings are required; efficient algorithms can be developed for most of these mappings.

The transfer ratio for a number of important data manipulations is shown in Table 2. The figures for large arrays correspond to the blocked data scheme. For large arrays the transfer ratio is normalized by the cost of an operation on the whole array.

This technique may be applied to almost any parallel architecture. It is expected that the results obtained for the MPP would be very similar to those obtained for other like architectures such as the Distributed Array Processor (DAP) [9] or NCR's GAPP processor chip which contains 72 PE's with local memory [4]. The results given in Table 2 could be used by programmers to predict the performance of algorithms on the MPP. A more detailed analysis of data mappings on the MPP is given in [10].

For the MPP, the results indicate that, although arbitrary data mappings may be very costly, some important data manipulations can be done very efficiently. The shift register, which has a 2 times speedup factor for multi-bit arithmetic also has a significant effect on the implementation of several of the multi-bit data manipulations studied. Especially interesting is the improvement of over 10 times for real data distribution. The shuffle cannot be implemented fast enough for efficient FFT implementation; however, other data mapping strategies for the FFT, such as butterfly permutations, are well known which have a much more efficient implementation on the MPP.

On the DAP row and column distribution is implemented directly by special hardware buses. For the MPP we can see from Table 2 that no advantage would be gained from this hardware for real data operations and possibly very little advantage for integer operations.

The MPP can effectively implement algorithms on pyramid data structures. Horizontal operations are done with near neighbor operations; i.e. single element shifts. Vertical operations require data mappings similar to the shuffle permutation; the times for these operations are given in Table 2. These times for numeric data are quite reasonable for many pyramid algorithms. A detailed analysis of pyramid operations on the MPP is given in [11].

Sorting has been proposed as one technique for doing arbitrary data permutations on the MPP; the cost of bitonic sorting on the MPP is given in Table 2. This cost is very high although not as high as the arbitrary data mapping algorithm.

The last row in Table 2. shows the transfer ratio for swapping a matrix with the staging memory. In this case the transfer ratio indicates the number of operations which

Table 2: Transfer Ratios for Different Data Manipulations and Array Sizes

Data Manipulation	Array Size								
	128 x 128			256 x 256			512 x 512		
	Boolean	integer	real	Boolean	integer	real	Boolean	integer	real
Data Shift									
a) 1 element	1.0	0.32	0.16	2.0	0.5	0.24	2.0	0.5	0.24
b) worst case	33	10.2	5.2	33	10	5.2	33	10	5.2
Broadcast									
a) Global	2	0.64	0.32	0.88	0.28	0.14	0.59	0.20	0.09
c) Row (or column)	68	3.2	0.52	35	1.68	0.30	18.2	0.92	0.19
Shuffle (2-dimensional)	640	90	42	640	90	42	640	90	42
Transpose	840	110	44	840	110	44	840	110	44
Flip	190	43	21	190	43	21	190	43	21
Pyramid Up (sum reduce)	330	45	21	110	15	7.5	28	4	2.3
Pyramid down	330	44	19	110	15	6.4	28	3.9	1.7
Sort	19000	1100	280	12000	870	230	14000	790	210
Swap	256	20	10	256	20	10	256	20	10

must be performed on each swapped matrix for there to be no significant overhead due to swapping. Since the transfer ratio does not change with array size, this suggests that spooling with large matrices for near neighbor operations would be more efficient than spooling with 128 x 128 blocks.

APPLICATIONS

The MPP was originally designed for processing multispectral satellite imagery and synthetic aperture radar imagery. Both of these applications have now been demonstrated on the MPP. Other uses of the MPP are now being explored. There are now 36 active projects on the MPP; these can grouped as follows: physics (10), earth sciences (5) signal and image processing (7), and computer science (14). A full list of these projects is given in Appendix A.

CONCLUSION

The Massively Parallel Processor is a highly effective computer for a large range of scientific applications. It is representative of the class of highly parallel SIMD mesh connected computers. Novel features of the MPP design are the staging memory and the PE shift register. The MPP has demonstrated its capability to implement the image processing algorithms for which it was originally designed; however, the current system lacks the very high speed peripheral devices needed to optimize its performance. It is also being used for a much broader range of scientific applications. The main limitation with the MPP when used for other applications is the limited amount of local memory (1024 bits/PE). This should not be a problem with future systems especially since the recent advances in memory technology. This problem has been offset on the MPP by judicious use of the staging memory.

The problem frequently cited for mesh connected SIMD architectures is the inefficiency of the mesh interconnection scheme when used for other than near neighbor tasks. However, this is not a problem for many practical applications on the MPP; for example, FFT and pyramid operations can be effectively implemented especially for very large data structures.

The MPP is more cost effective for suitable applications than supercomputers of a similar age. Future SIMD mesh connected computers may be anticipated which will take advantage of recent VLSI technology and will be much more powerful than the MPP. These systems can be expected to be much more cost effective than more conventional supercomputers for suitable applications such as low level image processing. These architectures can also be effectively implemented at smaller scales; for example, as attached processors to microprocessor systems.

The cost of using a highly parallel computer is the change of programming style and the need to reformulate existing programs. However, programming in an appropriate high level language is often not conceptually more difficult than programming a conventional computer. In fact, in some respects, it is simpler since arrays are manipulated without the multiple do loops required in conventional serial programming languages.

REFERENCES

1. K. E. Batcher, "Design of a Massively Parallel Processor," *IEEE Transactions on Computers* C-29(9) pp. 836-840 (September 1981).
2. J. L. Potter, *The Massively Parallel Processor*, MIT Press (1985).
3. A. P. Reeves, "Parallel Pascal: An Extended Pascal for Parallel Computers," *Journal of Parallel and Distributed Computing* 1 pp. 64-80 (1984).
4. NCR Corporation, *Geometric Arithmetic Parallel Processor*, NCR, Dayton, Ohio (1984).
5. S. G. Morton, E. Abreau, and F. Tse, "ITT CAP-Toward a Personal Supercomputer," *IEEE Micro*, pp. 37-49 (December 1985).
6. A. P. Reeves, "Fault Tolerance in Highly Parallel Mesh Connected Processors," in *Computing Structures for Image Processing*, ed. M. J. B. Duff, Academic Press (1983).
7. A. P. Reeves, "Languages for Parallel Processors," *International Workshsop on Data Analaysis in Astronomy,* , Erice, Italy(April 1986).
8. A. P. Reeves, "The Anatomy of VLSI Binary Array Processors," in *Languages and Architectures for Image Processing*, ed. M. J. B. Duff and S. Levialdi, Academic Press (1981).
9. R. W. Gostick, "Software and Algorithms for the Distributed-Array Processor," *ICL Technical Journal*, pp. 116-135 (May 1979).
10. A. P. Reeves and C. H. Moura, "Data Manipulations on the Massively Parallel Processor," *Proceedings of the Nineteenth Hawaii International Conference on System Sciences*, pp. 222-229 (January, 1986).
11. A. P. Reeves, "Pyramid Algorithms on Processor Arrays," *Proceedings of the NATO Advanced Research Workshop on Pyramidal Systems for Image Processing and Computer Vision,* , Maratea, Italy(May 1986).

APPENDIX A. Current Research Using the MPP

Dr. John A. Barnden Indiana University	Diagramtic Information-Processing in Neural Arrays
Dr. Richard S. Bucy Univ. of Southern California	Fixed Point Optimal Nonlinear Phase Demodulation
Dr. Gregory R. Carmichael University of Iowa	Tropospheric Trace Gas Modeling on the MPP

Dr. Tara Prasad Das State University of New York at Albany	Investigations on Electronic Structures and Associated Hyperfine Systems Using the MPP
Dr. Edward W. Davis North Carolina State University	Graphic Applications of the MPP
Dr. Howard B. Demuth University of Idaho	Sorting and Signal Processing Algorithms: A Comparison of Parallel Architectures
Dr. James A. Earl University of Maryland	Numerical Calculations of Charges Particle Transport
Mr. Eugene W. Greenstadt TRW	Space Plasma Graphics Animation
Dr. Chester E. Grosch NASA-Langley Research Center	Adapting a Navier-Stokes Code to the MPP
Dr. Robert J. Gurney Goddard Space Flight Center	A Physically-Based Numerical Hillslope Hydrological Model with Remote Sensing Calibration
Dr. Martin Hagan University of Tulsa	Sorting and Signal Processing Algorithms: A Comparison of Parallel Architectures
Dr. Harold M. Hastings Hofstra University	Applications of Stochastic and Reaction — Diffusion Cellular Automata
Dr. Sara Ridgway Heap Goddard Space Flight Center	Automatic Detection and Classification of Galaxies on "Deep-Sky" Pictures
Dr. Nathan Ida The University of Akron	Solution of Complex, Linear Systems of Equations
Dr. Robert V. Kenyon MIT Man Vehicle Laboratory	Application of Parallel Computers to Biomedical Image Analysis
Dr. Daniel A. Klinglesmith, III Goddard Space Flight Center	Comet Haley Large-Scale Image Analysis
Dr. Daniel A. Klinglesmith, III Goddard Space Flight Center	FORTH, an Interactive Language for Controlling the MPP
Dr. Chin S. Lin Southwest Research Institute	Simulation of Beam Plasma Interactions Utilizing the MPP
Dr. Stephen A. Mango Naval Research Laboratory	Synthetic Aperture Radar Processor System Improvements
Dr. Michael A. McAnulty University of Alabama	Algorithmic Commonalities in The Parallel Environment
Dr. A.P. Mulhaupt University of New Mexico	Kalman Filtering and Boolean Delay Equations on an MPP

Dr. John T. O'Donnell Indiana University	Simulating an Applicative Programming Storage Architecture Using the NASA MPP
Mr. Martin Ozga USDA-Statistical Reporting Service (SRS)	A Comparison of the MPP with Other Supercomputers for Landsat Data Processing
Dr. H.K. Ramapriyan Goddard Space Flight Center	Development of Automatic Techniques for Detection of Geological Fracture Patterns
Dr. John Reif Harvard University Computer Science	Parallel Solution of Very Large Sparse Linear Systems
Dr. L.R. Owen Storey Stanford University	Particle Simulation of Plasmas on the MPP
Dr. James P. Strong Goddard Space Flight Center	Development of Improved Techniques for Generating Topographic Maps from Spacecraft Imagery
Dr. Francis Sullivan National Bureau of Standards	Phase Separation by Ising Spin Simulations
Dr. Peter Suranyi University of Cincinnati	A Study of Phase Transitions in Lattice Field Theories on the MPP
Dr. James C. Tilton Goddard Space Flight Center	Use of Spatial Information for Accurate Information Extraction
Dr. William Tobocman Case Western Reserve University	Wave Scattering by Arbitrarily Shaped Targets Direct and Inverse
Mr. Lloyd A. Treinish Goddard Space Flight Center	Animated Computer Graphics Models of Space and Earth Sciences Data Generated via the MPP
Dr. Scott Von Laven KMS Fusion, Inc.	Free-Electron Laser Simulations on the MPP
Dr. Elden C. Whipple, Jr. UCSD/CASS/C-001	A Magnetospheric Interactive Model Incorporating Current Sheets (MIMICS)
Dr. Richard L. White Space Telescope Science Institute	The Dynamics of Collisionless Stellar Systems
Dr. Lo I. Yin Goddard Space Flight Center	Reconstruction of Coded-Aperature X-ray Images

PARALLEL PROCESSING:

FROM LOW- TO HIGH-LEVEL VISION[1]

Steven L. Tanimoto

Dept. of Computer Science, FR-35
University of Washington
Seattle, WA 98195
U. S. A.

SUMMARY

While parallel processors have been constructed to support image processing, these systems are not suitable for kinds of image analysis where pictorial information is encoded in non-raster format, e.g., as object-boundary chains, topological graph structures, or tree data structures. Studies of more general image-analysis systems can be based on a trichotomy of image analysis operations (low-level, intermediate-level, and high-level), in which the operations are characterized by the data structures they manipulate. Two computer systems are discussed each of which offers a means for handling all three categories of operations. The first system, the "PIPE" is a commercially available image processor. The second is one proposed by the author. The possible impact of systems such as these on data analysis in astronomy is discussed.

INTRODUCTION

Parallel processing gives us a hope of being able to make data analysis keep up with the increasing rates at which astronomical data is being acquired. This trend toward ever-higher data-acquisition rates was described nicely by Disney (1985). Much of this data consists of digital images, and consequently, much of astronomical data analysis involves image processing. Fortunately, a great many of the operations that astronomers wish to perform on images (e.g., filtering) can readily be mapped onto a variety of parallel computer architectures, resulting in efficient implementations. Parallel computers, while not yet in the hands of every astronomer, are well on their way to common usage.

---

[1]Research supported in part by NSF Grant DCR-8310410

Some of the operations on images, however, do not map so nicely onto existing parallel computers. For example, algorithms that construct contour maps from images do not execute efficiently on processor-per-pixel architectures. This paper explores some of the ways in which efficient processing may be brought to the full "iconic to symbolic" spectrum of image-analysis operations. Particular attention is paid to the "intermediate-level" operations and computer architectures that support it.

LOW-LEVEL AND HIGH-LEVEL VISION

Over the past ten years, substantial experience has been gained in the development of pixel-oriented parallel computers and their application to image processing. Good examples of such systems are the Massively Parallel Processor or "MPP", described by Batcher (1980) and the cellular-logic image processor CLIP4, described by Duff (1978). The MPP contains an array of processing elements of size 128 by 128, and the CLIP4 has an an array of size 96 by 96. These systems may be called "iconic" because they provide one processing element for each image pixel, and the processing elements are connected together in a mesh, just as the pixels of the image are connected together (e.g., in an 8-connected fashion, where each cell has eight neighbors). These iconic systems have been very successful in achieving highly efficient image processing. Unlike general multiprocessing systems in which several distinct processes may run concurrently, the iconic systems use a single-instruction-stream/multiple-data-stream approach, which avoids the runtime inefficiencies and the programming problems of synchronization.

The CLIP4 and the MPP have some additional characteristics which distinguish them, in practical terms, from other multiprocessors: First, their processing-elements' arithmetic-logic units are only 1-bit wide, necessitating bit-serial implementations of arithmetic operations. Second, the addresses that the processing elements use to address their local memories are all the same at any given time; they are broadcast by the centralized program-interpretation unit.

Some iconic processors achieve high throughput with a pipelined architecture, in which only a single neighborhood of pixels is processed at a time, although the computation on all the pixels of the neighborhood may be done in parallel.

Typical uses of iconic processing systems include the following: contrast enhancement, low-pass filtering, detecting edges, identifying the groups of pixels that belong to separate objects, and computing statistics of an image such as the mean pixel value, standard deviation, minimum and maximum.

Along with the development of powerful computers for low-level image processing, there has been significant progress in architecture for symbolic computation. Exemplifying such development are the Symbolics 3600-series workstations, Lisp Machine Inc.'s workstations, the Xerox Dorado and its descendants, and Texas Instruments' Explorer workstations. "High-level vision," which is generally regarded as a symbol-manipulation process rather than a pixel-manipulation one, therefore shares with low-level vision the status of being a beneficiary of architectural advancement.

Within the computer vision community, there is usually agreement that processing at all three levels (low, intermediate and high) of analysis is necessary for the "general" vision capability that is desired.

Although current computer architectures provide well for low-level and high-level vision, they do not do as well for the intermediate level. For example, architectures that are suitable for rapidly transforming image data into symbolic form are largely missing. The pixel-oriented architectures, are efficient for image processing, but they become very inefficient when they are confronted with symbols and lists. Symbol processors such as LISP machines, on the other hand, do not have the high degree of parallelism that is needed for real-time image processing.

In order to clarify the term "intermediate-level vision," I define it to be any process which takes as input an image represented as a two-dimensional array, and outputs a structure which is not a two-dimensional array. Although this definition permits some strange transformations to be classified as intermediate-level vision operations, it does serve the purpose of distinguishing transformations such as chain encoding, shape measurement, and constructing a region-adjacency graph from image processing operations such as median filtering and convolution. Furthermore, it is a simple definition, and it is usually easy to use it to classify operations as either intermediate-level or not.

With existing hardware there are a number of possible ways to perform intermediate-level vision operations. The "bottom-up" approach is to attempt to use pixel-oriented image processing hardware to handle such things as chain encoding. Needless to say, this is usually very inefficient; a CLIP4 or MPP would usefully employ only a few of its many processing elements at a time.

The top-down approach involves using a powerful symbol processor such as a LISP machine to do the intermediate-level operation. While there are no lost cycles in such an arrangement, the fact that there is little or no parallelism means that the operation is not as rapidly done as it could be with more appropriate hardware.

Certain computers currently under study are called "reconfigurable." These may be well-suited for intermediate-level operations; however, they are not generally as highly parallel as machines like the CLIP and MPP, and so they are relatively weak on low-level vision. Also, a variety of computer networks have been proposed that overcome particular problems of existing systems (such as the large network diameter of the mesh), and many of these are described by Uhr (1984).

Because of the these limitations of existing hardware, I believe that researchers should explore special hardware to facilitate the computations of intermediate-level vision (in the context of systems which must handle all three levels).

Two distinct approaches to solving the hardware problem for intermediate-level vision can be identified. The first is to provide special devices to compute iconic-to-symbolic and/or symbolic-to-iconic transformations; the second is to provide a high-speed interface between the low-level (pixel-oriented) computing hardware and the high-level (symbol-oriented) hardware. These two approaches do not exhaust the possibilities, since a combination of the two is possible; one could build hardware that would perform both the transformation and interface functions. Nonetheless, these two approaches give us a framework in which to explore possible designs at the intermediate level of processing.

POSSIBLE FUNCTIONS OF SPECIAL HARDWARE

A number of researchers in computer vision are attempting to build systems to understand outdoor scenes. These systems are to handle tasks such as navigating visually along a road or over open terrain. These systems may use such techniques as binocular stereo for depth, Hough transforms for finding lines (e.g., road boundaries), and segmentation of the scene into regions according to pixel brightness or color. To perform the desired tasks in real time seems to require special hardware not only for the pixel-oriented operations and the symbol-oriented operations, but also for iconic-to-symbolic operations such as the Hough transform, edge linking, raster-to-vector conversion and the extraction of connected-component information.

While these tasks may not be those needed in current approaches to astronomy image analysis, I think that they suggest a range of problems needing special hardware, some of which probably are relevant to astronomy. It therefore seems worthwhile to discuss these image-analysis problems and their hardware solutions, since these solutions may also provide or suggest solutions to the astronomy problems.

Three classes of iconic-to-symbolic transformations are the following: (1) those which determine the locations of special points in an image (e.g., centers of stars, vertices of a polygonal contour, centers of mass of galaxies) and output their coordinates, (2) those which identify prominent lines and curves in an image and output descriptions of them, and (3) those which determine the extents and properties of regions in an image and output such information. Let us discuss two iconic-to-symbolic problems: extracting straight lines from an image and determining the connected regions in an image. The problem of finding lines is in the second class, and it is often approached with some form of the Hough transform. The connected-components problem is in the third class, and it is often solved using a labelling algorithm that searches through the pixels of an image and then outputs summary information such as the areas and positions (e.g., centers of mass) of the regions found.

HOUGH TRANSFORMS

The Hough transform, in its classical form, is a method for finding the parameters for the straight lines in an image. As can be seen from the description which follows, the Hough transform is a good candidate for implementation in special hardware; it is computationally costly for a conventional computer, yet the essential operations involved are simple. Not only that, there are useful variations of the original technique which can be implemented inexpensively.

The Hough transform of an image $F(x, y)$ may be defined as the array of values $A(\rho, \theta)$ where

$$A(\rho, \theta) = \sum_{x=1}^{N} \sum_{y=1}^{N} F(x, y) \, h(x, y, \rho, \theta)$$

where $h(x, y, \rho, \theta)$ is a "hit function" or incidence function defined as

$$h(x, y, \rho, \theta) = \begin{cases} 1, & \text{if } x \cos \theta + y \sin \theta - \rho < C; \\ 0, & \text{otherwise.} \end{cases}$$

In other words, the point (x, y) is said to "hit" the line $\rho = x \cos \theta + y \sin \theta$ if its distance from the line is less than C.

It is usual to refer to $A(\rho, \theta)$ as an "array of accumulators" since each sum is normally computed iteratively by accumulating "votes" at each array position. If the input image F is a binary image, then each of its nonzero pixels casts a set of votes where each vote is for one of the pairs of parameters ρ and θ that specify a line that the pixel hits. If the input image F is not binary but has integer values, then votes are normally weighted by the value of the pixel that casts them.

Once the transform has been computed, it is inspected to see whether any particular pair of parameters has been "elected." If $A(\rho_0, \theta_0)$ has a relatively large value, then this indicates that the line $\rho_0 = x \cos \theta_0 + y \sin \theta_0$ has received strong voting and is one of the prominent lines in the image. The process of finding the parameter pairs that have high values is usually called "peak detection." In practice, the peaks may be slightly smeared across several array positions in A because of the fact that both the image space and the parameter space are quantized into discrete cells. Smearing can also result from either curvature or thickness in the line. Thus, in peak detection, groups of adjacent array positions are usually considered.

The need for each image pixel to cast many votes and to have these votes organized and counted means that the Hough transform is expensive to compute. By modifying the original Hough transform, the computational requirements can be lowered. Here are some of the modifications that are possible:

1. the use of a coarse partition of parameter space. Thus one may reduce the number of distinct values of ρ and θ which are considered in the parameter space. This has the effect of lowering the number of different lines that can possibly be found in the image. These lines then become representatives for sets of lines. That is, each pair (ρ, θ) for which a vote total is accumulated represents a rectangular region in the ρ-θ parameter space.

2. reduction of the resolution of the image. One way of accomplishing this is to not permit individual pixels to vote but to permit only blocks of pixels to vote. This also reduces the number of votes that are cast and which must be counted.

3. application of the Hough transform locally instead of globally in the image. For example, instead of computing one big Hough transform, one may compute many small ones, e.g.,

over 8 by 8 neighborhoods. These small Hough transforms should use a reduced parameter space as well as the small image space.

4. assumption that an orientation is computed at each pixel of the image; this is the orientation of the dominant edge or line passing through the pixel. One then restricts the Hough transform to only accumulate votes from pixels with a particular orientation. The parameter space may then be collapsed to a one-dimensional space and much less voting and vote-counting need be done. In order to obtain lines in various directions, this procedure must be repeated a number of times. However, the total voting and vote counting is reduced when only a relatively small number of directions are of interest.

Special hardware for the Hough transform can be designed to speed up the computation at the bottleneck which is in distributing and accumulating the votes. It has been suggested by Ballard and Brown (1985) that a special VLSI chip be fabricated to handle this function. This chip would contain a large number of wires to interconnect the pixels (vote sources) to the cells of the parameter space (vote destinations).

Each form of the Hough transform might have its own special hardware. Research on parallel hardware for the Hough transform is only beginning; but it appears that much can and will be done in this direction.

CONNECTED-COMPONENTS DETERMINATION

Let us now consider an example of a region-oriented operation which falls into the realm of intermediate-level vision. The problem we now consider is that of constructing a description for each of the regions in an image, such that each description is based upon all of the pixels in the region and none of those outside the region. For such an operation, the input is an image. The output, however, is not an image. The output is typically a list of attributes of each connected region. Such attributes may be a region's area, perimeter, average gray-value or color, or any of a large number of shape descriptions.

While the Hough transform permits a large degree of parallelism in its computation, computing region connectivity seems to require a significant sequential component. For example, to determine that a point X is in the same region as a point Y, it is necessary to verify that there exists a path from X to Y that is entirely contained within the region. Finding such a path is normally accomplished by a labelling process that must make its way from one point to the other. However, there are some region-growing methods that can achieve some parallelism by growing many small regions in parallel in the image and then merging ones that belong together.

Some region-connectivity algorithms for computers with various architectures have been presented by Danielsson and Tanimoto (1983). Whether much parallelism can be brought to the problem of determining connected components of image in an economical way is a problem open for research.

THE PIPE SYSTEM

Let us now examine a state-of-the-art image processing system whose designers have given some thought to the problems of intermediate-level vision.

The PIPE ("Pipelined Image Processing Engine") is a commercially-available image processing system. Its architecture is described by Kent, Shneier and Lumia (1985) and it was originally developed at the National Bureau of Standards. It is manufactured and sold by Digital/Analog Design, Inc., of New York.

The PIPE consists primarily of from three to eight "modular processing stages" each of which comprises two frame buffers (each 256 by 256 by 8 bits). The stages can be configured in a variety of different ways, including as a pipeline and as independent image processors. In addition to these stages there is an input stage and an output stage which also contain pairs of frame buffers. Within each modular processing stage are several comparators, arithmetic/logic units and lookup tables which can be programmed to perform a wide variety of functions. Figure 1 illustrates how the modular processing stages are interconnected.

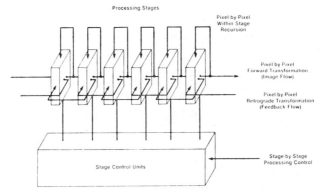

Figure 1. Interconnections among PIPE processing stages. (courtesy of E. Kent).

Dr. E. Kent, in conjunction with the staff of Digital/Analog Design, Inc., has designed a device called the "ISMAP" (Iconic-to-Symbolic MAPper). The ISMAP is intended for use with the PIPE. It can compute a histogram and it can use the histogram to construct a table in which the coordinates of pixels of the original image are indexed by pixel value. With this table, it is easy to have a list of the coordinates for all the pixels with a particular value, e.g., 54. With the ISMAP, such a table can be produced in three video-frame times, or approximately one-hundred milliseconds. This capability allows a programmer considerable flexibility in writing algorithms for operations such as finding lines. For example, an algorithm that uses the PIPE, ISMAP and host to find the dominant lines in an image could run efficiently by allocating an appropriate part of the job to each hardware component.

Since the ISMAP works with a single stream of image data, it does not incorporate significant parallelism itself. However, it runs concurrently with the PIPE, thus achieving some degree of parallel processing that way. The ISMAP is limited to operations which can be accomplished using a fixed image-scanning order. It cannot, for example, trace arbitrary contours and output a list of the contour segments all in order.

THE BIMODAL MEMORY

Whereas the ISMAP provides hardware to compute a class of iconic-to-symbolic transformations, the other approach is to provide an interface that will allow many general-purpose processors (that would otherwise perform high-level tasks) to access the pixels of an image stored in the iconic hardware with a suitably high bandwidth. If there are enough processors, then the iconic-to-symbolic transformations may be computed quickly enough.

Let us imagine a parallel processing system for computer vision which comprises M symbolic processors interfaced to an iconic processor which itself includes an N by N array of processing elements. If M is a power of 4 and N is a power of 2, and $M < N^2$, then each symbolic processor may be assigned to a distinct block of iconic processing elements. Such an assignment implies that only the members of the block are capable of communicating with their assigned symbolic processor at the maximum speed. Other processing elements would have to route their information into another block to reach the symbolic processor, or alternatively route it through other symbolic processors. This restriction is employed in some proposals, e.g., that of Weems et al (1985). However, it would seem to create havoc for the programmer responsible for intermediate-level algorithms.

Another approach is to give each of the M symbolic processors direct access to all N^2 iconic elements. This approach has two problems. First, the cost of the system is higher because of the increased number of connections needed. Second, there can be contention, among the M symbolic processors, for access to the iconic memory. If cost is not a barrier, one need only be concerned with the second problem. This problem can be resolved by replicating the image data and allocating to each symbolic processor its own copy or version. In order to store these extra copies, additional memory is required. An interface based around memory for storing such images can be called "memory-based".

A proposal by the author (1985a) for an efficient interface between the iconic and symbolic processing levels centers around a large memory buffer which supports two kinds of access: binary-image-at-a-time access and pixel-at-a-time access. The binary-image-at-a-time access allows a bit plane to be stored or read from the buffer, while pixel-at-a-time access allows several 8-bit pixels to be stored or read from particular points in an image. Such a memory may be called a "bimodal memory". The low-level, iconic hardware system may read or write a binary image in the bimodal memory in one step. Meanwhile, several general purpose microprocessors may be reading or writing 8-bit pixels in several different layers of the bimodal memory.

In order to efficiently construct a bimodal memory, an unusual component is required: a memory chip that provides access both in terms of binary images and in terms of pixel coordinates. The VRAMs which have recently come onto the market support raster-graphic displays well, but their image-access mode is serial rather than parallel; this would greatly slow down the interface, and what may be worse, require the parallel image processor to spend a great deal of its time in input and output operations.

APPLICATIONS IN ASTRONOMY

While the majority of the data-analysis problems with astronomy images are strictly image-processing problems (in which both the input and output forms are two-dimensional arrays), with further automation of the analysis process, we can expect an increase in the amount of structural processing done on astronomy images. Whereas the identification of star groups in images and subsequent measurement of their properties may be done manually or interactively at present, such tasks could conceivably be automated. While the algorithms for image segmentation and feature extraction are computationally expensive, the development of new kinds of hardware will reduce the computing time.

Although it may be some time before such boards are widely available, I don't think it unreasonable for astronomers to dream of plug-in boards for their image processing systems that accelerate (for example) the detection of stars in particular classes, where such classes may be defined in terms of spatial properties rather than spectral ones.

Generally, it will become easier to perform analyses of the shapes of astronomical phenomena than it has been, and this may be useful in automatically classifying such phenomena as well as deriving descriptions of them.

FUTURE PROSPECTS

Several manufacturers of image processing systems are currently studying ways to provide better support for intermediate-level vision computations. We can expect a variety of solutions to the problem. Coordinated families of boards for various image-analysis functions are one direction for the future. There is commercial appeal to the idea of having plug-in modules for particular iconic-to-symbolic transformations, e.g., chain-encoding, connected component description, Hough transforms, region-adjacency graphs, etc.

But there is also appeal to the idea of coupling the iconic and symbolic processing levels through high-bandwidth interfaces. This approach offers the possibility of high performance and generality; this kind of interface would permit almost any iconic-to-symbolic transformation or symbolic-to-iconic transformation to be computed more rapidly than is possible with a conventional (e.g., 32 bit bus) interface.

CONCLUSION

The computations of machine vision may be classified as low-level, intermediate-level, or high-level, accordingly as they are image-to-image, image-to-symbol, or symbol-to-symbol operations. Many advances in computer hardware support either image-to-image or symbol-to-symbol computations, but relatively little supports the image-to-symbol variety. However, both researchers and manufacturers are currently giving attention to the problem. Two approaches have been identified, one in which hardware is provided to compute particular iconic-to-symbolic transformations, and the other in which a high-speed interface is created between an iconic subsystem and a symbolic subsystem.

In either case the users of the resulting systems are likely to have new flexibility in designing algorithms. Whereas today's vision algorithms are often distributed over two components, future algorithms may contain parts on three or more very different processing components.

REFERENCES

1. Ballard, D.H. and Brown, C. M. 1985. Vision. Byte, Volume 10, Number 4 (April), pp245-261.

2. Batcher, K. E. 1980. Design of a massively parallel processor. IEEE Trans. Computers, Volume C-29, pp836-840.

3. Danielsson, P.-E., and Tanimoto, S. L. 1983. Time complexity for serial and parallel propagation in images, Architecture and Algorithms for Digital Image Processing, A. Oosterlinck and P.-E. Danielsson (eds.), Proc. SPIE 435, August, pp. 60-67.

4. Disney, M. 1985. The data analysis facilities that astronomers want. In Di Gesu, V., Scarsi, L., Crane, P., Friedman, J. H., and Levialdi, S. (eds.) Data Analysis in Astronomy. NY: Plenum, pp. 3-14.

5. Duff, M. J. B. 1978. A review of the CLIP image processing system. Proceedings of the National Computer Conference, pp. 1055-1060.

6. Kent, E.W., Shneier, M.O., and Lumia, R.L. 1985. PIPE (Pipelined image processing engine). Journal of Parallel and Distributed Computing, Volume 2, pp50-78.

7. Levialdi, S. (ed). 1985. Integrated Technology for Parallel Image Processing, London: Academic Press.

8. Tanimoto, S. L. 1985a. An approach to the iconic/symbolic interface. In Levialdi 1985 , pp.31-38.

9. Tanimoto, S. L. 1985b. Data Structures and Languages in Support of Parallel Processing for Astronomy. In Di Gesu, V., Scarsi, L., Crane, P., Friedman, J. H., and Levialdi, S. (eds.) Data Analysis in Astronomy. NY: Plenum, pp. 497-507.

10. Uhr, L. 1984. Algorithm-Structured Computer Arrays and Networks: Architectures and Processes for Images, Percepts, Models, Information, New York: Academic Press.

11. Weems, C., Lawton, D., Levitan, S., Riseman, E., Hanson, A., and Callahan, M. 1985. Iconic and symbolic processing using a content addressable array parallel processor. Proc. IEEE Comp. Soc. Conf. on Computer Vision and Pattern Recognition, San Francisco, June 19-23, pp.598-607.

LOW LEVEL LANGUAGES FOR THE PAPIA MACHINE

O. Catalano[1], G. De Gaetano[1,3], V. Di Gesu'[1,2],
G. Gerardi[1], A. Machi'[1], and D. Tegolo[1]

[1]I.F.C.A.I./CNR Palermo
[2]Dipartimento Matematica, Universita' di Palermo
[3]Instituto di Fisica, Universita' di Palermo

ABSTRACT

The paper presents the low-level languages implemented up to date to program the PAPIA machine.

The parallel assembly-level P-MACRO package, the microcode level instruction set and a machine simulating environment are described.

1. INTRODUCTION

PAPIA is a project to build a multiprocessor system for image analysis based on a pyramidal cellular processing unit (pyramid).

The actual machine architecture and project status have been reported in [1],[2].

In the following we refer to a 'virtual' view of the machine that we found useful in the development of both software and hardware of the machine prototype.

Four levels or shells are defined in the virtual machine. Each layer includes the functions of the lower ones, and offers to the upper level higher services.

A different language codifies the functions of the level and its instructions flow from it to the lower level. Adequate data and status information flow in the opposite direction.

The four levels are defined as the 'PYRAMID CELLULAR AUTOMA', the 'SIMD PROCESSORS', the 'P-MACHINE' and the 'WORKSTATION' , see fig. 1.

The PYRAMID CELLULAR AUTOMA is constituted by the pyramid and executes the instruction set provided by the E.P. implemented in VLSI technology

The SIMD PROCESSORS level includes a Microcoded Control Unit and an Image Memory Managenment Unit. At this level the machine operates as several SIMD cellular processors (pyramid sections) independent from each other, performs bit-serial arithmetic and image processing primitives and manages image buffering. Its instruction set , microprogrammed on the Control unit will be referred in the following as M-CODE. A formal model for M-CODE is described in [3] and a description of the Image Processing library in [4].

The P-MACHINE is the first level able to run a stand-alone user program. It adds to the lower level a Global Control Unit using a scalar processor to perform scalar data processing. An executive kernel performs machine control. Interaction with the user is provided via a system console.

At the uppermost level complete machine operation is finally provided through the facilities of the computer in which the P-MACHINE is hosted.

2. LANGUAGES FOR THE PAPIA MACHINE

Program languages, in particular languages for parallel machines, have to cope with several conflicting requirements as execution efficiency and self-explication and inner parallelism. Hardware dependency has to be hidden while exploiting machine possibilities. Portability and generality are still desirable features conflicting with previous others.

Attempting to obtain a satisfactory compromise among the various requests in the design of languages for the PAPIA machine, a multilayer approach has been adopted.

Several levels of languages , from microcode to problem oriented ones have been defined and assigned to the levels of the virtual machine.

Each language enbodies the facilities of the corresponding virtual machine in its instruction set and/or constructs, hiddens the lower level dependencies and defines a set of primitives in which the higher level language may expand its constructs.

So, for istance, an artificial intelligence language may use subroutines written in a medium-level parallel-C language to perform its operations

2.1 THE P-MACRO LANGUAGE

The P-MACRO language and program development package is the first tool developed to build complete user application on PAPIA at the P-MACHINE virtual level.

It is addressed to program the machine scalar processor, the cellular processor(s) and the Image Memory buffer (Active Memory Unit)

The scalar processor executes the MC68000 microprocessor instruction set, the parallel processors a set of instructions microcoded in the pyramid and image buffer controllers (M-CODE).

P-MACRO is a hybrid language mixing instructions directed to different processors in a same source file. In particular M-CODE instructions are inserted into a 'C' source file directed to the the scalar processor.

A complete user task is built in various steps (see fig. 2):

Firstly the user writes its application program on the host computer environment using a 'C' syntax for the serial instructions and M-CODE syntax on for the parallel ones. Symbols and macro routine definitions are allowed for the parallel sections and 'C' control flow constructs may surround them.

Then the source file undergoes a series of processing steps which translates it in a complete 'C' compatible source.

In the first step a preprocessor performs symbol and macro substitutions. In the second step a processor identifies m-code statements, translates each section of parallel statements into a block of opcodes for the M-CODE machine, saves it into a temporary file and inserts into the source file a 'C' compatible I/O call to the scalar processor firmware implementing the machine executive functions. The third step completes the job including in the source file a section which preinitializes a 'C' array variable (buffer) with the M-opcodes already saved into the temporary file.

An executable module is obtained compiling the file on the host and linking it with the scalar processor executive routines. The module may be then loaded on the Global Control Unit and executed.

Up to date a prototype version of the P-MACRO package has been implemented. 'C' language in 'UNIX' environment have been chosen for portability purposes. A VMS version of P-MACRO and its evolution towards a completely parallel 'C' is foreseen in the near future.

on parallel data while the medium-level parallel-C language expands in turn its constructs in parallel MACRO-instructions .

High level languages based on A.I. will be available at the 'WORKSTATION' level , a Parallel-C language and a parallel-MACRO assembler language will program the 'P-MACHINE' operations, a Controller Micro-assembler will finally program the 'SIMD PROCESSORS'.

Up to date the MACRO and micro-assembler languages and programming software has been implemented and are available in an UNIX operating system environment the first and VMS the latter.

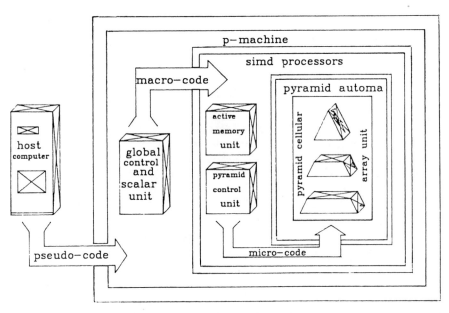

Figure I. Papia Workstation

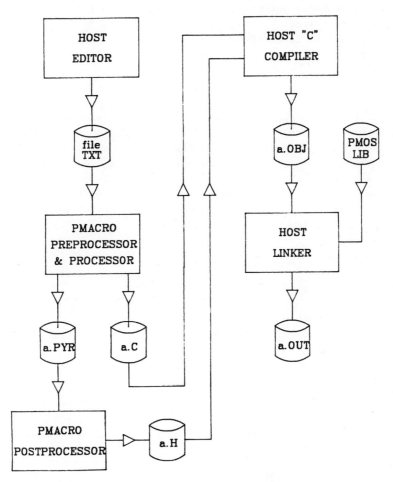

Figure II. PMACRO user TAsk Building

2.2 THE CONTROLLER MICROASSEMBLER

PAPIA sequences of instructions for the execution of Image Processing primitives are provided by the Microcode Control Unit.
Its assembler is described in the following.

The microassembler intruction is formatted by one line divided in three different fields.
The first is the label field. The mnemonic label must be terminated with the special character ":".
The second is the instruction field. In this, different instructions can be interlinked putting the special character & in between. Each subinstruction is composed by the operation mnemonic followed by a number of related arguments. There are four subinstruction categories, each related with a special part of the Microcoded Contro Unit. They are listed in tables 1-4: their meaning is evident. Only the last subinstruction in table 1 needs some comments; It represents the all pyramid set intructions here not presented. In the case its binary opcode has a variable part, like memory o n.n. addressing, it can be choosed to be the output of an address counter. The number of bits requested in overlay are indicated in the same intruction. The advantage of using such an implementation is evident; parametric microroutines can be used.
The third field is the comment one. It must be started with the special character ";".

The assembler text program is processed by an general purpose meta assembler, commercially available from Microtec Research Company, (META29R). It has three fundamental sections:
1)Definition section; It allows intructions and related bit-map definition, in order to customize the assembler instructions.
2)Relocatable meta assembler section; Its output is the assembled program as binary codes with relocatable addresses and cross references tables.
3)Linker loader section; Several assembled programs can be linked together and an absolute coded binary program is produced ready to be loaded and executed by the MCU.

Table 1

SEQUENCER INSTRUCTIONS

REGISTER AND PYRAMID INSTRUCTIONS

MNEMON.	DESCRIPTION
INHRP	INHIBIT REGISTERS AND PYRAMID OPERATIONS
LDR	LOAD SPECIFIED REGISTER FROM FIFO
INTS	INTERRUPT GLOBAL CONTROL SCALAR UNIT
(PYR.IN STRUCT)	OUTPUT SPECIFIED PYRAMID INSTRUCTION OVERLAYING 0,8,6 BITS OF SPECIFIED ADDRESS COUNTER

Table 2

COUNTERS INSTRUCTIONS

MNEMON.	DESCRIPTION
JZ	JUMP ZERO
CJS	CONDITIONAL JUMP TO SUBROUTINE
JMAP	JUMP MAP
CJP	CONDITIONAL JUMP
PUSH	PUSH AND CONDITIONALLY LOAD INTERNAL COUNTER
JSRP	JUMP TO SUBROUTINE, ADDR. FROM INT.REG./BR. ADDR.
JRP	CONDITIONAL JUMP TO INT.REG./BRANCH ADDRESS
RFCT	REPEAT LOOP IF INT. CNT.<> 0 AND DECR. INT. COUNT.
RPCT	JUMP IF INT.CNT.<>0 TO BRANCH ADDRESS AND DECR.CNT
CRTN	CONDITIONAL RETURN FROM SUBROUTINE
CJPP	CONDITIONAL JUMP AND POP
LDCT	LOAD INT. REG./COUNTER AND CONTINUE
LOOP	TEST THE END OF THE LOOP
CONT	CONTINUE
TWB	TREE WAY BRANCH

Table 3

COUNTERS FLAG OUTPUT INSTRUCTIONS

MNEMON.	DESCRIPTION
INHC	INHIBIT ALL COUNTERS OPERATIONS
CLRU	CLEAR SPECIFIED COUNTER IN UP MODE
CLRD	CLEAR SPECIFIED COUNTER IN DOWN MODE
LDC	LOAD SPECIFIED COUNTER AND INHIBIT COUNT. FLAG OUT
LDCU	LOAD SPECIFIED COUNTER IN UP MODE
LDCD	LOAD SPECIFIED COUNTER IN DOWN MODE
INC	INCREMENT SPECIFIED COUNTER
DEC	DECREMENT SPECIFIED COUNTER
INDE	INHIBIT AND DECREMENT SPECIFIED SETS OF COUNTERS
SHIFT	SHIFT PARALLEN-IN SERIAL-OUT SHIFT REGISTER

Table 4

MNEMON.	DESCRIPTION
INHCO	INHIBIT COUNTERS OPERATIONS AND FLAG OUTPUT
EOUT	ANABLE OUTPUT OF SPECIFIED COUNTER

4. THE PAPIA SIMULATOR

In this section is given a short description of the System for the Analysis of Pyramid Algorithms (SAPA) developed as an efficient tool for the implementation and analysis of pyramid algorithms for the pyramid PAPIA.

SAPA allows the user to write pyramid algorithms in PASCAL environment calling a set of pyramid instructions and image processing primitives as library procedures.

The simulator has been implemented by a multidimensional linked list, where each node is a processing element (PE) and the links are the interconections beetwen the PEs.

SAPA allows to develop primitives of the pyramid cellular arrays and their flow is displayed on the computer monitor.

Some debug facilities allow the user to stop the program execution, control the result or/and to have a trace of all the PAPIA-instructions under execution.

Results may be displayed on a pictorial device in order to visualize the contents of the all registers (containig) masks and images for all the pyramid-planes.

SAPA is written in Pascal under VAX/VMS operating-system. the display section is written in Fortran77, for the DEC/VS11 pictorial terminal, and is called externally by the main program in order to isolate nonstandard I/O calls and to enhance its portability.

SAPA allows to perform time evaluations in terms of elementary instructions. In fact at the end of the run a table is computed containing the frequency of occourrence of each elementary instruction.

SAPA has been used in order to implement part of the software primitives of the PAPIA processor and to develop some algorithms for image analysis and processing.

REFERENCES

[1] V. Cantoni, L. Carrioli, O. Catalano, L. Cinque, V. Di Gesu', M. Ferretti, G. Gerardi, S. Levialdi, R. lombardi, A. Machi', R. Stefanelli: "The PAPIA image Analysis System"
II ANRT/SPIE Technical Symposium on Optical and Electro-Optical Applied Science and Engineering. Cannes, December 1985

[2] S. Levialdi: "The PAPIA Project"
Intern. Workshop on Data Analysis in Astronomy. Erice, April 1986.

[3] A. Machi': "M-CODE An Instruction Set for a Pyramid Machine".
IFCAI Technical Report 1/85

[4] V. Cantoni, L. Carrioli, L. Cinque, V. Di Gesu', M. Ferretti, S. Levialdi: "Parallel Image Processing Primitives"
III Int. Conf. on Image Analysis and processing, Rapallo, Italy 1985

NEW DEVELOPMENTS

Chairman : S. Levialdi

EXPERT SYSTEM FOR DATA ANALYSIS

J.M. Chassery

Laboratory TIM3 UA CNRS 397 USTMG
C.E.R.M.O. - BP n° 68 - 38402 ST. Martin D'Heres
Cedex, France

INTRODUCTION

At every final step of feature evaluation in any physical domain, the user is faced with a problem of data manipulation and data interpretation. How can a coherent and significant interpretation of the volume of data be obtained. Moreover this volume of data is rapidly increasing because of the progress of technology for feature evaluation.

A first question concerns the user's desiderata. Generally, the information has been modeled on a qualitative formulation to give a quantitative representation. For example, temperature is measured in a numerical form with a predefined precision, but the user interprets it in semantical terms such as warm or cold. So the first question concerns the problem of interface between the quantitative formulation and the qualitative one with which the user is familiar.

A second question concerns the data volume. Such a large volume is justified when many runs of an experiment are realized in specified initial condition. Initial conditions can be related to a modification of a physical constraint or a modification of the physical support on which featuring is performed.

For example, during the evaluation of some features on a predefined population, we have to take a very large sample of individus if we want to obtain a representative sample

A third question concerns the data interpretation. To what type of information do we need to access ? Do we wish to separate the population of individus for whom we have obtained the featuring phase ? Do we wish to identify each individu in a set of predefined classes ? Do we wish to explain the modele of some specific phenomenon. For example explain the evolution of some feature in the population set ? Is it important to classify the population elements in a same order than that obtained by human reasoning ?

Such characterizations characterize the input parameters of data analysis.

Two major methods have been investigated to obtain to the solution.

The first concerns the use of the classification domain in statistical theory. A major reason for the success of this approach is essentially related to the concept of modeling. The formulation of the problem in terms of population analysis is in accord with constraints of every statistical method. Moreover the constant progress of computer technology, particularly in computation time and geometrical representation, have contributed to the implementation of classification methods.

A second orientation concerns the use of methods directly derived from Artificial Intelligence. Such methods are currently refered to an Expert Systems. The main objective of such method is essentially the modelling of human reasoning. For Data Analysis the objective is to identify, accumulate knowledge and experiments, and to copy the reasoning of an human expert who is confronted with same problem. As is the case for the statistical approach, expert systems has benefited from important developments in computer languages and recently in specific hardware for the Artificial Intelligence domain.

Each one of these two orientations is related to the same prospective research : the assistance to the decision process. In this paper we shall discuss the different approaches to the notion of assistance in terms of a more or less direct mechanism compared with human reasoning. Moreover we shall present the dual contribution of Expert Systems and the classification domain, particular in the case of algorithms and methods selection.

I. DATA ANALYSIS - STATISTICAL APPROACH - ADVANTAGES AND LIMITATIONS

We shall first discuss the performance and advantages of the statistical approach for data analysis.

When we use classification methods for diagnosis, a first step is related to the choice of a learning set. Such set is constructed by sampling and accumulating data from a representative subset of the population. Then each new observation is analyzed in the context of such learning set. The learning set is generally organized into classes [1].

Different strategies are used for the decision step. They are grouped in two major groups : parametric and non parametric methods [2]

In non-parametric methods each new observation is assigned to a predefined class only by geometrical considerations : linear or non linear separations between classes are generated in a geometrical representation.

Parametric methods are related to BAYES Theory. Probability measures are computed and the new observation is assigned to the class in which it has the greatest probability value.

Statistical methods have limitations.

- The formulation of statistical theory is itself based on the choice of the learning set.

- The elementary rules of the theory of probability have to be satisfied. For example if we want to use Bayes formula it is necessary to verify conditions of exhaustivity and mutual exclusion between the different observations.

- Generally such a theory is not well integrated in every domain to which it is applied : for example, the medical domain. A major reason concerns the mathematical aspect of hypothesis and formulation. The theoretical environment cannot be expressed in a current human language. Terms in current use are related to notions of system or mathematical formulations.

- The computationnal technics which are used to access results are far from human reasoning.

Taking into account these points, we can justify the introduction of Artificial Intelligence methods by considerating the following constraints :

- the environment of the machine must be in close relationship with human reasoning.
- the performance of the developped product must be in accord with the performances of an expert in the same considered domain.

To satisfy these conditions, the notion of Expert Systems has been developped and we can list its principal characteristics.

- Possibility of interaction in a natural language with the user.
- Explanation and justification of its reasoning at every step of processing.
- Integration of the knowledge of the human expert.
- Manipulation of knowledge by accumulation, deletion and modification.
- Use of knowledge in an evolutive environment by accumulation of data.

In opposition to the statistical approach, the Expert System approach is adapted to the domains in which the human expertise is currently used. Moreover, the Expert System integration is really justified if data are insufficiently structured and if the data base is evolving with new information.

In addition the above considerations, the notion of Expert Systems is differentiated from other techniques by its algorithmic environment. In a mechanism of classification related to the statistical domain, data and algorithms are imbricated. Algorithms are built to resolve a specific problem in a specific data environment. With the conception of Expert Systems, and essentially to satisfy the similarity of the machine evolution with human reasoning, a new concept of processing has been developed. Data are isolated from the mechanism of knowledge manipulation. On one hand data are stored in a set currently defined by a "knowledge base" and on another hand we have a mechanism defined by the term : "inference engine".The knowledge base is modified and completed by use of production rules and user interaction [3].

It is essentially the knowledge base that we have to correctly define for the problem to be solved. As for the inference engine, different models exist which take into account various modes of knowledge manipulation and reasoning evolution.

Before characterizing the approach to classification by Expert Systems we shall briefly present the general outline of such a notion.

II. EXPERT SYSTEM - GENERAL PRESENTATION [3]

Two major modules constitute the notion of expert system : the knowledge base and the inference engine.

- Knowledge base

The knowledge base stores the set of elements corresponding to the initial problem. Generally, the notion of knowledge is characterized in terms of information and meta-information.

In comparison with the notion of information, meta-information is related to the use and the significance which is given to each unit information.

The information set, also called set of facts, is related to the initial data, definitions, static knowledge and objects supplied by the human Expert. The notion of meta-information is related to deductive knowledge, to rules, to actions on the information set.

Facts and rules are introduced to modele the knowledge representation. A fact is expressed by a logical proposition and a rule is expressed by an implication rule represented by two logical propositions in the following relation :

"if hypothesis then conclusion".

Without entering into detail, it is known that different logical or non logical models may be used to represent knowledge. In the case of meta-information, knowledge may be classified in different typologies such as heuristic knowledge, strategic knowledge or procedural knowledge.

Example : heuristic knowledge.
   if hypothesis 1 then conclusion 1 is plausible
        strategic knowledge.
   if hypothesis 1 then try production rules which satisfy conclusion 1
        procedural knowledge
   if hypothesis 1 then do rule 1 and rule 2.

- the inference engine

The inference engine is the heart of the system. It works on the knowledge base of which it knows only the syntax and must explore the knowledge base and interpret it.

Such exploration is defined in order to infer new knowledge by using into account specific elements of the problem formulation and initial data of the knowledge base. The notion of inference is generally related to the notion of deduction in a system based on production rules.

The system "reasons" by successive cycles, each one being represented by an evaluation phase and an execution phase. In the evaluation phase, the system tests hypothesis and selects proposed rules, and in the execution phase the system executes and tests the plausibility of the conclusion.

We will not give the description of the inference engine. Its characteristics are essentially included in the procedures which are used to activate the cycles.

More details can be currently found in the litterature describing the organisation of Expert Systems.

III. EXPERT SYSTEM AND CLASSIFICATION

At the first level, the conception of an expert system for classification can be related to the elaboration of a system of production rules. In such a system each knowledge unit may be associated with a fact or a production rule [4].

A fact is defined by the sequence "object-predicate-value-attribute". Predicates which are employed are of logical type (=, ,) or non logical type (is, has). The notion of rule is related to an expression given in the following format if((fact1, fact2, ..., factn) then (action1, action2, action p). To associate such a formulation to the classification problem, we have to specify the rule interpreter in the environment of the inference engine. The interpreter of rules acts in 3 steps :

- search for the rules to be activated
- selection of the rule to be activated
- execution of actions related to the activated rule.

An advantage of the first approach concerns the Artificial Intelligence formulation as comparised to a classical algorithmic approach.

Rules are formulated by the user and such a formulation is essentially modular. Moreover such a formalism facilitates the manipulation of the knowledge and permits access to subjective criteria. The elaboration of control structures introduces the possibility of eliminating some non productive rules.

Nevertheless, this approach is not sufficiently structured to construct some models. All the knowledge units are situated on the same level. Such a notion of models is important in the classification domain in comparison with the statistical concept of class.

Therefore to develop an Expert System oriented to the problem of classification it is necessary to take in consideration that knowledge is not described by rules only. We have to integrate the notion of modules in order to use global concepts which are considered to be models of class in which the different elements are present. These global concepts are defined in the terms of structured objects, also named by prototypes. The Expert system is then composed of the following elements :
- a knowledge base
- a control module also called an inference engine
- an interface module to dialog with the user
- an explanation module to give explanations to the user of the reasoning of the system.

The knowledge base is composed of the essential structures
- prototypes which constitute the class descriptors.
- rules which describe the operating knowledge
- data which are related to the descriptors of elements to be classified.

The prototypes are organized in a network architecture in which the arcs constitute the intra-class relations.

The rules correspond to procedural relations between the different components of the prototype. They assume different tasks such as access to data. Prototypes are essentially used to orient the system and order the activation of specific rules and dialog with the user.

III.1. Prototype - Description [4, 5]

The prototype is a structured object to which different descriptors are associated in order to identify it. Some procedures, essentially in the control environment, are related to that identification. Different relations may exist between the prototypes in order to approach the solution during reasoning.

Different types of descriptors are associated with the notion of prototype
- parameters in which we have numerical information (scalar value or interval)
- attributes in which we have symbolic information (symbol value or a set of symbol values)
- prototypes which are used to characterize complex prototypes.

The implementation of a prototype notion is carried out in an object oriented language. A prototype is an object which is composed by some specific subsets, each one corresponding to the descriptors of the prototype. In that same language, the different objects we have to classify are also introduced into the system. These objects constitute the base of facts of the Expert System.

At the beginning of the classification the user introduces the parameters of the object to be classified and the inference engine establishes a correspondance between object and prototypes by successive filtering operations.

To perform such operations we have to define rules of production which concern the operating center of the system.

A rule of production is described by the sequence IF PREMISS then ACTION COEFF in which we define :

PREMISS :   <DESCRIPTOR>   <PREDICAT>   <VALUE>

ACTION :   <DESCRIPTOR>   <OPERATOR>   <VALUE>

COEFF : Numerical value belonging to [0,1] which corresponds to a plausibility degree.

The rule of production is implemented in order to complete the knowledge units on the objects to be classified.

III.2. Control structure and prototypes

In the particular case of classification process, the control structure of the Expert System has to satisfy the following conditions :

- adaptability to a hierarchical organization in order to work at different levels (from the general to the particular).
- possibility of considering objects which are inserted between two prototypes. Such condition requires the implementation of a distance function between elements.

The control structure has to integrate the two notions of representation of knowledge : prototypes and rules.

As in case of prototypes, the integration of such types of knowledge acts at the control level by elaboration of strategies in order to orient the searching process. The hierarchical organization of the prototypes is a network which allows the exploration of the prototypes set to be oriented. At each node of the network the following steps are applied :

- selection of the prototype to be activated. The selection is made in accord with a current hypothesis. We select every prototype which agrees with the formulated hypothesis which is present in the descriptor of the object we have to analyze.. The order of selection is not of great importance because the knowledge units are accumulated in the knowledge base, then each selected prototype will be activated.

- activation of the prototype. During this phase the prototype activates the production rules with which it is associated.

- confirmation of some prototypes. For this step a distance between object descriptors and selected prototypes descriptors is calculated. A prototype is selected according to the distance value. If every selected prototype is rejected then we explore another node of the graph of prototypes.

As for the control structure and the rules of production, when a prototype is selected its base of production rules is activated. The rules are used following a basis cycle :

- selection
- choice of the rule to be activated
- activation of the rule
- memorisation

The selection step is performed by filtering between the set of premisses of rules and the description of the object. The choice of the rule to activate is related to the coefficient of degree of plausibility.

With the activation of the rule we complete the set of descriptors of the object to be classified. Memorisation is necessary for the explanation phase.

III.3. Reasoning and incertitude [6]

One of major characteristics of the Expert Systems is the manipulation of incertitude. Such notion could also be integrated in statistical methods by consideration of BAYES theorem but its implementation requires a large number of data and that the data selection is independant. Different models have been proposed for the use of BAYES theorem and the most satisfying are found in PROSPECTOR [7] and MYCIN [8] For example, in the case of PROSPECTOR let us consider the rule "if E then H" where E is an observed fact and H is an hypothesis to be established.

BAYES theorem gives the formula :

$$P(H/E) = \frac{P(E/H).P(H)}{P(E)}$$

Let us define $\overline{H}$ = not H.

Then we can write :

$$\frac{P(H/E)}{P(\overline{H}/E)} = \frac{P(E/H)}{P(E/\overline{H})} \cdot \frac{P(H)}{P(\overline{H})}$$

Initially the expert gives the value of the factor of plausibility for each hypothesis.
We also quantify the measure of quality $\frac{P(E/H)}{P(E/\bar{H})}$ of the rule "IF E then H".
Then by such a formulation the Expert System acceeds to the coefficient of plausibility of the activated rule.

Such a presentation of the use of prototypes to approach the problem of classification by expert system has been implemented in the specific domain of classification of galaxies [4]
The goal of such an application is the identification of a galaxy by morphological considerations. The classification process has been expressed in terms of a continuous coding to characterize the degree of inhomogeneity of the galaxy and in term of a hierarchical coding to characterize the structures which are encountered inside the galaxy (spiral, ring ...).

Such a problem does not have a direct solution by statistical methods because there is no probabilistic information about the occurrence of the types of galaxies.

Thus we use the knowledge of an expert (astronomer) to construct prototypes and formulate the knowledge base with specified rules.

This system has been connected to a digital image processing system in order to obtain measurements of the morphological characteristics on the observed galaxies.

IV. EXPERT SYSTEM. ASSISTANCE TO CLASSIFICATION METHODS

In the first part, we presented the implementation of an expert system in data analysis environment. At a first level of the discussion we can make the following observations :
- The elaboration of an Expert System needs to access the information that we wish to use. Such information is given by an Expert.
- Actually existing Expert System are limited in the number of knowledge units and applicable rules. Each domain needs a specific elaboration of the knowledge base [9]
- The notion of Expert System is particularly well adapted to classification problems in which notion of incertitude and notions of symbolic knowledge are to be considered. Such domains are charaterized by perceptual evolutive knowledge and it is of great interest to develop methods which are easily adapted to such a modification of diagnosis rules.
- Faced with the problem of a large volume of data, it is necessary to use methods of data reduction. Data reduction can be implemented by the concept of filtering on the set of initial measurements. It must also be implemented in the representation domain to avoid combinatorial methods. For this the notion of prototypes must be implemented to structure the set of "objects" to be recognized. The notion of heuristics must also be taken into account.

For the second part of the discussion we can make the following observations :
We dont know to address a knowledge base associated to a data analysis problem when there is no expert. It may be possible to use classification methods to elaborate the initial elements of the knowledge base and to define the concept of prototype.

A second remark concerns the elaboration of an Expert System in order to assist the user in the classification process. How can an Intelligent System be defined so that it select satisfactory methods in a classification process.

These remarks introduce a new concept of using Expert Systems in data Analysis. We must define a tool to assist the user who is faced with data and algorithms [10]

Such a concept of using an Expert System is also integrated into the Image Analysis domain. A large number of elementary algorithms are defined to perform a specific process, and the notion of expert systems is introduced to assist the user in the elaboration of a strategy to access to image interpretation.

IV. 1. Expert knowledge elaboration

At an initial step of the data analysis problem, the data set is unorganized and it would be of great interest to access a structured knowledge base in order to use it in an environment of Expert System. Such problem is generally encountered in situations where there is no Expert related to the domain. It is also necessary in the case where quantitative measurements are not readily interpretated by human expert.

In such a situation the expert waits for assistance for the elaboration of the knowledge base.

Let us consider a number of facts issuing directly from a featuring phase.

Let us also consider a set of diagnoses which are on the set of accessible interpretations. Data analysis must explain the observations included in the set of facts by a diagnosis or a limited list of diagnosis. The first limitation to the use of classification methods concerns the qualitative aspect of some facts. A quantitative representation by binary or ternary formulations may resolve such problems, but the number of levels is not sufficient when compared with the continuous aspect of quantitative data.

A solution to this problem of data reduction is the use of Analysis in Principal components method.

New diagnoses are obtained as a linear combination of initial one. In general, we consider the two first components which are the most significant in order to represent the set of facts. Another interest of such transformations is the extended dynamics of the new variables in order to use them in other techniques of classification.

IV.2. Expert system and method selection [10]

It is well known that misure of statistical software may lead to incorrect analysis. Thus it is necessary to help the user. It is that aspect of assistance which has to be formulated in the concept of Expert System in order to select an adapted method.

A first important task concerns the knowledge acquisition. The system has to inform the user of problems occurring from a wrong sample set.

The knowledge acquisition is dependant on the domain of analysis and the problem definition.

An extended relational analysis can be used to capture the global aspect of the data. At this level we have only a static notion of knowledge.

No dynamic performance of the expert or strategic knowledge is introduced at this time.

The elaboration of the interpretation model is closely related to the level at which knowledge analysis take place. In order to take into account strategic knowledge it is necessary to have recourse to information from the human expert. Two elements have to be considered to help the user : the constraints to the statistical methods and the methods themselves. The work of expert system is to specify to each constraint a set of methods which can be applied profitably [10].

An example of such occurrence concerns the problem of modelling the evolution of a parameter. To do that we can use a supervised analysis method.

CONCLUSION

Data Analysis can be viewed on different aspects such data manipulation data interpretation and data explanation. Two strategies are actually offered to take into account these considerations in a classification problem. In the actual developments statistical methods are performed in the domain of numerical data manipulation and interpretation. These methods are used in a mathematical context to access to data explanation by data volume reduction and geometrical technique. An actual problem with such an approach concerns the misuse of the offered abilities.

In counterpart , in case of existing expertise, a new approach has been defined with the integration of Expert Systems. These methods are used in a human reasoning context to favorise the modelling of human reasoning and to understand the evolution of such activity.

These approaches, statistical one and Expert Systems, are effectively different in their initial conception but in their own finality they concern the assistance to decision process. Such difference is essentially a consequence of traditionnal approach to expert system building.

Another domain, would concern the statistical aid by expert systems in order to access to knowledge acquisition, implementation and representation in an environment of processing rules derived from actual classification domain.

REFERENCES

-1- HOOKE, R. Getting people to use statistics properly. The American statistician 3, 4 1 , 1980.

-2- BENZECRI, J.P. Pratique de l'Analyse de Données. Vol 1-2-3 Dunod 1980-1981.

-3- FARRENY, H. Les Systèmes Experts - Principes et Exemples - Techniques Avancées de l'Informatique. Cepadues Editions - Toulouse - 1985.

-4- GRANGER, C. Reconnaissance d'objets par mise en correspondance en vision par ordinateur. Ph. Diss. in Informatics - Nice University 1985.

-5- AIKINS, J.S. Prototypes and Production rules : a knowledge representation for computer consultations. Ph. Diss. Computer Science Department. Stanford University 1980.

-6- CAYROL, M., FARRENY, H., PRADE, H., Fuzzing Pattern Matching - Kybernetes 11 pp 103-116, 1982.

-7- DUDA, R.O., HART, P.E., NILSSON, N.J. Subjective Bayesian methods for rule-based inference systems. Proc. of the AFIPS 1976 - National Computer Conference. Vol 45 - pp 1075-1082.

-8- CLANCEY, W.J. ,SHORTLIFFE, E.H. Reading in Medical Artificial Intelligence : The first decade Addison Wesley Eds. 1984.

-9- HAYES-ROTH, F., WATERMAN, D.A., LENAT, D.B. Building Expert Systems - Reading Mass : Addison Wesley 1983.

-10- HA-KONG, L. Expert Systems Techniques for statistical Data Analysis. Progress Report, April, Polytechnic of South Bank, London - Also in Computers and computing. Etudes et Recherches en Mathématiques. pp 149-162, Wiley-Masson 1986.

AN UNIFIED VIEW OF

ARTIFICIAL INTELLIGENCE AND COMPUTER VISION

D. Dutta Majumder

Electronics & Communication Sciences Unit
Indian Statistical Institute
Calcutta 700 035, India

ABSTRACT

After introducing natural and intrinsic link between the evolving subjects of Artificial Intelligence and Computer Vision research, particularly in the context of next generation of computer system research, the paper presents an overview of the framework of current image understanding research from the points of view of knowledge level, information level and complexity. Because a general purpose computer vision system must be capable of recognizing 3-D objects, the paper attempts to define the 3-D object recognition problem, and discusses basic concepts associated with this problem. The major application area often mentioned an industrial vision system and scene analysis in aerial photography.

KEY WORDS

Artificial Intelligence, Computer Vision, Image Understanding System, Knowledge Organisation, 3-D Object Recognition, Spatial Reasoning.

INTRODUCTION AND PERSPECTIVE

Without entering into the philosophical issues involved in an attempt to define the meaning of artificial or intelligence I intend to attempt an working definition delineating the approximate boundary of the evolving concept of Artificial Intelligence (AI) which will be automatically and intrinsically linked with the ideas inherent in the development of Computer Vision Systems (C.V.S.). AI is the study of how to make machines to do some kind of mental and associated activities, which at the moment man can do better than computers. Such tasks to mention a few are writing computer programs, perceiving and understanding languages, pictures, photographs and visual environments, game playing and theorem proving, medical diagnosis, chemical analysis and engineering design, doing mathematics and problem solving, engaging in common sense reasoning etc. We might say that the systems that can perform such tasks possess some degree of AI.

Perception of the world around had been crucial to the survival of living beings. Animals with much less intelligence than man are capable of very sophisticated visual perception. Early effort at simple static visual perception by machines led in two directions. Firstly pattern recognition and machine learning, and secondly image processing and understanding systems. First group of activities being based on strong mathematical foundation are yet to fully collaborate with AI which from a loosely structured and empirical orientation is improving very fast. Whereas, because of inherent flexibility latter group is typically regarded as falling within the perview of AI.

During the past two decades, the field of Computer Vision (CV) including its subfields of image processing and image understanding or scene analysis, has developed from the seminal work performed by a small number of researchers at the few centres of AI research into a major subfield of AI with widespread involvement. The intellectual climate for progress and theoretical basis for IUS & CVS has improved with the work conducted under the US DARPA IU program at CMU, University of Maryland, MIT, SRI, University of Rochester, Stanford University, The Virginia Polytechnic and State University and University of Southern California. The goals and motivations of these researches in the last two decades were varied in nature, such as understanding and modelling of human vision system, development of comprehensive theories of perception, solution of some fundamental problems in AI, most of the others were engaged in solving practical problems in applications of Computer Vision Systems. Research in designing computer systems to 'see' continues to be fascinating, challenging, exciting and to some extent bewildering. Bewildering, because the construction of effective general purpose CVS has proven to be exceedingly difficult, though vertebrates carry out this task with very high level of sophistication easily. Though Human Visual System (HVS) need not be considered as the best possible vision system, but it is definitely the best known one, so we shall often try to understand our perceptual mechanism, in course of our discussion.

1.1 Motivations for Research in CVS-Diverse Applications

The field of CVS now contacts such diverse disciplines and areas as cognitive psychology, pattern recognition, image processing, computer systems hardware and software, geometical optics, computer graphics, electrical engineering, neurophysiology, psychophysics, and mathematics, and shares common problems from areas in automatic speech recognition, knowledge base management systems, robotics and artificial intelligence. The boundaries of this research are rather amorphous, particularly when we consider the important application domains in the context of designing next generation (commonly called fifth generation) of computer systems (FGCS). In order to present the status and an added motivation of research in AI and CV a very brief overview of FGCS may be useful at the introduction itself (16)(17)(18).

The main functions of FGCS as envisaged by Japanese task force (17) are : (1) problem solving and inference making functions, (2) knowledge based management functions and (3) intelligent man machine interface functions (Fig.1). In this diagram the upper half of the modelling software system circle corresponds to the function (1), lower half to the function (2), and the portion that overlaps the human system circle to function (3) above. Function (3) is very heavily dependent on IUS and CVS along with ASR and NLP. It is more or less agreed that about 80% of the information that needs to processed are visual in nature and so in order to provide flexible interactive intelligent man-machine interface which will function as a front-end processor IUS and CVS will have to play a vital role (16) (18).

1.2 Prospective Application Areas of CVS and IUS (18)

1. Man-Machine Communication - (a) OCR Systems, (b) Cursive Script Recognition System, (c) Image understanding, (d) Document reading and Text Processing, (e) Speaker identification and recognition by Sonagram Analysis, (f) Aids for partially sighted such as a system that read a document and say what was read, or automatic guide-dog navigation.

2. Bio-Medical Applications - (a) ECG, EEG, EMG Analysis, (b) Cytological, Histological and other steriological Application,(c) X-Ray Ultrasound, Isotopic and Tomographic Analysis, (d) Diagnostics, (e) Mass screening of medical images such as protein chains, cells and chromosome slides for detection of various diseases, cancer smears, X-ray and ultrasound images and tomography, (f) Routine screening of plant samples, (g) Analysis of chemical compositions, (h) Determination of spatial orientation of Neurons.

3. Application in Physics - (a) High energy Physics, (b) Bubble chamber and other forms of Track Analysis, (c) Astronomical applications in having improved images of stars and planets and their chemical composition.

4. Crime and Criminal Detection - (a) Fingerprint, (b) Hand writing, (c) Speech Sound (Sonagrams) and (d) Photographs.

5. Remote Sensing and Natural Resources Study and Estimation - (a) Agriculture, (b) Hydrology, (c) Forestry, (d) Geology, (e) Environment, (f) Cloud pattern, (g) Urban quality, (h) Cartography, the automatic generation of hill shaded maps, and the registration of satellite images with terrain maps, (i) Monitoring traffic along roads, docks, and at airfields, (j) Exploration of remote or hostile regions for fossil fuels and mineral ore deposits.

6. Stereological Applications - (a) Metal Processing, (b) Mineral processing, (c) Biology and (d) Mineral detection from microphotographs of ore sections.

7. Military Applications - All the above six areas of applications PLUS (a) Detection of Nuclear Explosions, (b) Missile Guidance and Detection, (c) Radar and Sonar Signal Detection, (d) Target identification, (e) Naval Sub-marine Detection, (f) Reconnaissance Application, (g) Automatic navigation based on passive sensing, (h) Tracking moving objects, (i) Target acquisition and range finding, (j) Tactical Analysis.

8. Industrial Applications - (a) Computer Assisted Design and Manufacture, (b) Computer Graphic Simulation in Product Testing, (c) Automatic Inspection in Factories, (d) Non-Destructive Testing, (e) Object acquisition by robot arms, for example by "pin picking", (f) Automatic guidance of seam welders and cutting tools, (g) Very Large Scale Integrated related processes, such as lead bonding, chip alignment, and packaging, (h) Monitoring, filtering and thereby containing the flood of data from oil drill sites or from seismographs, (i) Providing visual feedback for automatic assembly and repair, (j) The inspection of printed circuit boards for spurs, shorts, and bad connections, (k) Checking the results of casting processes for impurities and fractures.

9. Robotics - (a) Intelligent sensor technology, (b) Three-dimensional outdoor and indoor scenes, (c) Object acquisition and placement

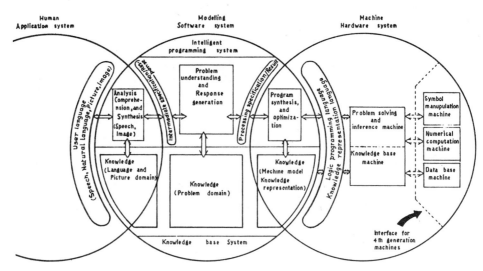

Figure 1. Conceptual Diagram of the Fifth Generation Computer System.

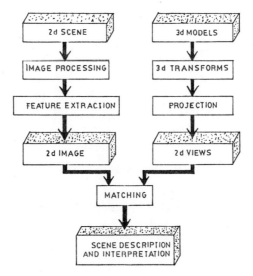

Figure 2. Machine Vision Principle.

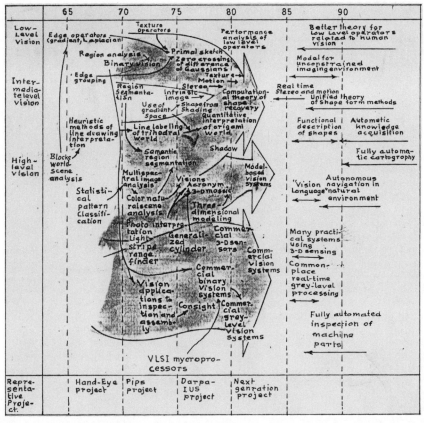

Figure 3. Computer Vision Efforts have Advanced over the Past 20 Years along Three Fronts: Low-Level vision, the Extraction of Basic Features such as Edges from an Image, Intermediate-Level Vision, the Deduction of the Three Dimensional Shape of Objects from the Images; and High-Level Vision, the Recognition of Objects and their Relationships. Some Representative Research Projects Include the Hand-Eye Robotic Vision Project Initiated at the Massachusetts Institute of Technology in Cambridge and at Stanford University in Palo Alto, Calif; the Pattern-Information Processing System (PIPS) Project in Japan, one of the Earliest Focussed Research Programs Sponsored by the Ministry of International Trade and Industry; the U.S. Defense Advanced Research Projects Agency's Image-Understanding System (IUS) Project; and the Current Data Next-Generation Project.

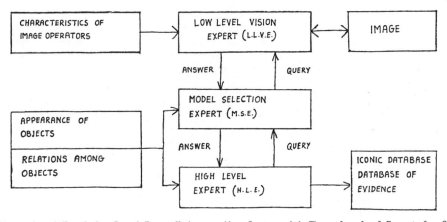

Figure 4. A Knowledge Based Image Understanding System with Three Level of Expert for Combining Evidence.

by robots, (d) Designing expert systems for specific Robot vision applications.

10. Management Applications - (a) Management information systems that have a communication channel considerably wider than current systems that are addressed by typing or printing, (b) Document reading and other office automation works, (c) Design aids for architects and mechanical engineers.

As major motivation for developing computer vision was to develop application oriented tool for solution of some contemporary problems, most of the successful scene analysis systems were based on ad-hoc working principles (1)(2)(3), with a limited domain of specialized applications. In the last decade there were several proposals to obviate these limitations (4)(5)(6), aimed at developing competent re-usable, extensible but general at the system level. Although concern for generality would appear natural in the context of biological vision or abstract vision theory, it is not necessarily a desirable characteristic of a methodology directed towards application oriented vision system (7). This realisation has resulted in gradual transition in AI from general purpose problem solvers to knowledge specific systems. A general CVS comparable to HVS implies large range of objects and background with invariant system performance to large changes in viewing angle, illumination angle, contexts and obscured areas, along with ability to withstand rapid contextual changes such as indoor and outdoor environment.

It seems very difficult to achieve any of these characteristics, in the present state-of-the-art, and we should look at the necessary system characteristics in terms of a range of real problems from several application domains. We should also understand that the human vision and reasoning cannot be so neatly subdivided as (a) sensing, (b) segmentation (c) recognition, (d) description and (e) interpretation as in computer vision. An elimentary machine vision principle is illustrated in Fig.2, which is self explanatory. As for example recognition and interpretation are very much interrelated in HVS but is not understood to the point that they can be analytically modelled. These five subdivision of functions we should look at for limited practical implementation of the state-of-the-art CVS.

1.3 Levels of Vision

Taking into account the above developments we see that in some sense we may divide the general purpose CVS arbitrarily into several and at least three basic levels of vision. Levels proposed by Tennenbaum (19) are : Level 0-Original image; Level 1-Intrinsic surface characteristics; Level 2-3-D Surface descriptions; Level 3-3-D object descriptions; Level 4-Symbolic description of scene. But most CVS are based on three step process. The computational problems involved in deriving Level 1 from Level 0 are fairly well understood now. The next step from Level 1 to Level 2 is being extensively studied. But the choice of object representation at Level 3 will influence surface extraction and so the Level 2.

In Fig.3 we present the historical development in CVS research efforts over the past 25 years or so along the three fronts in some very significant projects in USA and Japan. At the lowest level (LLV) (sensor information - usually an intensity image) pictures are segmented into regions of similar primary features to extract 'primitive' information from a scene ranging from modelling the characteristics of incident and reflected light properties of a body to the detection of edge segments

(20)(21)(22)(23) and connecting them into lines or curves or regions with uniform properties (8) (Fig.4). The next intermediate level of vision (ILV) refers to the procedures that use the results from LLV to produce structures in the picture or portions of the picture where complete knowledge regarding model features and topological structure are available. Techniques are edge linking, segmenting, shape analysis (24)(25)(26) description and recognition of objects. Techniques such as local graph search and global optimization using dynamic programming as developed in AI can be employed to merge regions and to assign label sets to them (9). Highest level vision (HLV) may be viewed as the process that attempts to emulate cognition, encompassing a broader spectrum of processing functions. HLV may use a relational data base to store knowledge and a vision strategy akin to production systems, which has to be based on knowledge directed or goal oriented analysis. Although this three-level process applies to many vision, systems, several systems omit or add one or more steps depending on complexity of environments.

There are several competing paradigms to achieve the goal in this rapidly evolving field. It may not be possible for me to discuss in depth the paradigms and research issues facing the field. Rather I intent to provide with state-of-the-art overview of the breadth of problems which must be considered in the development of general computer vision systems.

The overview will include the framework of current image understanding research from the point of knowledge level, information level and complexity level along with knowledge organisation and control structure in image understanding system (IUS). Different computational approaches to IUS will also be discussed briefly. We shall try to present examples of problems in designing knowledge based computer vision systems (15)(16) for applications such as organization of aerial image analysis and industrial inspection system.

2. FRAMEWORK OF IMAGE UNDERSTANDING RESEARCH

Binford (10) gave a good survey of the different IUSs developed during late seventies as feasibility studies. Some of them were proved to be good in some application areas as indicated in the earlier section of this paper, but several crucial problems became clear (11). These are : (a) view point dependent image model, (b) weak segmentation ability and (c) limited number of object classes in restricted environments. Though the scenes were essentially 3-D, the systems model scenes by 2-D image features, and weak segmentation produced erroneous results.

It was pointed out by Takeo Kanade (12) that the discrimination between 2D-image features and 3D scene features is essential in IUS, and the interpretation must be based on 3D features and relations. Michael Brady (13) indicated the extensive researches that are conducted to extract 3D features from 2D imagery. Barrow and Tanenbaum (14) proposed to use the photo-geometry as the theoretical basis to recover intrinsic properties of 3D objects such as range (depth), orientation, reflectance and incident illumination of the surface element visible at each point in the image. The idea find very good support as these are useful for higher level scene analysis, humans can determine these characteristics irrespective of viewing conditions, and such a description is obtainable from noncognitive process. It has been shown that 3D shape of object surface can be recovered from 2D image features such as shading, texture, and contour shape.

David Marr (5) advocated segmentation methods based on HVS with symbolic representation of pictorial information known as primal sketch. Haralick proposed a functional approximation of local gray level distribution to capture more informative pictorial characteristics. Chanda, Chowdhury and Dutta Majumder recently suggested some preprocessing techniques (20)(21)(22) useful for improved segmentation work. The importance of segmentation based on 3D scene characteristics rather than 2D image features was also indicated (14). Paul Besl and Ramesh Jain (27) proposed an effective utilization of all the information present in range images as according to them range image understanding problem is a well-posed problem in contrast with the ill-posed intensity image understanding problem. Most segmentation work for single intensity images is based on thresholding, correlation, histograms, filtering, edge detection, region growing, texture discrimination or some combination of the above. The key issues in range image processing are planar region segmentation, quadratic surface region segmentation, roof edge detection etc.

Methods and techniques of Artificial Intelligence can be used in this problem (of segmentation) above which is a central issue in realising intelligent computer vision systems. Intelligence often implies smart selection from a huge number of alternatives. In the sense that if the number of alternatives is small not much intelligence is required for the system to work well. The problem now is how to increase the level of intelligence of IUS by using different AI ideas.

2.1 Levels of Knowledge for IUS

Problems (a),(b),(c) mentioned above in this section are closely related to the levels of knowledge required in IUS and CVS :

2.1.1 <u>Physical Knowledge</u> : The physical laws governing imaging process in the multidimensional physical world along with the geometry among camera, light source and object, and spectral properties of light source, sensor and material of the object provides powerful knowledge sources. Shape from X (X : shading, texture, motion, object contour) and stereo vision can use this knowledge to recover 3D shape from projected 2D image features.

2.1.2 <u>Visual Perception Knowledge</u> : Gestalt laws of proximity, similarity, continuity, smoothness, symmetry etc. are used for the grouping of primitive pictorial entities into more global ones. This knowledge plays an important role in segmentation and also to group primitive 3D features into global characteristics.

2.1.3 <u>Semantic Knowledge</u> : For recognition of objects knowledge about properties and relations between them is essential. First two types of knowledge are general and domain-independent, semantic knowledge is domain specific.

2.2 Levels of Information

Fig.5 shows information levels in IUS and the processes developed so far to transform information across the levels. Here also we observe three levels of analytic processes. In the low level process (LLP) physical and neurophysiological knowledges are to be utilized to define and extract the most informative image features (primal sketch).

In the middle level process (MLP) the local features of LLP are to be grouped into global image features using the perceptual knowledge, again the image features are to be transformed to scene features (2-D features) using the physical knowledge so that matching can be performed with the 3D object model. There are many possibilities in the grouping and also 3D interpretations of a projected 2D image feature which calls for use of AI techniques. Probabilistic relaxation labelling (28) is a useful computational scheme to reduce such ambiguities.

The major task of the high level process (HLP) is to find the object model which matches with the information extracted from the input image. Problems in this are : (a) Depending on the viewing angle and the time of observation 2-D appearance of 3-D and moving objects change very much, (b) If an object is occluded by others it is difficult to predict its appearance, and (c) An abstract object can have widely varying appearances. These are the problems of underconstraint and has to be solved by sophisticated model representation and utilization of the semantic knowledge.

It should be understood that knowledge representation and control structures are key issues in the HLP in both IUS and AI and so also in CVS.

2.3 Levels of Complexity of a Scene

Depending on several environmental and other factors the levels of complexity of a scene can be assessed. These are, to mention a few : (a) Natural VS Artificial; (b) 3D VS 2D; (c) Flat VS Curved Surface; (d) Non-isolated VS Isolated object, (e) Generic VS Specific Model; (f) Uncontrolled VS Controlled Imaging environment.

Important factors in assessing complexity levels in motion understanding are : (a) Solid VS Deformable object; (b) Constrained VS Unconstrained Motion; (c) Physical VS Semantic Description.

It is well known that because geometric relations and shapes of man-made artificial objects are often composed of analytically well defined open and closed curves such as (24)(25)(26) line segments and disks, it is easier to recognise and group them by such knowledge. Homogenity and texture are also usual characteristics of artificial and natural scenes respectively. Hough transformation (25)(29) is an effective method to extract well defined global image features such as straight lines and ellipses (2D appearance of a flat disk). Some scenes are essentially 2D such as maps, design charts, documents etc. It should be noted that partial matching is inevitable in 3D object recognition. In 3D scene analysis flat VS curved surface can be used as a measure of complexity. In the case of non-isolated occluded (overlapped) object, local property measurement is to be performed and partial matching is a must.

Most of the CVS developed so far are for specific models, such as for recognition of industrial parts with specific properties of shape, material, colour, texture etc. Generic models are abstract objects such as, airplane, boat, table, house etc. If the imaging environment is under control as in industrial CVS the S/N ration and information level can be increased. Active sensing using a laser range finder and a structured pattern projector greatly facilitates the feature extraction process. (56)(57)(58)(59)

Regarding the factors of complexity in motion understanding it is obvious that if the motion of the camera can be constrained the analysis is facilitated. Description of the motion of deformable objects such as cloads is difficult because the shapes can change during the motion. It is also to be understood that the exact physical description of the motion is to be interpreted to obtain the semantic description.

3.0 FROM IMAGES TO OBJECT MODELS

There is a wide gap between raw images and an understanding of what is seen. It is too difficult to bridge this wide gap for CVS design. To identify, describe and localize objects, we need intermediate representations that make various kinds of knowledge explicit and that expose various kinds of constraint. Visual interpretation of completely unconstrained scene is far beyond the current state of the art of IUS and CVS. This view has led many researchers to the development of general-mainly 3D feature extraction methods. The other aspect of understanding is of course recognition, which again requires feature measurement. The difference between recognition and measurement is that, the former is in terms of generic objects and the later is of a specific object instance.

The principle of recent IUS researches toward 3D object recognition is based on the proposition that 3D objects are generic models to understand a scene, and the features measured from an image are their specific appearances.

3.1 3-D Object Recognition

P.J. Besl and Ramesh Jain (27) reviewed the object recognition problem in the following subject areas : (1) 3-D object representation schemes, (2) 3-D surface representation schemes, (3) 3-D object and surface rendering algorithms, (4) Intensity and range image formation, (5) Intensity and Range image processing, (6) 3-D surface characterization, (7) 3-D object reconstruction algorithms, (8) 3-D object recognition systems using intensity images, (9) 3-D object recognition systems using range images.

There are several overview papers on computer vision treating 3-D issues using intensity images as inputs [(19)(13)(30)(10); Barrow and Tenenanm 1981; Brady 1982; Rosenfeld 1984; Binford 1982].

3.1.1 <u>3-D Object Representation</u> : In the area of Computer-Aided-Design (CAD) geometric solid-object-modelling systems several representations are commonly used. I shall mention them without any explanation for the sake of completeness. These are (1) Wire-frame representation, (2) Constructive solid geometry representation (CSG), (3) Spatial-Occupancy representation consisting of (a) Voxel, (b) Octree, (c) Tetrahedral or (d) Hyperpath representations, (4) Surface boundary representation.

Most 3-D object representations in CVS literatures can be categorized as one of the above mentioned schemes or as one of the schemes mentioned subsequently.

3.1.2 <u>Generalized Cylinders or Sweep Representation</u> : Generalized cones or generalized cylinders are often called sweep representations because object shape is represented by a 3-D space curve that acts as the spine or axis of the cone, a 2-D cross-sectional figure, and a

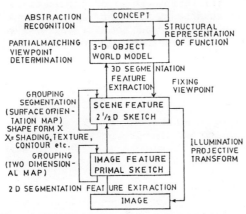

Figure 5. Information Levels in IUS (Facts about Brightness Value are Explicit in the Image; Brightness Changes Groups of Similar Changes, Blobs, and Texture are Explicit in the Primal Sketch; Surfaces are Explicit in the 2½ D Sketch; Volumes are Explicit in the World Model).

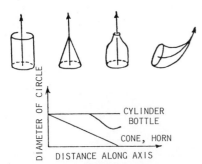

Figure 6a. The Generalized Cylinder Representation is good for a Large Class of Objects. The Simplest Generalized Cylinders are Fixed, Two-Dimensional Shapes Projected along Straight Axes. In General, the Size of the Two-Dimensional Shape need not Remain Constant, and the Axis need not be Straight. Also, the Two-Dimensional Shape may be Arbitrarily Complex.

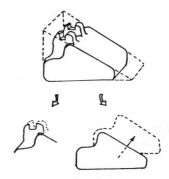

Figure 6b. Complicated Shapes can be Described as Combinations of Simple Generalized Cylinders. A Telephone is a Vaguely Wedge-Shaped Cylinder with U-Shaped Protrusions.

295

sweeping rule that defines how the cross section is to be swept and possibly modified along the space curve. Fig.6 (a) and (b) illustrates the idea, which like many great ideas is quite simple. An ordinary cylinder can be described as a circle moved along a straight line through its center. A wedge can be described as a triangle moved along a straight line through its center. The shape is kept at a constant angle with respect to the line. The shape may be any shape. The shape may vary in size as it is moved. The line need not be straight. For some objects with varying cross-sections, the circle shrinks or expands linearly as it moves.

Though this is most suitable for many real world problems, is not very general as it is almost impossible to describe an automobile or human face by this technique. But despite its limitations this is most suitable for vision purposes.

3.1.3 <u>Multiple 2-D Projection Representation</u> : In this method 3-D objects are represented by 2-D silhouette projections. Silhouettes have also been used to recognize aircraft in any orientation against the well-lit sky background. A more detailed approach of a similar nature is the characteristic-views technique described in Chakravorti and Freeman (31).

3.1.4 <u>Skeleton Representation</u> : A skeleton can be considered (32) an abstraction of the generalised cylinder description and consists of only the spines or axis curves, the idea of which is similar to the medial axis or symmetric axis transform of Blum (33).

3.1.5 <u>Generalised Blob Representation</u> : Generalised blobs have been used as a 3-D object shape description scheme in Mulgaonkar et al (34) by sticks (lines), plates (areas), and blobs (volumes).

3.1.6 <u>Spherical Harmonic Representation</u> : For convex objects and a restricted class of non convex objects shapes can be represented by specifying the radius from a point as a function of latitude and longitude angles around that point.

3.1.7 <u>Overlapping Sphere Representation</u> : In this scheme (35) many spheres are required to represent a relatively smooth surface. Though it is a general purpose technique, it is rather awkward for precisely representing most man-made objects.

The object recognition problem requires a representation that can model arbitrary solid objects to any desired level of detail and can provide abstract shape properties for matching purposes, which none of the existing schemes are capable. But whatever representations are used, it will be necessary to evaluate surfaces explicitly in at least one module of a vision system because (a) range images consist of sampled object surfaces and (b) intensity images are strongly dependent on object surface geometry. Object recognition is largely dependent on surface perception.

Both intensity and range image formation and their processing has been studied by researchers in detail. The book by Ballard and Brown (1982) (36) provides a thorough treatment of these and also object reconstruction aspects of vision and graphics, and in order to save space and time we have to avoid these aspects in this paper.

3.1.8 Some Distance Measures for Shape Discrimination and Recognition : Several authors suggested distance measures (51)(52)(53) for 2-D shape matching and understanding in addition to the usual Fourier and other descriptors which are computationally complex. In the recent past Dutta Majumder and Parui suggested six new shape distance measures (24(26)(54)(55) out of which five were information preserving and satisfies all the metric properties (None of the previous shape distance measures satisfies all the metric properties). The formal approach of Dutta Majumder and Parui are mathematically rigorous. Two distance functions are for simple curves and four are for regions without holes.

Another originality of this approach is the use of the major axis in normalizing the orientation of a region in order to construct the shape distance functions explicitly as a result of which they can deal with almost any shape.

The directional codes used to construct some of the shape distances are also a generalization of Freeman's Chain Codes. There have been several extensions to higher order (16,24 etc.) chain codes. But in our case the codes are much more general in the sense that they can take any real value 0 and 8 which has not been used before.

In order to extend some of the shape definitions and algorithms to 3-d, we intend to define 3-d continuous directional codes in 3 dimensions. Some of the shape distances can be extended to 3d cases in a straight forward manner. The 2-d shape distance based on shape vector can be extended to 3-d by considering concentric spheres instead of concentric circles. Similarly other shape distances are also extendable - in some cases one has to consider skeletal voxels instead of pixels. Similarly theoretically speaking some of the definitions of measure of degree of symmetry and antisymmetry can also be extended. The approach of Dutta Majumder and Parui along with the approach of generalised cone/Cylinder will lead to a more meaningful solution to the shape recognition problem.

4.0 MODEL-BASED 3-D OBJECT RECOGNITION USING AI TECHNIQUES

We have already mentioned about several 3-D object recognition schemes based on intensity images. Consistency among local features and ambiguity in data and knowledge are essential problems in CVS and IUS. The role of control strategy in recognition process is to resolve such ambiguity and to identify global objects by examining the consistency among local image features.

4.1 Control Structure

In order to control the recognition process knowledge is crucial to reduce the necessity for "search". On the other hand search can compensate for lack of knowledge Nagao (37) gave a survey of control strategy in IUS. At this point it may be worthwhile if we look at how model-based 3-D interpretations are possible using an actual rule-based system such as ACRONYM (Brooks et al 1979 and Brooks 1983) (38)(39), which is often mentioned in CVS literature. This is probably because of the flexibility and modularity of its design, its use of view-independent volumetric object models, its domain independent qualities, and its complex, large scale nature. Fig.7 shows a block diagram of the ACRONYM system and its hierarchical geometrical reasoning process. The system based on prediction - hypothesis - verification paradigm has three main data structures namely object graph, restriction graph and prediction graph, which are

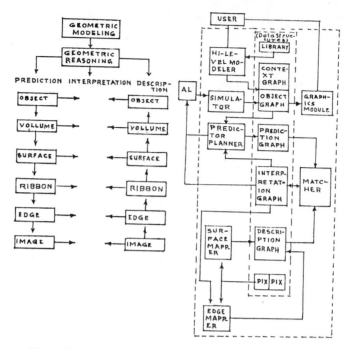

Figure 7. The ACRONYM System. From Brooks et al. (1979).

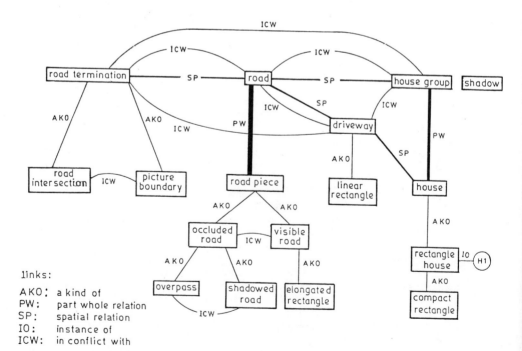

links:

AKO: a kind of
PW: part whole relation
SP: spatial relation
IO: instance of
ICW: in conflict with

Figure 8. Knowledge Organization about Suburban Scenes.

formed on the basis of the world model and a set of production rules. Nodes of the object graphs are generalized cone object models, arcs are spatial relationships among the nodes and the subpart relations (e.g., is-a-part-of). Nodes of the restriction graph are constraints on the object models, and directed arcs are subclass inclusions. Nodes of the prediction graph are invariant and quasi-invariant observable image features of objects and arc are image relationship among the invariant features - which are of the types : must-be, should-be and exclusive.

Every data 'unit' of the object has 'slots', such as a cylinder has a length slot and a radius slot which accept fillers or quantifier expressions. The image is processed in two steps. First, an edge operator is applied to the image. Second, an edge linker is applied to the output of the edge operator and is directed to look for ribbons and elipses, which are 2-D image projections of the elongated bodies and the ends of the generalized cone models. The higher level 3-D geometric reasoning and searches in ACRONYM is based entirely on 2-D ribbons and ellipse symbolic scene descriptions. The heart of the system is a nonlinier constraint Manipulation System (CMS) that generalizes the linear SUP-INF methods of Presburger arithmetic (Bledsoe 1974) (40). Constraint implications are propagated Top-down during prediction and Bottom-up during interpretation. ACRONYM system is implemented in MACLISP. Its prediction subsystem consists of approximately 280 production rules and in a typical prediction phase approximately 6000 rule firings occur. But we have not yet come accross any published results of 3-D interpretation using ACRONYM accept that of some jets on runways.

In the recent past, as we have already mentioned some, there are several other 3-D object recognition schemes based on intensity images have been developed such as Mulgaonkar et. al.(1982) (34) using generalized blobs. Fisher (1983) (41) has implemented a data-driven object recognition program called IMAGINE, in which surfaces are used as geometric primitives. Though there are several criticism of this system, the program did achieve its goal of recognizing and locating a robot and "understanding" its 3-D structure in a test image. Valuable ideas concerning occlusion are also presented in the paper. In all these and in several others including in automatic speech recognition system, unification of bottom-up and top-down process is very important.

4.2 Control Strategy For Unification of Bottom-up and Top-down Processes in Spatial Reasoning

It should be noted as above, that geometric relations are used for consistency verification in bottom-up analysis and hypothesis generation in top-down analysis. Matsuyama, Hwang Davis, and Rosenfeld (1983) proposed a control scheme (42) named evidence accumulation for spatial Reasoning in Aerial Image Understanding" an important characteristic of which is that it integrates both bottom-up and top down processes into a single flexible spatial reasoning process. There are three levels of representation and control in that system as discussed earlier.

A binary geometric relation between two classes of objects, O1 and O2 is denoted by REL (O1,O2), and is used as a constraint to recognize objects from these two classes, at first by extracting pictorial entities satisfying the intrinsic properties of O1 and O2, and then checking that the geometric relation is satisfied by these candidate objects (Fig.8). In this bottom-up recognition scheme, analysis based on geometric relations cannot be performed until pictorial entities corresponding to objects are extracted. In general, however, some of the correct pictorial entities often fail to be extracted by initial image segmentation. So one must additionally incorporate top-down control to find pictorial entities missed by the initial segmentation as described by Selfridge (43) (1982). At this point it may be noted that ACRONYM does not have any top-down goal-oriented segmentation for detecting missing image features.

The above relation can be functionally expressed as

$$O1 = f(O2) \text{ and } O2 = g(O1).$$

Given an instance of O2, say r, function f maps it into a description of an instance of O1, f(r), which satisfies the geometric relation, REL, with r. The analogous interpretation holds for the other function, g.

In this system knowledge about a class of objects is represented using the frame theory as enanciated by Minsky (1975) (2), and a slot in that frame is used to store a function such as f or g. Whenever an instance of an object is created, and the conditions are satisfied, the function is applied to the instance to generate a hypothesis or expectation for another object which would, if found, satisfy the geometric relation with the original instance. A hypothesis is associated with a prediction area (locational constraint) where the related object instance may be located   In addition to this area specification, a set of constraints on the target instance is associated with the hypothesis. In the case of a road hypothesis the frame name is : Road, and Slot names are : Length, Direction, Left-adjacent-road-piece, Right-adjacent-road-piece, Left connecting-road-terminator, Right-connecting-road-terminator; Left-neighbouring-house-group, Right-neighbouring-house-group etc. All hypothesis and instances are stored in a common database, the iconic data-base (Fig. 4) where accumulation of evidence i.e. recognition of overlapping sets of consistent hypotheses and instances is performed. Similar ideas have been proposed by Haar (44) and McDermott (45) to solve spatial layout problems and to answer queries about map information.

Two types of geometric relations "spatial relation" (SP) and part-whole relation (PW) are used. SP represent geometric and topological relations and PW represent AND/OR hierarchies. "A-kind-of" (AKO) relations are used to construct object specialization hierarchies. There are restrictions to avoid redundant hypothesis generation. Fig.4 shows the organization of the entire system in which HLE undertakes the following iterative step : (1) Each instance of an object generates hypotheses about related objects using functions stored in the object model (frame). (2) All pieces of evidence - both instances and hypotheses are stored in the common data-base-called iconoic database. They are represented using an iconic data structure which associates highly structured symbolic descriptions of the instances and hypotheses with regions in a 2-dimensional array. (3) Pieces of evidence are combined to establish "situations", consisting of consistent evidences. (4) Most reliable situation is selected. (5) The selected situation is "resolved" which results either in verification of predictions on the basis of previously detected/constructed image structures or in top-down image processing to detect missing objects. (6) Instantiation of objects at the very beginning of interpretation is performed by the MSE which searches for object models that have simple appearances, and directs the LLVE to detect pictorial entities which satisfy the appearances. The instances thus constructed are seeds for reasoning by the HLE. (7) The HLE maintains all possible interpretations and maximal consistent interpretation is selected.

In order to resolve a situation one of two actions are taken : confirm relations between instances or activate top-down analysis. In the paper (46) mentioned earlier the MSE analysed the partial knowledge structure of a suberban scene detecting visible road, occluded road, overpass, shadowed road etc. (Fig. 8).

Some of the problems that needs to be solved are as follows : knowledge organization should have the knowledge of how to reason about failures depending on their causes. Secondly, some sort of meta knowledge about the dependency among geometric relations should be established, so that which one should be examined first, which one is prohibited, which one cannot be done unless some others are established etc. can be coped with. Thirdly, how to manage mutually conflicting interpretations and perform reasoning on them.

To cope with the problems of ambiguity in data and knowledge because of partial information - all attempts should be made to increase the amount of information. Range sensing is a typical example. The Bayesian probabilistic model has been widely used to compute reliability values, but there are some basic problems in them. The concept of dependency graph as enunciated by Lowrance (47) seem to be a useful method in IUS.

Lee and Fu (48) proposed a design for a general purpose CVS that allows for the proper interaction of top-down (model-guided) analysis and bottom-up (data-driven) analysis. Chakravorti and Freeman also developed an interesting technique using characteristic views as a basis for intensity image 3-D object recognition.

Before concluding this section, for the sake of completeness I have to mention about object recognition using range images, which for lack of space and time, I am not dealing with in this paper. Range image understanding is quickly becoming an important and recognised branch of CVS, as these contain a wealth of explicit information that is obscured in intensity images. In certain environments range-image CVS will be more suited - and this research will perhaps give us new insights into the whole problem of general purpose CVS. Some relevant references for this are Nevatia and Binford (1977) (49), Birbhanu (1984) (50) and Besl and Jain (27).

5.0 CONCLUSION

In this paper we have tried to explain how the research in the fields of Artificial Intelligence, (AI) Image Understanding Systems(IUS) and some aspects of Pattern Recognition(PR) are unified in the CVS research largely motimated by galaxy of applications.

The development of a general purpase CVS that can approach the abilities of the human eye and brain is remote at present, despite recent progress in understanding the nature of HVS.

There are many factors that are confounded in the image. A surface may look dark because of low reflectance, shallow angle of illumination, insufficient illumination or unfavourable viewing angle. The objects such as houses, cars, ships, roads, trees, ponds etc. to be interpreted required large body of knowledge-not only about them-but also how they fit in together.

Though we have not dealt with architectural aspects in this paper at all, it should be noted that CVS involves large amount of memory and many computations. For an image of 1000 by 1000 pixels some of the simplest procedures require 10^8 operations. The human retina with 10^8 cells operating at roughly 100 hertz, performs at leest 10 billion operations per second, and the visual cortex of the brain has undoubtedly higher capacities.

We have reviewed in depth the status of research in different levels of vision and problems of determining 3-D shapes from 2-D images which in my view is on its way to systematic solution. The progress at the higher level problem of recognizing the shapes deduced as objects and identifying them is limited. So it can be concluded that much research remains to be done. To develop generic systems, much more knowledge at the world has to be incorporated into the program. There must be a mechanism to store large-scale spatial information about an area, from which relevant data can be extracted and into which newly acquired information can be fed. Finally there must be dramatic rise in the speed of CVS processors. Once such high speed processors are available highly computationally intensive methods may be attempted that have not been tried so far, leading to more versatile systems. Next generation of computing systems with non-VonNeumann architectur will provide a greater opportunity.

6.0 ACKNOWLEDGEMENT

I wish to acknowledge with thanks the discussion I had with my colleagues M.K. Kundu, S.N. Biswas, K.S. Ray, B.B. Chowdhury and S.K. Pal in preparing this paper and to N. Chatterjee, S.De Bhowmick and S. Chakravorti in preparing the manuscript.

7.0 REFERENCES

1. P.Winston, The Psychology of Computer Vision (MeGraw-Holl,1975).
2. M.Minsky "A Framework for Representing knowledge "In the Psychology of Computer Vision, P. Winerton, Ed. (McGraw Hill 1975).
3. D.Marr, "Artificial Intelligence - A Personal View", Artificial Intelligence, Sept. 1977.
4. S.Zucker, A Rosenfeld and L.Davis,"General-Purpose Models : Expectations About the unexpected", RT-347, Computer Science Center, U of Maryland, June 1975.
5. D.Marr "Analyzing Natural Images", A.I Memo 334, AI Lab, M.I.T., June 1975.
6. P. Winston "Proposal to ARPA", AI Memo 366, AI Lab. M.I.T. May 1976.
7. Bruce L. Bullock "The necessity for a theory of specialized vision" A.P.Hauson and E.M. Riseman, Ed., Vision Systems (Academic Pres 1978).
8. Takeo Kanade and Raj Reddy "Computer vision : The challenge of imperfect inputs", IEEE Spectrum November 1983.
9. Martin D. Levin "A knowledge Based computer vision system", same as (7).
10. T.O.Binford "Survey of model-based image analysis systems", The Int. Journal of Robotics Research, Vol.1,No.1 pp.18-64, 1982.
11. Takashi Matsuyama "Knowledge organisation and Control Structure in image understanding" Proceedings 8th ICPR, IEEE 1984, pp.1118-1127.
12. Takeo Kanade "Region segmentation signal vs semantics" CGIP, vol.13, No.4, pp.279-297, 1980.
13. Michael Brady "Computational approaches in image understanding",ACM computing surveys, vol.14,No.1, March 1982.
14. H.G. Barrow and J.M. Tanenbaum" Recovering intrinsic scene characteristics from images" in Computer Vision Systems (A.R.Hauson and E.M. Riseman eds.) Academic Press, pp.3-26, 1978.
15. D.Dutta Majumder "Pattern Recognition and Artificial Intelligence Techniques in Intelligent Robotic System", Proc. National Convention of Production Engineering Division of Institute of Engineers(India), August 17-18, 1986.

16. D.Dutta Majumder "Pattern Recognition Image ←Processing Artificial Intelligence and Computer Vision in Fifth Generation Computer Systems", Sadhana, Proc. The Indian Academy of Sciences, Bangalore 1986.
17. T.Moto-Oka et al, "Challenge for knowledge information processing systems" (Preliminary Report on FGCS) Proc. Int. Conf. on F.G.C.S. Oct.19-22, 1981, pp.1-85.
18. D.Dutta Majumder "Impact of Pattern Recognition and Computer Vision Research in FGCS Framework", Proc. Int. Conf. on Advances in Pattern Recognition and Digital Techniques, Calcutta, 6-10 Jan. 1986.
19. J.M. Tanenbaum and H.G. Barrow "Experiments in Interpretation guided segmentation". Artificial Intelligence, 8,3,1977.
20 B.Chanda and D.Dutta Majumder, "A Hybrid edge detector and its properties", Int. J. Syst. Sc. Vol.16, No.1, 1985.
21. B.Chanda and D Dutta Majumder "On image enhancement and threshold selection using grey level co-occurance matrix" Pattern Recognition Letters, Vol.3,No.4, 1985.
22. M. Kundu, B.B.Chowdhuri and D.Dutta Majumder "A generalized digital contour coding scheme". CVGIP, 30(3) '85 (July 1985).
23. S.N.Biswas, B.B.Chowdhury and D.Dutta Majumder "An Interactive Curve Designb Method Through Circular Areas and Straight Line Segments"1986 Fall Joint Computer Conference, Dalla, Texas (Communicated).
24. S.K. Parui and D.Dutta Majumder "A New Definition of Shape Sinilarity PRL, Vol. 1982.
25. D.Dutta Majumder and B.B.Choudhuri "Recognition and Fuzzy Description of sides and symmetries of figures by computers" Int. J. Syst. Sc., Vol.11,1980.
26. D.Dutta Majumder and S.K. Parui "How to quantify Shape Distance for 2-D Regions" Proc. 7th ICPR, 1982.
27. P.B.Besl and R.C.Jain "Three Dimensional object Recognition" Computing Surveys, Vol.17, No.1, 1985.
28. A.Rosenfeld, R.A. Hummel and S.W. Zucker "Scene Labelling by Relaxation operations" IEEE. Trans. SMC, Vol.10, No.2, Feb. 1980.
29. R.O. Duda and P.E.Hart "Use of the Hough Transformation to Detect lines and curves in Pictures" Communications of the ACM, Vol.15, January 1972.
30. A.Rosenfeld "Image Analysis : Problems, Progress and Prospects" Pattern Recognition, 17,1(Jan). 1984.
31. I.Chakravorti and H.Freeman "Characteristic views as a basis for 3-D object recognition" IPL-TR-034, Rensselar Polytechnic Inst. Troy. N.Y.1982.
32. K.J.Udupa and I.S.N.MUrthy "New concepts for 3-D Shape Analysis" IEEE Trans. Comp., C-26, 10,Oct.1977.
33. H.A.Blum transformation for extracting new Descriptors of Shape" In Models for the perception of speech and visual form. W.Wathan-Dunn Ed., MIT Press, Cambridge, 1967.
34. P.G. MUlgaonkar, L.G.Shapiro, R.M. Haralick "Recognizing 3-D objects single perspective views using geometric and relational reasoning" Proc. PR & IP Conf. IEEE, Lasvegus, 1982.
35. J.O'Rpurke and N. Badler 1979 "Decomposition of 3-D objects into spheres" IEEE Trans. PAMI, 3(July) 1979.
36. D.H. Ballard and C.M. Brown, Computer Vision, Prentice Hall Inc. 1982.
37. M. Nagao "Control Strategies in Pattern Analysis" Pattern Recognition Vol.17, No.1, 1984.
38. R.A.Brooks R.Greiner and T.O. Binford "The ACRONYM model-based vision system" 6th Int. Jt.Conf.AI, TOKYO, IJCAI,1979.
39. R.A.Brooks "Model-based 3-D interpretation of 2-D images",IEEE Trans. PAMI, 5,2, (March) 1983.

40. W.W. Bledsoe, " The Sup-inf method in Presburger arithmatic", Dept. of Math. and CS Memo ATP-18, Univ. of Texas, Austin 1974.
41. R.B. Fisher 1983, "Using surfaces and object models to recognize partially obscured objects" 8th IJCAI, 1983.
42. T. Matsuyama V. Hwang and L.S. Davis, "Evidence Accumulation for Spatial Reasoning" CAR-TR-54, Univ. of Maryland, 1984.
43. P.G. Selfridge, "Reasoning about Success and Failure in Aerial Image Understanding", Ph.D. Thesis University of Rochester, 1982.
44. R.L. Harr, "The Representation and Manipulation of Position Information Using Spatial Relations", TR-923, CVL, University of Maryland, 1980.
45. D. McDormitt, "A Theory of Metric Spatial Inference", Proc. of Natl. Artificial Intelligence Conf. Aug. 1980.
46. V. Hwang, T. Matsuyama, L.S. Davis, and A. Rosenfeld, "Evidence Accumulation for Spatial Reasoning in Aerial Image Understanding", CAR-TR-28, University of Maryland, 1983.
47. J.D. Lowrance, "Dependency-Graph Models of evidential support" Coins Technical Report, Univ. of Mass, 1982.
48. H.C. Lee and K.S. Fu, "Generating object Descriptions for Model Retrieval", IEEE Trans. PAMI-5, 5(Sept) 1983.
49. R. Nevatia and T.O. Binford, 1977, "Description and Recognition of curved objects", Artificial Intelligence, 8, 1.
50. Bir Bhanu, "Representation and Shape Matching of 3-D Objects", IEEE Trans. PAMI-6, 3(May), 1984.
51. E. Bribiesca and A. Guzman, "How to describe pure form and how to measure differences in shapes using shape numbers" Pattern Recognition, Vol.12, No.2, 1980.
52. L.S. Davis, "Understanding Shape : Symmetry", IEEE Trans. SMC-7, 1977.
53. R.L. Kashyap and B.J. Oommen, "A geometrical approach to polygonal dissimilarity and shape matching", IEEE Trans. PAMI-4, 1982.
54. S.K. Parui and D. Dutta Majumder, "Symmetry Analysis by Computer", Pattern Recognition, Vol.16, 1983A.
55. S.K. Parui and D. Dutta Majumder, "Shape Similarity Measures for Open Curves", Pattern Recognition Letters, Vol.1, 1983.
56. B.R. Suresh, R.A. Fundakowski, T.S. Levitt and J.E. Overland, "A Real-time Automated Visual Inspection System for Hot Steel Slabs", IEEE Trans. PAMI-5, No.6, Nov. 1983.
57. G.J. Agin, "Computer Vision Systems for Industrial Inspection and Assembly", IEEE Computer, May 1980.
58. W.A. Parkins, "INSPECTOR : A Computer vision system that learns to Inspect posts", IEEE Trans. PAMI-5, No.6, Nov.1983.
59. Michael Brady, "Artificial Intelligence and Robotics", Artificial Intelligence, 26, (1985) (North Holland).

DATA STORAGE AND RETRIEVAL IN ASTRONOMY

François Ochsenbein

European Southern Observatory

Karl-Schwarzschild-Straße 2, D-8046 Garching bei München (FRG)

Summary: This review paper describes the available sources of stored astronomical data: archives, catalogues and bibliography. After a short review of the present status of these data, reliability requirements are discussed, including quality estimators. The organisation of the different sources in data bases are summarized. Problems and limitations in accessing and maintaining these data are discussed from a user's point of view; long and short term solutions are proposed to facilitate the access and maintenance of stored astronomical data.

Introduction

Astronomical observations produce a large amount and a large variety of data. It is probably not worth mentioning here the importance of data in astronomy and astrophysics; a good way to understand what is underneath "astronomical data" is to follow the different elements of data management exposed by Vette (1985).

Astronomical data are first acquired from a receptor installed on an astronomical instrument, via a command and control process; the output of the detector is called *raw* data (*e.g.* the photographic plate). The next process consists in detecting anomalies and removing known instrumental biases by means of calibration adjustments, and leads to *calibrated* or *edited* data; the information contents is not modified by this step. A further process, controlled by the observer for ground-based observations, leads to *reduced* or *analyzed* data expressed in physical units (*e.g.* fluxes as a function of wavelength); this step makes use of more sophisticated image or data processing (filtering, contrast enhancement, model fitting, etc), and frequently implies a significant reduction of the data size. The publication of the results and their interpretation — sometimes in the form of catalogues — is the last element of this data management.

Storing selected results coming out of these processes is meaningful if a future use of old data can be foreseen. For example, even if some region of the sky is later reobserved with a much more accurate instrument, the old observation may still be of high interest, *e.g.* when a new source is later discovered, or to estimate slowly variable phenomena.

1 Storage of astronomical data

Let us suppose that, some years later, an astronomer wishes to test some hypothesis about, say, a possible long-period variability of a class of objects. He needs then an access to previously stored data; three typical sources of stored data can be used for this purpose:

1. **Archives**: an Archive can be defined as a collection of observations, each observation consisting of raw, edited or reduced data, generally of large size (say, more than 100 pixels). Reduced data are more useful for the astronomer who wishes to use directly the results, while the expert in data processing would prefer calibrated data; keeping both calibrated and reduced data seems therefore to be worthwhile. Archived raw data would likely be rarely reused; a temporary (a few years) archival would however help to recover possible calibration errors.

   Each observation included in an Archive must provide several elements besides the values of the pixels, known as a *header* (cf Albrecht, 1982), including:

   - means to locate the observation in time and space (telescope position, date and time, target designation)
   - means to identify the receptor and the instrument characteristics to allow an accurate interpretation of the pixels; a way to locate the necessary calibration data (dark or flat fields, wavelength calibrations) is an implicit requirement for raw data.
   - means to identify the owner of the data (observer), and the purpose of the observation
   - engineering data, like CCD temperature or events on a spacecraft

   The contents of this header differs if raw, calibrated or reduced data are stored: engineering data for instance are only meaningful for raw data, but units and accuracies are required for reduced data.

   An observation may also include a *trailer* made of data added after the data acquisition or reduction, like quality remarks derived from some later data analysis or a list of publications making use of these data.

2. **Catalogues**: the definition of catalogues is widely discussed by Jaschek(1985a): he adopted the definition of "an ordered collection of a large number of observations of a certain kind made for a certain purpose and carried out at a certain place and time". The number of parameters listed in each catalogue entry is generally small (say, less than 100). Catalogues are further classified as *Observational Catalogues* made of reduced data, like the IRAS point-source catalogue (Beichman *et al.*, 1985), and *Compilation Catalogues* (Bibliographical, Critical or General), like the catalogue of quasars and active nuclei by Véron and Véron (1985). Within a catalogue, a distinction should also be made between *primary* and *secondary* data: the primary data, related to the prupose of the catalogue, were analyzed by the author, like the IRAS fluxes; the secondary data are added for convenience, generally copied from other sources, like the diameters of associated extragalactic sources listed in the IRAS catalogue.

   As for Archives, a catalogue must contain, besides the data themselves, a "header" which describes what is listed (column headings) and how the data were obtained (instruments, methods, averaging algorithms for critical compilation catalogues, etc), and a "trailer" providing additional individual notes (remarks) and references (for compilation catalogues).

3. **Bibliography**: many data are still only available from the astronomical literature: for instance, I am not aware of any catalogue or Archive providing the observed lightcurves for variable stars. Results which can be summarized by a few parameters are frequently

published as catalogues or gathered later on within compilation catalogues, like photometric observations or redshifts. The bibliography is also the main source for description of instruments, of methods and algorithms used for catalogue compilation or data processing.

These three sources of stored data are tightly connected: the "header" of an archived observation or of a catalogue generally refers to a published paper for detailed descriptions, whereas bibliography can be the source for the creation or the improvement of catalogues.

2 Present status of stored astronomical data

Photographic plates taken before the beginning of this century can still be used: recent measurements of some old Carte du Ciel plates were experimented with by Rousseau and Guibert (1984); many plates however show important defects. The lifetime of archived data is nowadays a problem, since magnetic tapes have only a short lifetime (a few years), and the large amount of data involved discourages regular duplications. Optical disks, with an expected lifetime of several decades, will improve the situation in the near future (see *e.g.* the ST Archive, Albrecht *et al.*, 1985); usage of optical disks is also planned for the ESO Archive. Data compression methods could be used to store more efficiently large size observations (say, more than 10^6 pixels); such procedures should however only be applied if no loss of information occurs, *i.e.* the original image can be exactly restored from the compressed data.

Similar remarks apply for some very large catalogues: for instance, the old Astrographic Catalogue is still a valuable source to derive proper motions, as discussed by Fresneau (1985). But for most catalogues, the lifetime is a less crucial point than for Archives: catalogues generally evolve, and a safe storage for one century or more is probably not a requirement. Centres like the CDS (Jaschek, 1982) or the Astronomical Data Center (ADC) (Warren, 1985) collect a large number of catalogues on magnetic tape and distribute copies on request: even if the magnetic tape lifetime is short, the existence of many copies acts as a security.

The bibliography is mainly kept on print-out form, which has a life-time of several centuries. The usage of microfiches does not show a trend towards a large increase, but we can foresee an emergence of some kind of electronic publications for the future; the printed form will likely still be in use for many years.

A comparison between the present status of the three sources of stored data may be summarized by the following table:

	Archived data	Catalogued data	Bibliography
Size (Gb)	$> 10^6$	~ 1	$\sim 10^3$
Digital	Recent (< 10 yrs)	$\sim$ all	No
Copies	unique	many	many
Usage	rare	very frequent	frequent
Access	not easy	easy	easy

3 Quality of astronomical data

Most scientific publications are submitted to a review procedure: assumptions and results are discussed and criticized before acceptance for publication. Such a procedure ensures some minimal quality to bibliographical data.

The situation is generally less satisfactory for published catalogues; if the referee can criticize or suggest improvements about the assumptions and the methods used by the author (the "header" data), he is generally unable to check the reliability of a large amount of data. Such a check would obviously be extremely useful, in view of the number of errors that are regularly detected and (rarely) published (the *Bulletin d'Information du Centre de Données Stellaires* is an exception). With the use of a data base of computerized catalogues, checking procedures could be as easy to perform as the verification of an algorithm on some trivial examples. Such procedures could also, in some cases, provide estimators about the quality of data.

For Archives, the situation is generally worse: apart from space experiments, the "header" information which was in the past reported in the log-book is frequently inaccurate or incomplete; the usage of computers connected to large instruments ensures a more reliable acquisition of some important parameters (telescope position, time, detector parameters), and therefore improves the situation. But still only the observer is able to specify which observations are of good quality and therefore worth keeping or which can simply be thrown away. Another problem is the availability of reliable calibration data required for the reduction of raw data, or the lack of information about the exact procedures used to derive reduced or calibrated data.

An astronomer is frequently faced to the tricky problem of getting an estimation of the quality of the data he is using. It is generally hopeless to try to get figures about the accuracy of the data listed in bibliographical compilation catalogues without returning to the sources. In observational or critical compilation catalogues, the quality of the "primary" data (but not of the "secondary" data) is generally discussed in the "header"; sometimes uncertainties are listed in dedicated columns of the catalogue. But the estimation of the quality of an observation stored in an Archive is more difficult: many parameters interact (instrument, detector, weather conditions, calibration procedures), and the observer's judgement play a fundamental role. The meaning of a quality is also related to the purpose of an observation: an image may be unusable for the planned photometric analysis, but of high quality for astrometric studies.

In some very specific Archives, like the ESO Schmidt plate Archive, an accurate quality index was defined for the ESO/Uppsala Survey usage (Lauberts, 1982), including notes about seeing, weather conditions and plate aspects (elongated images, chemical fog, etc); this quality index is moreover homogeneous over the whole plate archive — but this homogeneity is mainly a result of one man's work. For general Archives, only some rough quality estimations could be made available as an output from some standard calibration procedure — perhaps performed by an expert system. However, even a quality index on a very rough scale (very good, good, poor, very poor) is quite instructive if it refers to a well-defined *purpose* of the observation.

4 Organisation of astronomical data

The main aim of facilitating access to astronomical data is to reduce the amount of unnecessary observing and computing that takes place because the astronomer has failed to find already existing data, or was unable to judge whether the data are of adequate quality for his purposes. This goal implies an organisation of astronomical data in easily accessible data-bases.

The present data-base capabilities allow the management of up to a typical Gbyte total size with good reliability and performances. The relational model is now considered as the best conceptual approach for a data-base management system (DBMS), and much software and even dedicated hardware is now available. Any DBMS must however be optimized to speed up frequent queries (creation of indexes, organisation based on some key parameter).

"Classical" bibliographical data-bases are organised to minimize the access time needed to retrieve published papers from some boolean combination of keywords related to a subject, an author or some words extracted from a summary. The result is generally a set of matching references, with on-line abstracts; the complete articles can presently only be retrieved in a library. In astronomy, besides the well known Astronomy & Astrophysics Abstracts which are not yet implemented as a data-base, the "Physics Briefs" data-base is an example described and illustrated by Lück and Behrens (1985).

A bibliography by astronomical object, allowing the retrieval of publications where some information about individual objects may be found, is somewhat different, and cannot easily be integrated in a "classical" bibliographical data-base: this requires the knowledge of a complicated object designation scheme — in which even trained astronomers have difficulties to find their way. A dedicated software was developed at the CDS to solve this problem (Ochsenbein, 1985a): an external table, easily (and frequently!) updated, containing the list of the acronyms associated to the existing formats (templates), is used as a "knowledge" data-base to parse and edit astronomical designations.

An organisation of astronomical catalogues may consist of a set of on-line catalogues sharing some common software like the SCAR (Starlink Catalogue Access and Reporting System, see Fairclough, 1985) or the Astronet DIRA (Distributed Information Retrieval system for Astronomical files, Hunt and Nanni 1985), or a complete integration of data extracted from many catalogues in a single data-base. The second approach has the advantage that the user is not assumed to know which catalogues he should investigate (which are the primary data, how reliable are *e.g.* the SAO magnitudes), and was adopted to build up the SIMBAD data-base detailed by Ochsenbein (1984, 1985a): the *primary data* were extracted from many catalogues, keeping full references to the sources of data; the largest known bibliography by object is an important part of this data-base and is regularly kept up-to-date. SIMBAD was designed to allow fast access to all presently known information about single objects from both astronomical designations and specifications about positions.

Due to the very large amount of data involved, an Archive is organised in two parts:

- the observations, with complete header, are stored on some permanent medium: magnetic tapes regularly copied, or better optical disks; "juke-boxes" of optical disks in the short-term future will have on-line storage capabilities up to a Tbyte (10^{12} bytes). This part does not need to be organised as a data-base, each observation being stored as a sequential file; note that a fundamental part of the presently existing archives is made of photographic plates.

- the catalogue of observations, comparable to the log-book, is organised as a data-base. Principles about the contents of such a catalogue may be stated as : "The information given should be sufficient to allow the astronomer to locate the data and to judge whether they will be suitable and of sufficiently high quality to justify the effort involved in obtaining them" (Wilkins,1982). Typically, such a catalogue includes:

    1. astronomical specifications extracted from the "header" (instrument, detector, purpose of the observation, telescope position, object designation ...)

    2. instrumental specifications extracted from the "header" (filters, grating ...)

    3. exposure specifications, *e.g.* exposure time, plate identification or file name, location of calibration data , observer's name ...

    4. secondary data extracted from other sources (*e.g.* SIMBAD), added to facilitate retrieval purposes, like a classification of the observed object

    5. evolving data: trailer additions like quality comments, or specifications about protection and availability. This part must be carefully managed, since it cannot be restored from a reading of "headers" and catalogues, and may include results involving a large amount of computing time or expert estimation.

Examples of accessible or developing Archives are IUE (Barylak, 1985), and the more sophisticated La Palma project (Harten and Lupton, 1985). The ESO Archive is under development, and will be tested on data from a recently implemented instrument (EFOSC). A description of the ST-ECF plans for the ST Archive management is detailed by Russo *et al.* (1986).

5 Access to astronomical data

The efficiency and reliability of computer networks have greatly improved during the recent years: even if it is not yet possible to transfer large sets of data, public networks allow a transfer of 10 Kbytes within 1 minute of connection time, which at least allow efficient query dialogues. A recent remote control test on a leased line between La Silla (ESO telescopes) and Garching showed an actual transfer capability of 34 Kbytes/min (G. Raffi, private communication).

For catalogued or bibliographical data, the situation is presently more or less satisfactory, due to the central data bases mentioned in the preceding section; small data sets can be transfered via the network, but large catalogues can only be exchanged through mailed magnetic tapes. The main problem in the exchange of astronomical catalogues is the lack of a standard output format including the "header" data; the FITS extension to catalogues proposed by Harten et al.(1985) could be a solution, but carefully applied such that the catalogue contents is not modified (Ochsenbein, 1985b).

Archived data cannot presently be easily accessed, with the important exception of some large plate vaults and space instruments like IUE. If many Archives exist, few are organised in data-bases (see e.g. Hauck, 1982), and very few have an on-line catalogue; it is however hoped that more and more organised Archives will be accessible to the astronomical community in the future. A centralized Archive similar to SIMBAD for catalogues seem unrealistic mainly for two reasons: this would imply transfer and processing of very large amounts of data, and the required expertise is located at the source observatories. A distributed data base is therefore a better approach.

Let us take an example to illustrate the problems an astronomer may encounter to locate and get existing information about, say, the bulge of M 104 (the Sombrero galaxy), for a surface brightness analysis. Published papers were previously located through SIMBAD, but nothing interesting was found in the recent literature. Getting more data from Archives implies the following steps:

1. Which are the Archives that *could* contain images of M 104?

2. The ESO Archive is a possibility. How to connect to this Archive?

3. The connection was successful. How to phrase my requirements?

4. How to understand what is listed on my screen, and how to select among several observations the best suited for my purpose?

5. How to get (order) a copy of the few observations I finally selected?

6. I might not be satisfied with the result: I know, for instance, than an observation was made, but I can't find it. How to communicate such errors and get an answer?

7. After all, other Archives may contain observations better suited for my purpose. How to get similar information from other Archives?

This example leads to some suggestions and recommendations about possible solutions:

1. A list of all available archives, with a description of their specificities, completeness, available services and connection facilities, must exist somewhere (a central node?) and *maintained*.

   A centralized data-base similar to SIMBAD quoting which archives include observations for specific targets would be another possibility (the location of possible archives would then be immediate). Such a solution would imply regular exchanges with all archiving centres; the contents of such a data-base would have to be defined, with a possible information ranging from a single bit (existence of data) to almost complete descriptions (used instrument, detector, wavelength range, etc).

2. A distributed data-base organisation linking archives is a possible solution; a pilot network project named ESIS (European Space Information System) is being set up by ESA to link a few astronomical archives and data bases.

3. The query language should be simple and adapted to astronomy; each commercial DBMS has presently its own query language, of course not oriented towards astronomical applications. The use of a different language for each system should be avoided: the occasional user would simply give up, a minimum of 4 hours of practice per month being generally considered as a minimum for an efficient use of any query language. Access to a specific data-base is performed through a "User Interface", a piece of software which can be made flexible enough to be usable on any terminal. Some basic commands like Help, Report, Select according to date of observation, to target astronomical designation or to position in classical coordinate frames could be standardized without too much effort. The "STARCAT" interface, developed as a collaboration between ST-ECF – ESO – ST ScI, is a good example of a User Interface (Russo et al., 1986).

4. If the definition of some standards can easily be adopted for a designation of some key parameters like RA and DEC (with which units?), it is a bit more difficult to agree on some scheme for object classification; the way of designating astronomical objects is also a problem (see e.g. Jaschek, 1985b) which can be solved with the help of the SIMBAD data-base. In any case, standard help facilities must provide detailed explanations about site-specific parameters, like instrumental specifications. But the most tricky problem is the estimation of qualities (see section 3): an expert's eye looking at the picture is still presently the most reliable procedure, but is still beyond the present network capabilities. The availability of a very rough quality index in the context of the observer's purpose would be quite helpful to decide which could be the best suited data.

5. A "Fetch" command could be the standard way to ask for a copy of archived data. Loading can be immediate for on-line data of small size (e.g. spectrograms) or for a local use, or delayed via standard-formatted (FITS) tapes. In any case, the command would be passed to an authorization procedure; minimal standards for remotely filling such requests are also desirable.

6. Communication with archive managers could include subsections related to: a) data errors (e.g. wrong object designation); b) bugs (e.g. the query provides a wrong result); c) encountered problems (e.g. obscure help phrasing); d) suggestions for improvements. How to route the answer to the remote user is likely the more difficult problem if public networks are used.

7. It is tedious to reformulate the same set of questions to many archives; once the archives are nodes of a network and make use of a common user interface, simple queries could be routed to several archives, returning either minimal information about matching observations, or only the number of matching observations if this number is large. This last facility is in fact an alternative to the possible "archive index" suggested in the first point.

This step by step example emphasizes the usefulness of minimal standards for at least the "User Interface". An easy connection between archived and catalogued data would also enhance the efficiency of astronomical data usage: the retrieval of catalogued data in the course of a reduction procedure is another illustration of the benefit of such "data sharing" possibilities.

6 Maintenance of astronomical data

Bibliography, catalogues and archives obviously evolve with time: new instruments are available, implying a rapidly increasing amount of accumulated material. Such an "astronomical maintenance" first requires routine work for the addition of newly acquired data, the correction of erroneous data, or the modification of obsolete results. The evolution of astronomical instruments also implies that detailed descriptions and calibrations of each used instrument are carefully stored — otherwise, the next generation of astronomers will be unable to understand properly observations, even carefully stored. A comparison of data from different sources would also help to improve the reliability of the stored material.

The "computer maintenance" implies an organisation which can easily be transported to computers of a next generation; this implies software architectures fully documented, clearly structured and layered, making use of well-defined interfaces to operating-system dependencies and to *data-base management systems* (DBMS). A cooperation in the development and implementation of several archives would obviously benefit both data producers and users, with a possibility of sharing software.

The existence of very large data sets is another aspect of this "computer maintenance": moving to a new computer should not result in an impossibility to read data stored on the presently used media. Photographic plates taken at the beginning of the century can still be used; it would be disastrous if the data now stored on optical disks will not be accessible in the future simply because the cost involved in a transfer to the new media is too high. This means that the usage of permanent media like optical disks is only meaningful if we are reasonably sure that we will be able to read data stored on such media in the future.

Concluding Remarks

The organisation of astronomical data (Archives, Catalogues, and bibliographical data) would greatly benefit from a cooperation between astronomers involved in data production, data processing, and data-base specialists. It is hoped that minimal standards will be adopted to enlarge the efforts involved in the developments of new astronomical instruments: this implies comprehensive descriptions from the observer's side (*e.g.* the target name), and a friendly "User Interface" to access stored data via networks from the user's side. An experiment of routine archiving the Faint Object Spectrograph and Camera (EFOSC) data is presently starting at ESO.

References

Albrecht R., 1982: in *Automated Data Retrieval in Astronomy*, C. Jaschek and W. Heintz (eds), D. Reidel, 87
Albrecht R., Kinsey J., Schreier E., Shames P., 1985: *Mem. Soc. Astron. It.* **56**, 371
Barylak M., 1985: *IUE ESA Newsletter* **24**, 9
Beichman C.A., Neugebauer G., Habing H.J., Clegg P.E., Chester T.J., 1985: *IRAS Catalogs and Atlases*, Joint IRAS Science Working Group
Fairclough J.H., 1985: in *Starlink User Note* **70.3**, 25 September 1985
Fresneau A., 1985: *Astron. J.* **90**, 892
Harten R.H., Lupton W., 1985: *Mem. Soc. Astron. It.* **56**, 415
Harten R.H., Grosbøl P., Tritton K.P., Greisen E.W., Wells D.C., 1985: *Mem. Soc. Astron. It.* **56**, 437
Hauck B., 1982: in *Automated Data Retrieval in Astronomy*, C. Jaschek and W. Heintz (eds), D. Reidel, 217

Hunt L.K., Nanni M., 1985: *Mem. Soc. Astron. It.* **56**, 565

Jaschek C., 1982: in *Automated Data Retrieval in Astronomy*, C. Jaschek and W. Heintz (eds), D. Reidel, 3

Jaschek C., 1985a: *Mem. Soc. Astron. It.* **56**, 259

Jaschek C., 1985b: *Mem. Soc. Astron. It.* **56**, 331

Lauberts A., 1982: *"The ESO/Uppsala Survey of the ESO(B) Atlas"*, European Southern Observatory

Lück W., Behrens H., 1985: *Bull. Inf. Centre de Données Stellaires* **28**, 133

Ochsenbein F., 1984: *Bull. Inf. Centre de Données Stellaires* **26**, 75

Ochsenbein F., 1985a: *Mem. Soc. Astron. It.* **56**, 293

Ochsenbein F., 1985b: *Mem. Soc. Astron. It.* **56**, 313

Rousseau M., Guibert J., 1984: *Bull. Inf. Centre de Données Stellaires* **27**, 43

Russo G., Richmond A., Albrecht R., 1986 *in this workshop*

Véron-Cetty M.P., Véron P., 1985: *E.S.O. Scientific Report n°4, April 1985*

Vette J.I., 1985: *Mem. Soc. Astron. It.* **56**, 269

Warren W.H. Jr, 1985: *Mem. Soc. Astron. It.* **56**, 285

Wilkins G.A., 1982: in *Automated Data Retrieval in Astronomy*, C. Jaschek and W. Heintz (eds), D. Reidel, 193

VECTOR COMPUTERS IN ASTRONOMICAL DATA ANALYSIS

Donald C. Wells

National Radio Astronomy Observatory [†]
Edgemont Road, Charlottesville, VA 22903-2475 USA

> "A scalar computer is one that provides instructions only for manipulating data items comprising single numbers, in contrast to the vector computer that also has instructions for manipulating data items comprising an ordered set of numbers (that is to say a vector)."[1]

INTRODUCTION

Three different astronomical observatories have installed vector computers during 1985-86. These machines have been purchased with astronomy money for use principally in astronomy data analysis applications, *not* principally for theoretical modelling. The three installations are:

Observatory	Installation Date	Model
NRAO (Charlottesville, VA, USA)	December 85	Convex C-1
DAO (Victoria, BC, CANADA)	April 86	Alliant FX/1
NRO (Nobeyama, JAPAN)	June 86	Fujitsu VP-50

This paper outlines the performance reasons why such vector machines are being chosen for astronomical data analysis applications—reasons which imply that more such vector machines will be installed in the future. The author makes several predictions for the period mid-1986 through 1988:

- Several more vendors will enter the vector computer market,
- Vector floating point workstations will be offered,
- Existing scalar computers will be given vector extensions, and
- More astronomical institutes will procure vector machines.

[†] NRAO is operated by Associated Universities, Inc., under contract with the U. S. National Science Foundation.

FLOATING POINT PIPELINING

A vector computer employs floating point pipelining to achieve a speed advantage over scalar computers built with a similar number of components and operating at a similar clock frequency. The concept of pipelining is based on the fact that floating point arithmetic hardware units generally perform their operations in a sequence of steps, each of which takes one clock period of the computer. It is typical for floating point operations to take five steps, or five clock periods (some designs have as few as three or as many as seven steps). A pipelined version of a floating point unit moves new data into the first stage of the pipeline as it moves the first partial result into the second stage; new data enter on each subsequent clock cycle. For a five-stage pipe, the pipeline is full on the fifth cycle, and the first result comes out. After this, the pipeline delivers a new result for each clock cycle, and the computation rate is five times higher than for a non-pipelined floating point unit built from similar components. The speedup of about five is the major reason for using vector computers. *This advantage is fundamental!*

ATTACHED ARRAY PROCESSORS ARE NOT VECTOR COMPUTERS

The performance superiority of floating point pipelining has been well known for more than a decade. During this period several vendors have offered such hardware in the form of specialized processors which are designed to be attached to conventional scalar computers. The basic strategy of the designers was to support signal processing and linear algebra applications with libraries of hand-coded algorithms, thus avoiding any requirement to construct a compiler capable of recognizing parallel constructs in original source programs. A user who wishes to execute an existing program on such an attached processor must change the program to call subroutines in the processor-specific library; in fact, it is usually necessary to call subroutines to move data from the scalar host into the processor before calling subroutines to initiate the processing in the array processor, and then to call other subroutines to bring the results back into the host. Programs which have been changed to function with one model of array processor are generally not compatible with models from other vendors, and are also not compatible with scalar processors unless a library of subroutines is provided to emulate the processor-specific library. These limitations and problems mean that array processors are generally used only in applications which are so compute-intensive or important that the extra costs of code changes and lack of portability are justified.

The principal distinction between attached array processors and vector computers is that *the computers have compilers*. Compilers hide almost all of the details of the hardware implementation from the programmer. They allow the high performance hardware to be used with all algorithms, not just those which are important enough to have been hand-coded for specific attached processors. Two forms of compiler support for attached array processors are likely to appear as products. First, compilers will appear which invoke attached array processors for scalar hosts, thereby converting the scalar machines into a true vector computers. Second, some existing array processors are intrinsically capable of becoming independent computers in their own right, and provision of operating system and compiler support, plus conventional peripheral device support, will convert these systems into true vector computers. While attached processors are likely to remain technically and economically competitive for specialized problems, the advantages of pipeline performance in general computing imply that vector computer implementations will be the main evolutionary path in the future.

FORMS OF PARALLELISM IN VECTOR COMPUTERS

Pipelining is a form of parallelism, of concurrent computation. It is the most important form for present-day high performance computers for engineering and scientific computation, but other forms are also important.

SCALAR PARALLELISM

Scalar parallelism is important because, for many practical applications, the scalar execution speed of a vector computer strongly influences its overall speed. This is because significant portions of most programs are not able to execute in the vector hardware. For example, if half of a program uses the vector hardware, and runs five times faster, the whole program will only execute 1.67× faster, not five times faster. But if the scalar half of the code can be speeded up by a factor of two by parallelism, then the whole program (scalar plus vector) can run 2.86× faster than the original scalar version (still not the full factor of five, of course).

The first form of scalar parallelism is multiple functional units, plus a "scoreboard" which records and interlocks register and functional unit reservations, allowing the execution of independent instructions to overlap in time. This form of parallelism was pioneered in the CDC 6600 in 1964, and it occurs today in Cray machines and in the Convex C-1. Cray machines and the C-1 also overlap the execution of many scalar instructions with floating point pipeline operations.

The second form of scalar parallelism is instruction pipelining, which appeared in several machines in the late 60's, and is used today in many computer implementations. In this case, the CPU starts the fetch of the next instruction from memory while the interpretation of the current instruction is still in progress. How can the CPU fetch the next instruction after a conditional branch until the condition has been evaluated? The answer is that it can't, but it can guess which way the branch is likely to go, and start the fetch for the most likely case. In fact, scalar loops contain branches which almost always branch back to the top of the loop, and so the heuristic trick almost always succeeds. The Alliant FX/8 is an example of a vector machine which uses instruction pipelining.

PARALLEL VECTOR OPERATIONS

The basic pipeline technique is subject to several forms of higher-order parallelism. First, two or more pipes can be operated in parallel for increased throughput (even elements into the first pipe, odd elements into the second). The CDC Cyber 205 and the Fujitsu VP-series machines are well-known examples of this form of parallelism, but many other machines also use it.

The second form of higher-order parallelism for pipes are two techniques for concatenating pipelines to execute two or more vector operations in parallel on data vectors. The first technique is usually called "linked triads"; it is used in the Cyber 205, the Fujitsu VP-series and the Alliant FX/8. In this case, two vectors and a scalar constant can be combined using a vector add and vector multiply to produce an answer vector; the two most common combinations are $\mathbf{A} + \mathbf{B} \times c$ and $\mathbf{A} \times \mathbf{B} + c$. These operations and several others are specific instructions in linked-triad machines.

The second technique is called "chaining"; it is used in the Cray-1 and the Convex C-1. The idea is that if two vector instructions are issued in quick succession, and if the input vector for the second is the same as the output vector of the first, then the two pipelines are *automatically* concatenated. In fact, in these machines, a vector load or store operation can be concatenated with the vector add and multiply operations, and the third operand can be a vector, not just a scalar constant. Although use of chaining produces a more versatile architecture, in practice linked triad machines often compete favorably with chaining machines (for example, the triad instructions usually include forms in which the vectors come from memory, just as in the Cray architecture). A technical point is that, unlike the Cray-1, the new Cray-2 cannot chain memory operations, although it can overlap them.

CONCURRENT VECTOR COMPUTERS

The final form of parallelism is concurrently executing CPUs, each with vector pipelines; the individual CPUs can each employ the forms of scalar and vector parallelism discussed above. We can classify these machines into two categories, architectures capable of "small granularity" parallel computing, and those only capable of medium to large granularity concurrency. The notable example in the first category is the Alliant FX/8, the first general purpose parallel computer with full compiler support. Its CPUs are able to synchronize their operations on a time scale of a few clock periods, thus permitting efficient parallel execution of inner loops (small "granules" of code). The Cray X-MP is the notable example of the second category; another machine in this class is the IBM 3090/VP. Synchronization delays in these machines are much longer, and they are best used for parallel execution of large outer loops (medium to large granules). The author believes that the programming techniques used to synchronize the CPUs of the X-MP are likely to be used with many other vector/parallel CPUs during the next five years.

Table 1. Architectural Parameters for Selected Vector Computers

Machine	Clock: nsec	Mhz	Pipes: Add	Mult	$n_{\frac{1}{2}}$	MFLOPs	
FPS-120B	167	6	1	1	2	12	
Sky Warrior	200	5	2	1	50:	15	
Numerix 432	100	10	2	1	2	30	
FPS 264		19	1	1	2	38	
Convex C-1		10	2	2	10	40	(32 bit)
Alliant FX/1		5.7	1	1	10	11.4	(32 bit)
Alliant FX/8		5.7	8	8	80:	94	(32 bit)
Cray-1	12.5	80	1	1	10	160	
Cray X-MP	9.5	110	1	1	20	220	(single CPU)
Cyber 205	10	100	2	2	200	400	(32 bit)

VECTOR COMPUTER PERFORMANCE COMPARISONS

It is common to compare vector machines using the unit "MFLOPs" (millions of floating operations per second). For this purpose, only addition and multiplication operations are counted. This single criterion is generally an oversimplification, but it has proven to be useful for approximate comparisons of competing architectures. Table 1 gives architectural details for a number of vector computing systems, sorted in approximate order of increasing peak MFLOP rating. The machines are tabulated in two groupings; the first group contains examples of the attached array processors discussed above while the second group contains the true vector computers. The array processors are included in this listing so that their performance can be compared to the vector computers. The FPS-120B is the classic example of the attached array processor; the Numerix 432 is a higher performance example of the concept. The Sky "Warrior" is a special case; with a proper compiler it could probably be operated as a vector register extension to the architecture of a scalar host computer. The FPS 264, which is often operated as an attached processor, is included in the second group because it is capable of operating as an independent computer as well as an attached processor.

The MFLOP ratings in the table are *peak speed*. They are computable from three of the other columns in the table: $MFLOPs = F \times (A + M)$, where F is the clock frequency in MHz, A is the number of adder pipelines and M is the number of multiplier pipelines. Real programs only rarely execute at these peak speeds. For a mix of random programs, expect about 10% of this speed. For reasonably "vectorized" programs, vector machines typically can obtain perhaps 30% of the peak speed, while for carefully coded cases as much as 70%, or even more may be obtained. Also, the percentage of peak speed obtained for a random mix of programs depends on the architecture; some machines in the table do better than others. For example, the FPS 264 can execute scalar programs several times faster than the Convex C-1, and so partially vectorized programs will generally be faster on the 264 than on the C-1, in spite of the fact that the peak MFLOP ratings of the two machines are similar for 32-bit data. A final point: the Convex, like the Cyber 205 and Alliant FX/8, is about twice as fast for 32-bit data as it is for 64-bit.

Because there are a wide variety of subtle architectural and implementation differences between existing vector computers, it is difficult to predict relative performance *a priori*. An analyst who needs to make such predictions will want to consider not only the theoretical peak MFLOP ratings shown in the table, but also several other factors discussed in the next section of this paper. The analyst must also remember that, in the end, there is no substitute for empirical measurements of the performance of production codes.

THREE IMPORTANT ARCHITECTURAL FACTORS

The $n_{\frac{1}{2}}$ parameter in Table 1 is the effective "half-length" of the floating point pipelines; this is the vector length for which the pipeline reaches half of the peak speed. Consider the Cray X-MP and the Cyber 205. While the 205 has a higher peak pipeline speed, the X-MP is faster for scalar code and has a much smaller $n_{\frac{1}{2}}$. The scalar difference means that the Cray will be faster for partially vectorized code. The Cray will also perform much better than the Cyber on vector operations involving short vectors because of its smaller $n_{\frac{1}{2}}$, which is comparable to the actual physical length of the Cray pipes. The $n_{\frac{1}{2}}$ of the Cyber is much larger because this memory-to-memory architecture has a long setup time to begin pipe operations—about 50 clock periods (equivalent to processing 50 vector elements). However, for carefully designed, highly vectorized algorithms, operating on long vectors (lengths of 1000 or more), the Cyber is capable of achieving higher performance than the Cray.

Achieved performance depends on two other factors which do not appear in Table 1. The first is the CPU-to-memory bandwidth. Consider a simple addition of two vectors in a Convex C-1, with the sum vector to be stored in memory (*i.e.*, A(I)=B(I)+C(I)). The C-1 can do only one transfer between a vector register and memory at a time. So the first vector is loaded into a vector register at a rate of one clock tick per element, and then the second vector is loaded into another register, again at a rate of one clock tick per element. While the second load is in progress, *the addition operation is chained to the load, and almost completely overlaps.* The vector store operation can then occur, again at one tick per element. Thus, the complete operation consumes three clock ticks per element, and the effective MFLOP rating for the vector add operation is $\frac{1}{3}$ of the 20 MFLOP peak rating, or 6.7 MFLOPs. If the C-1 had two memory-access functional units, which could simultaneously address different banks of main memory, then the two vector loads could overlap in time, and the addition and vector store operations could also overlap, requiring only two clock ticks per element, not three, and giving an effective speed of 10 MFLOPs. This concept is not just theoretical. The principal difference between the Cray X-MP and the Cray-1 is not the 110 Mhz *vs.* 80 Mhz clock difference shown in Table 1; rather, it is the fact that the X-MP has more simultaneous paths to memory than the Cray-1. As a result, for a wide range of memory-limited algorithms the X-MP achieves a higher fraction of its peak pipeline power than does the Cray-1.

The second architectural factor which is not represented in Table 1 is the I/O capability. Each data processing problem has a characteristic requirement which can be approximated by a single number expressed in units of FLOPs per I/O byte. The question is: how many floating point operations does a program perform for each byte of data that it moves into or out of the system? If the number is large ($\gg 100$), then the program will be compute-limited; but if it is small ($\ll 10$), then the problem will be in danger of being I/O-limited. Consider the worst case: subtracting two single-precision floating point digital images which are on the disk. With two inputs and one output, each with four bytes per pixel, the program does 12 bytes of I/O per FLOP. A Convex C-1 would need $12 \times 6.7 = 80$ megabytes per second to keep its floating point pipelines busy on this problem; in fact, the C-1 can hardly achieve $\frac{1}{10}$-th of this rate—the only way to make this problem be compute-limited is to hold all of the data in main memory, not disk. The deconvolution of radio aperture synthesis imagery does ≈ 10 FLOPs per byte of I/O. A high performance I/O system is required if this problems is to be compute-limited in a high performance vector computer. In fact, individual disk drives are not fast enough; it is necessary to operate drives in parallel in order to synthesize a drive with sufficient performance; such a disk system is said to be "striped".

TWO TYPES OF VECTOR ARCHITECTURES

We can divide current vector machines into two architectural groups: memory-to-memory and vector-register. The memory-to-memory group has one notable member, the Cyber 205. In this machine all operand vectors flow from memory to the pipes and result vectors go back to memory; no fast registers are available to retain vectors.

Other current vector machines, *e.g.*, Cray-1, Cray X-MP, Cray-2, Convex C-1, Alliant FX/8, IBM 3090/VP and Fujitsu VP-series, all have vector-registers to keep temporary results; in fact, both the Alliant and the Fujitsu machines are also capable of operating in the memory-to-memory mode when it is advantageous. These machines have eight vector registers, with lengths of 32 (Alliant), 64 (Cray), 128 (Convex) or 1024 (Fujitsu) vector elements. The Fujitsu is a special case: it is dynamically reconfigurable to 8, 16, 32... registers, with a total of 8192 elements. In all of the vector-register machines, single instructions (load, add, multiply,...) perform operations on *entire* vector registers.

The technique called "strip mining" is used to process vectors longer than the vector register length. For example, the code

```
DO I = 1, N
  Z(I) = X(I) + Y(I)
ENDDO
```

becomes

```
DO J = 1, N, REGSIZ
  DO I = 1, MIN (REGSIZ, N-J+1)
    Z(J+I-1) = X(J+I-1) + Y(J+I-1)
  ENDDO
ENDDO
```

for a machine with vector registers of REGSIZ elements. This transformation is automatically generated by the vectorizing compiler on such a system.

VECTOR PROGRAMMING—PROBLEMS AND TACTICS

The first big problem for use of vector computers is that they need to be able to execute old programs in the vector mode. This is often called the problem of the "Dusty Deck". It is supremely important that a compiler be able to recognize vectorization opportunities in code which was not written with vector computers in mind. In fact, the actual delivered performance of vector machines generally depends as much on the software technology of compilers as it does on the hardware technology of scalar concurrency, pipelining and concurrent CPUs. Not all compilers are created equal—some are capable of greater degrees of vectorization than others. Some of the compiler strategies which can be used have been discussed by Kuck and Wolfe.[2]

The second big problem is a result of "Amdahl's Law": the overall speedup due to parallelism of any form is ultimately limited by the fraction of a program which executes in scalar mode. This problem was mentioned earlier under the heading of scalar parallelism; in general, a fast vector processor must be accompanied by a fast scalar processor if it is to achieve its full potential. Also, it is *very* important that compilers perform local and global scalar optimization and automatically recognize opportunities for scalar concurrency. The paradox of the vector compiler is that as its vectorization capability is improved, its scalar optimizations must also be improved to fully realize the vector performance.

It is typical that 20% of the code of a program does 80% of the work; this is often called the "80/20 Rule". In practice, even higher concentrations are frequently observed in highly vectorized code. The concentration of the vector computing means that it is often economically and technically feasible to hand-optimize the innermost loops of a program to gain extra performance. Therefore, it is desirable that compilers should support advanced, often machine-dependent, explicit vector constructs to facilitate the hand-optimization coding, and permit the programmer to avoid having to write assembly language.

THE PROBLEM OF VECTOR DEPENDENCY

The independent execution of several iterations of a loop in parallel is only possible if iteration i does not need values computed in iteration $i - 1$. For example, consider the following subroutine:

```
      SUBROUTINE VADD (R, IA, IB, IC, N)
      REAL R(*)
      INTEGER IA, IB, IC, N
CVD$DIR NO_RECURRENCE
      DO I = 1, N
        R(IA+I) = R(IB+I) + R(IC+I)
      ENDDO
      END
```

If the R(IA+1:IA+N) and R(IB+1:IB+N) or R(IC+1:IC+N) vectors *overlap*, results *may* be improper if concurrent operation is attempted. Vectorizing compilers are able to recognize that this possibility exists because the values of IA, IB, IC and N are not known at compile-time, and they will generally refuse to compile instructions which would execute the above loop in the vector pipelines. But suppose the programmer knows that the values of IA, IB and IC are such that the vectors do not overlap? In this case, the comment card compiler directive can be inserted, as shown above, to assert to the compiler that there is no possibility of vector dependency in the loop, and the compiler will then compile the higher performance code.

Most vector/parallel compilers support several types of these comment-card compiler directives. The purpose of the directives is to supply the compilers with information which is not implicit in Fortran syntax and semantics. Unfortunately there is not yet any industry standard for the format of the comment cards or for the names and meanings of the directives (*e.g.*, "NO_RECURRENCE" is the Convex name for the directive used above; Cray calls the directive "IVDEP" and has a slightly different format for it).

VECTOR OPERATORS

Vector computers support many operations other than addition, subtraction and multiplication. Integer arithmetic, division and Boolean operators (AND, OR and XOR) are generally available. Branching requires vector comparison operators (LT, LE, EQ,...); these produce a vector of logical values which is used for branching and as an operand of other operators described below. One frequently needs to know how many values of such a vector are true, and for this it is common to have a "population count" instruction. Often one also needs to know the subscript of the first or last of the elements of the logical vector which is true, and instructions are generally provided to facilitate these searches.

Most machines provide a selection of advanced vector instructions. The first group are "reduction operators"; these include the SUM, PRODUCT, MAX, and MIN of vectors. In each of these, a vector operand is used and a scalar answer is produced. The remaining advanced instructions implement sophisticated subscripting operations on vectors. Indirect addressing is generally called "gather" and "scatter" in vector machines. For the "compress" instruction, a data vector and a Boolean vector are the operands; the result vector contains only those data elements for which the Boolean elements were true. The "expand" operator is the inverse of "compress"; these operations are frequently used with sparse matrices to save storage space. The "merge" instruction takes two data vectors and a Boolean vector; the result vector contains values from the first data vector when the Boolean is true and from the second vector when the Boolean is false. These advanced operations facilitate data editing operations and the parallel execution of IF-statements inside DO-loops. The use of these vector operators to avoid a difficult vector dependency in a synthesis mapping problem has been analyzed by Wells and Cotton.[3]

BENCHMARKING NRAO'S AIPS ON VECTOR COMPUTERS

Although several observatories have used vector machines for the processing and analysis of astronomical data,[4] the only comparative trials of the same programs on several machines have been made by NRAO. These involve the installation and operation of NRAO's Astronomical Image Processing System (AIPS). Two cases occurred: first, during 1985 NRAO installed AIPS on a Cray X-MP under the COS operating system to run aperture synthesis production work on this high performance machine. Second, NRAO installed AIPS on the Convex C-1 and the Alliant FX/8 in order to test them for a proposed procurement in Charlottesville; NRAO ultimately procured a C-1. Detailed technical reports on these projects are available.[5] The actual measured data will not be reproduced here; instead, summary conclusions will be tabulated:

- The VAX-8600 is 4-5× faster than the VAX-780 in CPU-time, but I/O-limited tasks are only slightly faster in real-time (because the two machines use the same model of disk drive).

- The Convex C-1 and Alliant FX/6** are about equal to the VAX-8600 for scalar execution.

** Alliant's "FX/8" model is composed of 1-8 separate, identical "computational elements" (CEs), each of which is a vector register computer with scalar speed about 4.5× greater than a VAX-780. A proprietary concurrency bus is used to synchronize the CEs for true parallel computation. In this paper the notations FX/6, FX/8 and FX/n refer to the FX/8 model with 6, 8, or n CEs; Alliant's "FX/1" model is a special configuration which uses only one CE and which has a different arrangement of cache memory from the FX/8. For its 1985 procurement decision, NRAO considered both the Convex C-1 and a 6-CE Alliant, here called an FX/6.

321

- The VAX-780+AP is 8-14× faster than the 780 for AIPS tasks; the C-1 and FX/6 are both about 3× faster than the 780+120B for AIPS "AP"-tasks.

- The C-1 disks appear to be about 3× faster than those on the 8600, due to the use of "disk striping" (writing alternate sectors to separate disks simultaneously, using independent disk controllers, to double or quadruple the effective disk transfer rate).

- NRAO's measurements in the autumn of 1985 showed that the FX/6 generally was faster than the C-1 in CPU-time, but the C-1 was faster than the FX/6 in real time.

- The X-MP uniprocessor is 5-10× faster than the C-1 and FX/6 for AIPS computing.

- The C-1 and FX/6 are both typically 25-30× faster than a VAX-780 for general scientific computing.

- The AIPS task MXCLN is a good vector-processing performance index. For this task, the 8600+AP is 1.4× faster than the 780+AP, and the C-1 is 2.2× faster than the 8600+AP. Note that while the AP is computing, the scalar capability of the 8600 (or 780) host is available for overlapped timesharing performance. This extra scalar capability is an advantage for the 8600+AP combination in comparison to the vector-register computers.

- The total AIPS benchmarking problem is a reasonable approximation of a typical AIPS job mix. For this case, the 8600+AP is 1.4× faster than the 780+AP, and the C-1 is 2.0× faster than the 8600+AP.

- For the total benchmarking problem, the Alliant FX/1 has 85% of the speed of the 8600+AP; consideration of the relative pricing of these systems suggests that the Alliant FX/1 offers one of the most attractive combinations of price and performance in the current market.

THE FUTURE OF VECTOR COMPUTING

The performance advantages of pipelining and general parallelism are so great that the only thing which prevents the general usage of these techniques is the lack of suitable compilers. Those vendors who provide vector hardware with compiler support will have a special price/performance advantage over vendors who offer only scalar systems. These facts are obvious to computer architects and to compiler designers, and a substantial number of new companies have been formed to exploit variations on the theme. For example, two vendors, SCS and Culler, have recently announced new vector systems in the same price performance range as Convex and Alliant; two more vendors are expected to announce new systems in this same range during the next two years.

Supermicro workstations for scientific/engineering calculations will also gain in performance from vectorization. For this reason, it seems obvious that such systems will soon appear. In general, all forms of parallelism will be exploited by astronomers for data processing and analysis during the next few years.

REFERENCES

1. Hockney, R.W., and Jesshope, C.R. 1981, "Parallel Computers—Architecture, Programming and Algorithms", Adam Hilgar Ltd., Bristol. The quotation is taken from page 10.

2. Kuck, D.J., and Wolfe, M. 1984, "Supercomputers simply need supercompilers that automatically restructure programs for concurrent processing", Physics Today, **37**, 67 (May 1984). This paper, which gave the "no" side of a debate on the question "Retire FORTRAN?", is a good introduction to the concept of automatic restructuring for vectorization and concurrency.

3. Wells, D.C. and Cotton, W.D. 1985, "Gridding Synthesis Data on Vector Machines", AIPS Memo No. 33 (30 January 1985). [send orders to: AIPS Group, National Radio Astronomy Observatory, Edgemont Road, Charlottesville, VA 22903-2475 USA]

4. Cornwell, T.J., and Dickey, J. 1986 (in preparation), the proceedings of a workshop on "The Use of Supercomputers in Observational Astronomy" which was held 4-6 November 1985 at Minneapolis, MN. Papers on various aspects of the processing and analysis of astronomical data using vector

computers were given during this workshop; subjects included pulsar searching, synthesis imaging with various radio interferometers, optical solar astronomy and speckle imaging.

5. AIPS benchmarking results are available for the following machines:
   VAX-780, 780+120B, VAX-8600 (AIPS Memo No. 36, "Certification and Benchmarking of AIPS on the VAX-8600", 24 June 1985)
   IBM-4341, Masscomp MC-500, ModComp+120B (AIPSLETTER, Vol. V, No. 3, 15 July 1985, p. 4, "Portability Column, CPU/OS Combinations")
   Alliant FX/1 & FX/n, Convex C-1, Cray X-MP (AIPS Memo No. 38, "Certification and Benchmarking of AIPS on the Convex C-1 and Alliant FX/8", 24 December 1985)
   VAX-8600+120B (AIPS Memo No. 44, "Benchmarking AIPS on a VAX-8600 with FPS-120B Array Processor", 19 April 1986)
   [send orders for copies of the AIPSLETTER and AIPS Memo Series to the address given in reference 3]

BIBLIOGRAPHY—FOR FURTHER READING

Reference [1] above is good on comparisons between the architectures of the Cray-1, Cyber 205 and AP-120B, especially on the concept of the vector half-length of a pipeline.

Ibbett, R.N. 1982, "The Architecture of High Performance Computers", Springer-Verlag, New York. Good historical source on hardware developments; good discussion of Cray-1 architecture in section 6.4; especially good discussion of the STAR-100 and Cyber 205 in sections 7.3 and 7.4.

Kuck, D.J., Lawrie, D.H., and Sameh, A.H. (eds.) 1977, "High Speed Computer and Algorithm Organization", Academic Press, New York. A *superb* reference source. Contains both facts and food for thought on a variety of issues, machines, applications, etc. See especially pp. 71-84, "An Evaluation of the Cray-1 Computer", and pp. 287-298, "A Large Mathematical Model Implementation on the STAR-100 Computers", and don't overlook pp. 3-12, "It's Really Not as Much Fun Building a Supercomputer as it is Simply Inventing One".

Kuhn, R.H., and Padua, D.A. (eds.) 1981, "Tutorial on Parallel Processing", IEEE Computer Society. See especially pp. 464-472, "Sorting on STAR", by H.S. Stone.

Metcalf, M. 1982, "Fortran Optimization", Academic Press, New York. Mostly concerned with scalar optimization strategies; somewhat weak on details in places. Note Chapter 10 ("Fortran Portability"). See especially the Hitachi Integrated Array Processor discussion in Chapter 11 ("Vector Processors"), and the brief discussion in Chapter 12 ("Future Fortran") of the array processing language extensions proposed for Fortran "8X" by the ANSI X3J3 committee.

Peterson, W.P. 1983, "Vector Fortran for Numerical Problems on CRAY-1", Comm. of the A.C.M., vol. 26, pp. 1008-1021. Contains an excellent discussion of vectorization strategies appropriate for vector register machines such as Convex and Cray, plus much food for thought about other architectures.

Rodrigue, G. (ed.) 1982, "Parallel Computations", Academic Press, New York. An excellent tutorial source. See especially pp. 129-151, "Swimming Upstream: Calculating Table Lookups and Piecewise Functions" by P.F. Dubois, for a sophisticated discussion of the uses of advanced vector operators.

PANEL DISCUSSIONS

PANEL DISCUSSION: DATA ANALYSIS TRENDS IN OPTICAL AND RADIO ASTRONOMY

Chaired by I. King

Participants: P. Moore,
E. Schreier,
G. Sedmak, and
D. Wells

KING: I was obviously chosen to conduct this panel because I know nothing about the subject, so that I can let the other people talk. What we're going to do is to divide the hour in thirds: each of the four panelists will talk for about five minutes, then for about twenty minutes we'll have further discussion among the panel, and then we'll throw the discussion open to the audience. We have met briefly and identified some questions that might be worth discussing, and allocated them somewhat among the panelists, who will now go ahead one by one.

SCHREIER: In the course of my previous career as an X-ray astronomer, and in the work I've been doing developing Space Telescope systems, it has become very clear that astronomy now is moving in the direction of what I've started calling the distributed data system, which has three main components. One of them is software that's portable, another one is a good usable archive system, and a third is a network that allows people to get at these facilities. Going into a little bit more detail in the area of portable software, systems are getting too big for everybody to reproduce them. We started out in individual observatories with small systems, and we've now hit the point where people are by necessity having to share data analysis systems.

The software that we do ourselves at Space Telescope is modular, has isolated I/O, etc. You've heard all of these words. We're trying to move toward Unix-based systems which are transportable to a lot of machines. As far as the coordination goes, we started off looking at community software. We then made our arrangements with Kitt Peak National Observatory to deal with IRAF-SDAS coordination. We started the SDAS-MIDAS coordination so that the European Coordinating Facility and the Space Telescope Science Institute would have interchangeable software. There's a recent initiative in trying to go toward standard imaging interfaces. This is the last major area in data analysis systems that has not had any good standards. As an example of how much people jump on this, we drafted a document containing some trial standards for imaging-device interfaces, and two groups have already implemented these. All we were expecting were comments back. It's clearly very important. You've heard about the FITS standard that NRAO took the initiative in, which is now more or less a world-wide standard in astronomy. Another example is optical disc formats, where again we tried to coordinate the development of standards of how we should use optical discs for image data. We're interested in work stations where the same portable software can be implemented.

In the archive area we've gotten interested in trying to find how we can get not only a catalogue but an actual image data archive on optical discs, and there are several projects going on in this area. We are, at the Institute, now seriously considering a project that you heard about earlier in another context, of putting the Guide Star Selection System data on optical discs. We are considering first the catalogue of twenty to thirty million stars up to magnitude 14 or 14 1/2, but then also the raw image data itself. This will take several years, clearly, but the idea of having such a data base on optical discs is now feasible.

Finally, a network that would allow people to share software, to access archives, and to do remote observing is extremely important. You heard that the benefits of this kind of network are going to be sociological as much as technical. Getting people to talk to each other, to do collaborations remotely as opposed to just going to meetings, to exchange drafts of papers, to be able to send mail messages -- all of these help immensely. Getting a little bit further than that, the possibilities are endless as far as really bringing everyone up to the same level of sophistication in the various techniques that are being developed around the world. Let me stop there, and we can come back to some of these items later.

SEDMAK: What I want to focus on is the user-system interface evolution and the impact that this evolution has on the realization of algorithms, which are used not only by astronomers but by scientists in general.

Starting from the archaeology, let me recall the times when data processing was done by means of cards, batch programs, and line-printer outputs. Then we had the introduction of the alphanumeric video terminal, of the graphic video terminal, of interactive graphics, and of image-display devices. All this may be seen as the basic first milestone. The second milestone, in my opinion, was the realization of software systems of increasing complexity, flexibility, and power -- say the VICAR, IHAP, MIDAS, and IRAF operating environments. The main features of these systems are the use of a structured command language, the availabilty of help files, the automatic logging of the operations carried out, and the use of standard I/O and data structures, including an archive. Another key feature of the present level of functionality is networking -- now, in particular, geographical networking. I am talking of course of what is common in the astronomical community; it might be different in the other scientific communities.

We have now reached the present time, and I would like to refer to the fine talk by Don Wells in order to introduce the third milestone: the availability of local networks of workstations consisting of a microcomputer communicating efficiently with a supercomputer. In particular I would like to stress the integrated video functionality of state-of-the-art workstations whose video memory is shared with the CPU, the widespread utilization of UNIX, and the availability of very useful new software tools like multiwindowing, icon-based facilities, and so on. The facilities offered by supercomputers were thorougly discussed by Don Wells, so I refer to his talk for this matter. The important point that I identify in this multicoloured field is the level of the complexity as seen by the end user. I predict a future user-system interface based on Artificial Intelligence tools, like EXPERTS, as one reasonable approach to an effective use of the resources available in the near future as well as in some mature systems presently running for astronomy.

All this leads me to a strong desire to isolate the algorithms, themselves system-independent, from the computer environment. This is because astronomers do generally use only proven and widely known algorithms. I felt from the fine talk of Friedman that many new algorithms, in particular in statistics, could be used advantageously by astronomers; but this does not happen, because these algorithms are usually hidden in complex, integrated software packages. So I believe that the isolation of the system from the end user by means of a suitable interface could help tremendously in isolating the computer side, which is not necessarily within the user's expertise, from the algorithms that the user wants to run on the system, which are not necessarily included in the computer man's expertise. This specialization should generate a higher global efficiency.

WELLS: I chose several topics to talk about. The first one has already been mentioned by both of the previous speakers, networks which encourage collaboration and communication among astronomers. Now, it seems to me that science is done by people and not machines, and enhancing the ability of people to work together is very important in enhancing the scientific output. Furthermore, astronomers are kind of a rare breed, and we're spread rather thinly in order to find places to support all of us; communication means that we can work together and collaborate even across oceans and from northern to southern hemispheres. Therefore, I believe long-distance as well as local-area communications improvement is critical for improving both the scientific output and the way astronomers feel about each other and about their profession. The way I would summarize this in a little saying here is, "From what I have seen, from all I have read, just watching how the primitive networks I have seen installed have worked out, it's clear to me that in networking, sociology is much more important than technology." We always talk about the technology, but in the end, it's what people do in the network that's really most interesting, and one of the big phenomena of networks is that it's hard to predict what they'll do with it. It's usually new, interesting things, and people soon insist they have to have the network. Incidentally, one phenomenon that you should know about is that when the subject of networking comes up and you're installing something new, or it's being proposed, someone asks you how much it will cost. Whenever that question comes up, the network is in trouble, because usually when you start off the network, the accountant will claim it costs too much. Once the network is in place, the users will claim they don't care what it costs; they have to have it. I say sociology is more important than technology.

Now, a practical subject associated with networking which is going to become more and more important with every year, and which is an issue of technology, is remote access to observing, computing, and astronomical archives, and to data-based management systems and other kinds of services distributed world-wide. We're going to do it, that's clear; the only question is, in which year and for what purposes are we going to do it? This is a technological question as compared to a sociological question, although I believe it's also true that remote access to services will change how people do astronomy. One thing I want to say, just a technological gazing in the crystal ball: very-high-bandwidth networking is coming. Optical fibers are being pulled across continents. The bandwidth is going up in the satellites which are being launched. It is at the moment a little bit worrisome to me, because I'm not quite sure that we know how to properly use that bandwidth. So I'm a little nervous that the major centers have got to start doing their homework on this, and we are; I'm hoping we'll be ready.

I was asked to talk about languages for new hardware, and I wasn't particularly thrilled about it, but I'll say what my opinions are. First off, I assumed the subject was about things like parallel computers. I'd prefer not to talk about things like Lisp or Prolog because I don't know much about them; I'll let other people talk about that. Regarding parallel computers, there's the dusty-deck problem, which we could also call the "Fortran forever problem," meaning that you can't change those old programs very easily. There's a corollary, which I'll call the "dusty-programmer problem," which means you can't change the habits of programmers very easily; and then there's the argument about whether the compiler will do it all for us because the parallelism is inherent in the code, or whether we have to help the compiler. The current practice is to help the compiler with some of the larger parallel aspects and let it do the details.

Now, one final thing that wasn't on our list of subjects but which I remembered that I'd like to talk about is optical character recognition. I claim it's nearly practical, and I claim that there is now coming to us a duty to begin to use it to digitize and capture our historic literature from before the age of computer typesetting and things like that, and to capture old catalogues, data files from the 19th century and the first part of this century. I believe that literature search is a part of data analysis. It's true for all astronomers preparing papers that they must be aware of the prior literature, and give proper credit to their predecessors. After all, if we do not credit our predecessors, what hope have we that our successors will credit us?

MOORE: I want to talk about two points; one point must be important because all three speakers have mentioned it so far, and the other point is one which I haven't heard anywhere amongst you so far. I'm a little surprised that there's been nothing said about it. Well, you guessed it of course; what I want to talk about first is networks, and I'd just like to gaze into the future a bit and say what we could conceivably do with them. There's been a lot of talk about using networks for transmitting data, mail messages, remote observing proposals, and quite a few things. There's been little talk about using them for passing around the final information that comes out of this, the final papers that people publish; and I think that's something we may want to think about. I guess André Heck talked a little about some of this earlier on, but let me just go through some of the things that you could consider doing with a network or with some kind of data base that you distribute to different sites. Clearly you can set up a much better indexing scheme than is available inside printed books nowadays. You can much more quickly browse through a complicated cross-reference index using a computer system than you can hope to do using your eyes, and following down long lists of fairly meaningless items. I think that's a fairly easy thing that you might think about doing. Another fairly easy thing of course is to put in things like abstracts and key words, so that when you've got the task of doing a complicated literature search you can use some kind of data-based searching algorithm and fish out the things you want, relatively quickly. Well, it needn't stop there, of course. I see no technical reasons why we don't extend it to the whole publishing scheme. You could include text very easily, you can fairly easily come up with a scheme for transmitting graphics around, so that the diagrams can be flashed up on your screen, and we already have a scheme for passing images around so that we can get out nice halftone prints, which now we have some difficulty publishing in journals but can be flashed around on networks in relatively short spaces of time. I just put forward the suggestion that this is something which may well come in the not-so-distant future with high-bandwidth networks. I'm not sure we want it, but it's something that may come, and perhaps we should think about it.

The other subject I wanted to talk about was image displays. Displays for two dimensions are in common use now. We all, pretty well, have access to them, and we're used to fooling around with lookup tables, color display, zooming, roaming, and all the nice things we enjoy doing with them. What I'm talking about is the fact that a lot of the data that we now observe is not two-dimensional. The VLA, for example, regularly observes in a three-dimensional mode and produces data cubes. There are two major problems with visualising these data cubes. The first one is obvious and that is, how do you represent more than two dimensions? There are various tricks you can play with polarising spectacles and fancy projectors, or two colors and stereoscopic views. There are tricks you can play by using motion. You can have a cube and you can use the motion cue to give you the artificial idea that the thing is really three-dimensional by rapidly rotating it and viewing it from different perspectives. You can use these various tricks to try to visualise three dimensions, but I've yet to see any of them which really give a good impression of what's going on. The first problem, as I say, is how do we really see more than two dimensions; and the second problem is probably even more difficult, and that is that a number of the axes that we observe now are not spatial axes. The third dimension of our cube is very often a frequency axis, or a velocity axis. How do we represent that in some way in this three-dimensional cube so that it gets the right neurons in the brain stimulated to think in the right sort of way? How do we represent this as a three-dimensional solid object, when its axes are X, Y and velocity, in a way helpful for trying to figure out the dynamics of a galaxy? We really need to start thinking about how, after collecting a huge amount of data, we display it and think about it.

KING: Thank you very much. We are now going to move into the mode where the panelists are able to comment on each other's remarks, or else add things that they hadn't yet said but have now been stimulated to.

SCHREIER: What Pat Moore just mentioned about publishing is an interesting point. The concept of electronic publishing has been discussed; and I think you should take it a little bit seriously, because the first half of it exists already. I think most of us probably either have at our institutions or else know institutions that have computerized libraries. You can do abstract or literature searches fairly easily. But as we move toward observations which involve extremely large volumes of digital data, you cannot publish the data that you are writing about. I'm sure many of you have been frustrated reading an article and saying you'd like to look at this, you don't believe that, or whatever. It used to be that the data was published. It's not the case anymore. If you're lucky you will find a plate, but often not at a level that supports quantitative analysis. Being able to use a network to access a digitized file of the published data is potentially very important. I think it's something that people ought to be taking seriously. I don't think we're there yet, but the technology certainly does exist, and I think as the sociology changes we should push it.

WELLS: I've got a very practical matter about this electronic publishing business; it's high time for the astronomical journals to admit that computer typesetting is here, and it's here to stay, and it's practical. I know of a certain prestigious journal in astronomy which, as I have heard it, has refused in recent years to accept typeset preprints, and I think that's deplorable. One of the last acts that I will make as chairman of the Astronomical Software Working Group in North America is to go and protest. I'm considering writing a letter to the American Astronomical Society to protest these policies if they do not get changed.

KING: I hope you'll get the endorsement of the whole group, in writing that letter.

WELLS: I'm glad to hear that. I've learned in recent months that one particular typesetting language in particular is sweeping the world without anybody particularly pushing it, and that's Donald Knuth's TeX system. In case you've been wondering how some of my view graphs got typeset, it was TeX, of course. I've noticed that some of the papers here that have been passed around were clearly done with TeX, and it's time for journals to admit it. In fact, you can see on the shelves in the library that the preprints are typeset quite frequently now, and I would just say the world-wide astronomical journals ought to agree on this. Now, there's a corollary, of course. If we have networking properly arranged, clearly the journal editor's mailbox is the place where the paper ought to arrive. That's obvious. Likewise, the refereeing process can be handled that way.

SEDMAK: I want to comment on the implications of the possibility of electronic publishing of papers, data sets, and so on. I believe that it is a very common experience for an astronomer to have a lot of difficulties when looking for the documentation of algorithms in the literature. I suppose that this is due to the fact that our astronomical journals do not publish the algorithm implementations, i.e., the software sources, by themselves. What normally gets published is results from observations processed by means of algorithms and procedures, but these are usually not documented in the same paper or journal. This introduces a true ambiguity in the results, which in my opinion is normally underestimated. Moreover, this implies that the effective library of algorithms known to astronomers remains quite small, simply because very few of them are documented and diffused in the community. So I believe that the time has come to give more room to the circulation of valuable, possibly refereed documentation on data treatment. I think that the possibility of networking could be exremely useful for this purpose, even up to the possibility of an electronic journal on astronomical data processing.

Another comment that I want to make follows from what Pat Moore said here. Let us start now thinking about very advanced visual presentation devices and techniques. Thinking to the past (and present also) we consider it normal to waste our astronomer time in entering data, texts, etc., by keyboard, while voice input devices have been available on the market for at least two years. This for the input side. Regarding the output, we do not consider presently the possibility of using display devices operating in three dimensions. Such devices could be realized by relatively simple holographic techniques and might be very useful when dealing with the analysis of three-dimensional data, as in radio astronomy. I believe that this point could be worthy of some discussion.

MOORE: Could I respond to a point that Giorgio Sedmak made in his first presentation about image displays? He was suggesting that we start thinking about image displays which share address space with the processing engine. He obviously suggested that as a much more efficient way, a much more conceptually simple way of feeding data in and out of an image display. I would like to point out another very very clear advantage of such a scheme, and that is you have in effect an image debugger. You can look through your program and you can actually map an image display over a running program, and you can really see what that program is doing internally. That's possibly a very very powerful technique for debugging some obscure image-processing algorithms which we're hearing about more and more now. You'd clearly need to extend the technology a little. You'd need to handle the conversion of floating-point-format numbers to colors or intensities, but I think there's potentially a great deal of scope there for enhancing the understanding of the way some algorithms work. Certainly in some of the alogrithms that I talked about earlier, I hope I gave a flavor of some of their uncertainties, and that the more we pursue them the more uncertainties we discover in them!

WELLS: I want to speak to that point. In order to integrate an image display tightly into the heart of a computer, which is what you're talking about doing, you've got to talk to computer architects. In fact, I've tried to do this, because we've been in the midst of this problem. We were buried in cubes. You've got to understand that at radio-synthesis observatories like the VLA and the Westerbork telescope, the cube problem obsesses us. Naturally, whenever I get a chance to talk to a computer architect this is one of the subjects that come up. The first architect I tried wouldn't talk about it. The second one did suggest the idea of actually attaching a Sun work station directly to a high-bandwidth bus and driving images out at high speed right into the workstation, out of the main memory. He was willing to talk about it. He even brought up the idea in its full form. It will take a while more to persuade them, I think. Now, you should realize there are applications other than astronomy with a lot more money that have very similar needs. A particular example that had a lot of money until the price went down recently was the seismic petroleum prospecting industry.

KING: There's one topic I'd like to bring up now myself. We've heard a lot about these marvelous large systems, mainly from the representatives of the large rich institutions; but I have some concern for the small institutes, the people who can't afford all these things. We're being told that astronomers need to have large computer facilities for all their work, but what happens to the small institute in the face of this?

SCHREIER: Well, I think a large part of the whole concept of portable data analysis that we've been talking about has been to deal with exactly that issue. A data analysis system now costs twenty, thirty, or forty man-years to develop, a good system that's wrung out and tested. Clearly that cannot be repeated many times. Thus, to the extent that you can, you make it portable. And when we talk about portable we don't just mean to another major computer system or a Cray. We're talking about a twenty-five-thousand-dollar workstation that can have most of the same capabilities. That's an extremely important concept, that is very much in our minds as far as the kind of systems we've been developing for our Space Telescope. We have been trying to come up with a sort of canonical work station in that class, twenty-five or fifty thousand dollars, that essentially any astronomy department should be able to afford.

MOORE: I think the effect you're describing, actually, is one where the big centers just lead the smaller people by a small fraction of time. I think in a few years time even the small institutions are going to be buying computers of comparable power to the ones at the big institutions that we're using today.

WELLS: Yes, the Microvax that you have is comparable to the state-of-the-art machines in the mid-70's.

SCHREIER: Even if it is cheap.

SEDMAK: I have some local and national experience of small institutes facing the problems which are normally referred to as requiring big computers. My opinion, again, is that we must not forget the difference between the user interface and the algorithm. What is really costly is the realization of the system, that is, the user-computer interface, the graphics and image display subsystems, and the system-dependent software. A lot of effort is required which is related not to the science but to the implementation of your algorithms on the available hardware. So I think that we can and must standardize the system up to the user-system interface, leaving the user free to implement in the easiest way his data processing procedure. I believe that international cooperation may be extremely useful just in the realization of such systems, like MIDAS, IRAF, etc., which after all can then be run on even small computers like PC's. This should result in stimulating the work of research fellows towards what is really important on the science side, the algorithms and procedures. I hope that others will comment on this.

KING: Now let's let everybody into the discussion. I think it's legitimate either to make comments of your own or to direct questions at any of the panelists.

SHAMES: I think that one comment has to be made here about networks and workstations in software interfaces. One of the most essential things is to arrive at a set of standards that are functional, and one of the real tricks here and one of the real dichotomies is doing that in such a way that you don't stifle creativity. I wonder if anybody wants to comment on that further. [Chairman: apparently no one did.]

KING: I would like to hear some discussion about what to do about the macho programmer, because he exists everywhere, probably in every one of our institutions. How do you control him or her?

SEDMAK: Use him for algorithms.

KING: But the macho programmer will not document the algorithm.

SEDMAK: Right, but remember that the algorithm may consist of fifty lines of code and that you document it after he wrote them. There is a point hidden here. In this panel, no comment has been made until now about the implications of new languages on the development of algorithms. I would like to focus your attention on an example, the implementation of a recursive algorithm in FORTRAN (difficult) or in PASCAL (easy). Now imagine using other languages that are more and more powerful. We obtain nearly automatically a separation of the opportunities of the so-called "macho programmer" and the normal users, simply because the programmers know of the existence of such facilities, where to find the documentation about them, and so on. The astronomer does not normally know. So I think that this point should be examined with respect to the programmers on one side, and to the astronomers on the other side. The astronomers need documentation of the code *and* the related know-how; otherwise no real progress can be envisaged.

WELLS: When I introduced the "macho programmer" term, what I was trying to do was to make sure that all of you who may not have been around these big projects understood the kind of organizational problems you get into in the very large software projects, and that it was necessary to have some cooperation and some kind of team spirit in such projects. Pure individualism tends to undermine the big projects. Now, the other side, maybe many of you thought that I didn't like individualistic programs. Well, when I see programmers who are writing programs all on their own and who are allowed to do what they want, I have a terrible feeling of envy. Once I was allowed to do that, or felt it was possible to do it, and I wish I could go back to those years. In fact, one of the great dreams of computing has been to make the man-machine interface and the programming techniques so flexible and so efficient that people could make computers do what they want with hardly any effort at all. At that point, there's no real reason to feel bad about macho programming.

KING: I don't feel bad about the macho programmer, but I feel bad about the programmer who doesn't document things. I can remember receiving something a centimeter thick which had not a single comment in it. I threw it away.

WELLS: Of course, Ivan, one of the goals of programming-language design would be to make the program tell the story without other documentation.

SCHREIER: I wanted to take Giorgio's comment one step further. I think that you can make the same statement about the influence of new software languages on documentation as well, not just on algorithm development. In fact you find that when you give even macho programmers good tools that make it very easy to document, to record, to do the revision control and the software tracking, they use these tools. We have many programmers at Space Telescope; amongst those are some macho programmers who fit in remarkably well in the environment, and who have produced documented software. At the beginning it was hard, but after some time, if you set up the right tools you find that even the macho programmers have egos, they want to leave a record of their accomplishments.

WELLS: Let me inject one more remark at that point, just to show us the kind of innovation in documentation technique that is feasible today. When I referred to the typesetting system, Knuth's TeX, most of us see TeX as a program and we say "RUN TEX." Actually, if you look and see the text that's inside TeX, how it's built, what you find out was that Knuth not only innovated in typesetting, he innovated the programming style in that project, because in fact the documentation is interleaved with the code. He has programs to take it apart and produce it in program form or in book form to describe it, and his program actually is presented to the world as a book with embedded text in the code, and the programs that he does this with he calls TANGLE and WEAVE, suggesting the classic English phrase:

> O what a tangled web we weave,
> when first we practise to deceive.

KING: I like your poetry, but I wish I could share your enthusiasm about the book. I find it impossible to use for reference.

WELLS: No, I'm speaking about the documentation of the program, not what you've seen.

KING: Okay, but we badly need a good reference book for TeX.

MOORE: Can I just make another point about the effect of programming languages on macho programmers? I think not only does it help automatically generate documentation, but I think another advantage of getting a suitable programming language is that it takes away the incentive to write in a macho style. Macho programmers tend to like writing in assembly language or rubbing their nose in the guts of the machine. If the language has all the tools to handle things you want to do easily and efficiently, then I think the temptation to program in that style becomes much much less.

CRANE: I think most of what you have talked about have been things which have not necessarily related to advances in doing astronomy. I think one of the important things that we need to really do something about is the area of algorithm development and algorithm comparison, efficiency, and these types of questions as opposed to the systems. We've pretty well settled down on what we think systems are all about, and what we think about graphics and interfaces, and in all of these things we seem to come to some not universal but quasi-universal feeling. My feeling of where we ought to be going in data analysis in both radio and optical astronomy would be much more towards the development of algorithms -- efficient algorithms -- and comparing algorithms amongst each other. I wonder if the panel members might have a comment on that?

WELLS: I was just going to propose a comparison with the world of mathematics which came to my mind as you were speaking. Mathematicians are fond of posing challenge problems in their journals. It's really fun to read their journals and see some of these challenge problems, and then to see people come back with attempts to answer them. In terms of data-analysis algorithms, we know very well there are some challenging problems in analyzing particular kinds of astronomical data. It's a fact, and it's alluded to commonly in these lectures. One of the things I have often found about comparing different people's problem solutions, like photometry programs or spectroscopy reduction programs, is that you never see these programs run on the same pieces of data. I believe that it is important that certain pieces of data be analyzed by multiple programs, so that algorithm performance can be compared accurately. I would like to see test data sets published with new algorithms, so that later algorithms can be compared to earlier ones. I would also like to see well-chosen data sets being posed as challenge problems. I believe that these practices will tend to stimulate innovation, and I think we could profit by something like that.

MOORE: Let me just emphasize that. I think the way to categorize algorithms and compare them is not to run them on real data; it is to generate an artificial data set which is as realistic as you can make it. We heard a good example of this this morning for the X-ray satellites. It's been done in a number of different fields; you generate a picture of the sky, you distort in the way you believe your instrument does, then ideally you give that data set to somebody else, without telling them what it looks like, and say, "Tell me what's there." I think something like that is about the only way you can realistically compare algorithms.

SAIDIO: Does any of the panelists want to comment on the possibility of working toward what we might call a general astronomical data base, in which we could collect optical, radio, X-ray, and whatever you want in a unified format?

HECK: Well, as far as I remember there were discussions of this kind within the meeting of experts set up by the European Space Agency. That was not the whole of astronomy, but it was various fields of astronomy. I think a right approach was to leave a given field to specialists in that field, because the problems of setting up the best radio data base are not the same as the problems of keeping and seting up the best UV data base. That's what I think has now been definitely adopted for the eighties project: UV data will stay with UV astronomers in Vilspa, X-ray data will be maintained by X-ray astronomers in Estec, and basic astronomical data will be managed by CDS with cooperation of experts in the field. I think, to say it briefly, it is dangerous to be too ambitious.

TANIMOTO: I think there might be a middle ground, between having a network of totally heterogeneous data bases and one sort of general data base, and that might be to have some reasonably smart programs that a user could run to essentially construct for him a customized data base by knowing where the right kinds of files are, maybe having the catalogue locally resident, and then pulling these things together and perhaps actually doing some mapping on some of the images to bring these into registration or to make them otherwise compatible. Has anyone had some thoughts along these lines?

SCHREIER: There is an effort in the United States that NASA has started, to try to do just this kind of thing. It resulted out of some CODMAC recomendations a couple of years ago (a committee on data management and computation). There's an effort to try to set up a hierarchy of archives, essentially. There would be a central directory of directories which Goddard Space Flight Center would set up. There would then be pointers to where the various archives live. It's actually recognized that not every archive should be in exactly the same format. Different disciplines in astronomy will have different needs, but there have to be common protocols on getting to them, getting to the catalogues, getting to the archives -- not only having the data in the archive, but the description of the data, and how it's organized. In fact it's gone further, and I wanted to mention this before: the software that's useful for analysing that data should be stored in the archive with the data. So, the idea of a network that makes these kinds of capabilities possible is very strong; if I'd gone into a network discussion more, I would have gotten to that. There are also some fairly sophisticated projects going on as far as setting up catalogues of catalogues, distributed data bases that even present the same user interface.

SCARSI: I would like to see if it is possible to stir up some comment on data-analysis trends, which is the title of the panel. Up to now we have heard about more powerful tools, more sophisticated tools, better management, and so on. Is there any comment about new methodologies of data handling? Let's see if there are any things that we can talk about there.

SEDMAK: In my opinion the methods may turn out to be quite complex if we want to discuss new methodologies of processing, analysing, archiving, etc., applied to information rather than data. We must separate the relevant subsets within information theory, and separate applications based on information theory from the technology which supports this information processing. Of course, we can consider some parts of general information theory, identify what is of interest to astronomers, and then detail the implementation. However, I think it is nearly impossible to give a general answer to the question asked by Livio Scarsi. I believe, however, that it could be very useful to make an effort to identify the subset of interest to astronomy, and I ask all the audience and panelists to comment on this point, because the question posed was a very good one.

SCHREIER: I had a couple of small comments on that. One way of looking at it is first things first. My thesis in this has been that it would be very nice to start freeing up those tens and twenties and perhaps hundreds of software people, who are working in various astronomy institutions, setting up somewhat duplicate architectures and structures for data analysis systems, and let them start working on the algorithms. Let them not have to worry about reinventing the structures. The second idea I should point out is that I think the conference was organized this time a little bit more in the direction of major problems and the trends. It was not really concentrating on algorithms. I think we've only heard one significant topic that was not on the program. It has turned out that several people are using artificial intelligence and rule-based systems in data analysis; we can call that algorithms if we want, extending the term a little bit. I think that's important, and I think that by two years from now there will be an awful lot more of that, and I think perhaps we ought to start planning now a little bit on how to get the proper technologists in, to deal with those kinds of problems.

KING: I don't think that one simply sits down and says, I am going to develop new algorithms; nor does one make a general search of the literature. I think that historically, these things have come from some single astronomer being faced with a problem and finding a solution for it, and from his colleagues seeing that this was a good way of doing things. We have an American expression, "when a man builds a better mousetrap, the world will beat a path to his door," and I think that's where our new algorithms may come from.

WELLS: One remark I'd make about algorithm development for things like image processing, which I think is one of our more challenging areas in astronomy, is that those of us who are associated with algorithms, with hardware/software systems for handling large masses of data, need to pay attention to the literature in allied fields. I personally skim the journals very frequently; I often stumble over wonderful-looking papers with clever ideas, things like the IEEE Transactions on pattern recognition. Of course I almost never have time to follow up these ideas and see how they would apply in astronomy, but I would just simply encourage people to take a good hard look at some of the graphics, image-processing, artificial-intelligence applications which are covered in journals. Many of these are problems not very different from ours, and much better funded sometimes.

MURTAGH: A number of speakers have alluded to algorithms, new algorithmic developments in this area, and I'd just like to mention that about nine months ago a working group was put together to try to focus some efforts in the area of statistical and pattern-recognition algorithms, and not really image processing, because I think that's fairly well looked after already. André Heck and myself have had one newsletter produced before Christmas of last year, and a second one is due shortly. I would like to let people know that this newsletter is currently circulated to about sixty or seventy individuals. There has been a lot of interest in it, and it's a possible forum for discussion of algorithms in the area that we have focused on, and might in some way help to push forward this area and perhaps also give people a forum for communicating ideas.

KING: You say you're distributing it to individuals; what about institutions and libraries?

MURTAGH: Initially the idea was simply to get people who were particularly interested in the area, in other words it wasn't going to be a matter of a mass-circulation journal or the sort. However, one recent development was that in order not to have too much of a burden of administration, what we are doing now is to use the CDS bulletin and incorporate the newsletter in it. Therefore people who already get the CDS bulletin will find in the next issue details about the newsletter, and those who do not get the CDS bulletin can send in their names and get on the mailing list in the future.

CRANE: Fionn, you might also mention something which I found extremely useful. This was a bibliography that you put together of references to clustering algorithms in astronomy. I don't know what you're planning to do with that, but I found that just the list of references and the comments that you made to go with them was extremely useful. How do you plan to distribute that?

MURTAGH: Yes, this bibliography of applications in astronomy of methods in this area has been put together with some reccomendable reading material in the general non-astronomical literature. Basically the answer to Phil's question is that it is intended to publish it, but there are no precise details yet. Perhaps since this has been brought up, if anybody wants a copy of this as it is at present, I'd certainly be more than willing to give them it, and I'd also very much welcome any feedback as to what sort of areas have been in particular not given sufficient attention.

WELLS: I want to give a dissertation on a subject that I think goes right along with all this. It suddenly dawned on me that maybe a lot of you have no idea why it is that all of us lined up in a row here say we want networking. Maybe many of you don't realize what it is we really mean by that, and why it's so important. What's the sociology that makes it so incredibly effective? I want to explain what those of us from the U.S. mean about networking. We're talking about TCP/IP, which is a protocol of the ARPAnet. What's important about that? Well, the ARPAnet spawned a whole scheme for interpersonal communication, for the kind of problems we're discussing here, and for exchanging information between people who are widely dispersed. They invented this technology ten years ago, and they've been using it ever since, which is one of the reasons why the computer science field is advancing very rapidly these days. In fact, whole major research projects have been conducted over networks by subgroups, special interest groups. I know a major book that was written by a group over a network. Now, what do they mean? They're talking about bulletin boards, basically. It's the bulletin board technology, where people post comments, post documents, and responses are made; the entire community sees the whole transaction, because it's digital files you can search on strings and make retrievals from. In fact you could argue for a bulletin board on astronomical algorithms where people would pose problems or ask questions about availability, actually insert code -- fragments of code! -- into the files. Certain people might be assigned to be the editors of the bulletin board and might produce periodic compendiums of information and cross-reference indices. An analogous thing you can recall from the computer-science field twenty years ago was the famous collected algorithms of the Association for Computing Machinery, which ultimately became books. These were often challenge problems; they were also algorithms presented and then referees presenting referee reports which were published. These kinds of activities are the way to make real progress efficiently.

MOORE: Could I just bring together two things that Don Wells has mentioned? Firstly, although I thoroughly agree with his points about bulletin boards, I'd also like to bring in another point that he made earlier, and that is that we don't just want a bulletin board for astronomy. One of the main purposes of a bulletin board is that everyone can see it, and a lot of our problems are common problems to people in other fields in research. So, I think we don't just need an astronomy bulletin board, I think we need something much much wider than that.

KING: I think we have time for just a brief comment from each of the panelists and perhaps one more from the audience.

SCHREIER: I was going to say earlier what Don just said. When the question came up before about standard data sets and trying them, the exact thing that went through my mind was that with a network, you will have standard data sets available; in fact you'll have the different algorithms available, and you can compare them.

SEDMAK: I would like to add a quick comment. Please do not forget that the availability of a network is one thing, but the ability to use the software that could be exchanged through the network is based on the availability of a system which must be a standard one. So I believe that our thinking about networking must go parallel to standard systems like MIDAS and IRAF. Otherwise we will find out what exists, but in general we will not be able to use it for our purposes. So please do not forget the standard systems.

KING: On behalf of everyone present, I would like to thank the panel very much for sharing their expertise with us.

[Final comment added by chairman: The communications exchanged in editing this discussion have clearly highlighted the need for better networking.]

PANEL DISCUSSION ON

DATA ANALYSIS TRENDS IN X-RAY AND GAMMA-RAY ASTRONOMY

Date: 22 April 1986

Participants: Hans-Ulrich Zimmermann (MPE, Garching), Chairman
Rosolino Buccheri (IFCAI/CNR, Palermo)
Jean-Marc Chassery (TIM 3 Cermo, St. Martin)
Roland Diehl (MPE, Garching)
Livio Scarsi (IFCAI/CNR, Palermo)
Ethan Schreier (STScI, Baltimore)
Wolfgang Voges (MPE, Garching)

ZIMMERMANN: Let me begin the panel discussion on Data Analysis Trends in X-ray and Gamma-ray astronomy by first introducing the panel members. These are from the left to the right: Roland Diehl, Ethan Schreier, Rosolino Buccheri, Livio Scarsi, Jean-Marc Chassery and Wolfgang Voges. With the exception of Jean-Marc Chassery, who is an expert on image processing and statistical analysis, all the others have longer experience in the X-ray and Gamma-ray fields. May I ask the speakers to keep their contributions to not more than five minutes, in order to allow ample room for discussion with the audience.

X-ray and Gamma-ray astronomy are relatively young members of the astronomical zoo. It appears therefore natural that data analysis here is still in a very active state and follows many trends. We still have to learn for example, which methods from different branches of astronomy can be applied successfully in our field. Also we are often faced with quite different kinds of instrumentation and observation techniques that enforce us to search for new and feasible solutions in the data handling area.

In order to understand these difficulties better, I propose that we begin first with a short introduction about the basic problems in the instrumentation and what we can get at the end of the data collection process. So may I ask Ethan Schreier to give us his view on these specific items for the X-ray field.

SCHREIER: This will be a five minute tutorial on X-ray astronomy. I will say first what X-rays are, and what the implications are of how X-rays behave as far as data analysis is concerned. Then I will go a little to the instrumentation requirements and the particular problems of having to deal with satellite based observatories. In this area, X-ray astronomy has been somewhat ahead of some other disciplines in having used rockets, then satellites right from the start. Finally, I would like to say a few words about the special computing needs. This is clearly not in strict logical order. There is not a simple mapping of the computing needs from the other areas, because what X-rays are, affects the instrumentation and the requirements. The uniqueness is in quotes because the further we go, the less unique the discipline is, the more commonalities you may find. We will come back to that tomorrow.

So, what are X-rays? X-ray astronomers tend to talk in terms of electron volts or kilo electron volts. The general range from about a hundred electron volts to a hundred thousand electron volts is what would normally be regarded as X-ray astronomy. Below that is the extreme ultraviolet, and above are the Gamma-rays. There is no sharp distinction. For the optical astronomers among you that corresponds to say a hundred Angstroms to a tenth of an Angstrom. Within this, there are two more divisions. There is what is usually called soft X-rays, which is typically a quarter keV to a few keV, and then the hard X-rays above that.

X-rays arise from very energetic processes. A typical source is very hot gas which occurs in two extreme environments. One is in very large clusters of galaxies in the universe. Essentially the only way this gas is observed is via X-rays. The other extreme is very energetic processes in very small regions, like accretion flows onto neutron stars, white dwarfs or black holes, or in the nuclei of active galaxies where it is not clear whether it is really a thermal process or not. There are shocks and jets near active nuclei. We see the X-rays from jets. It turns out it is consistent with synchrotron radiation, but it could also be consistent with thermal shocks that are needed to accelerate the electrons, that then emit in the radio band. Also in Supernova remnants there are clearly shock phenomena going on.

X-rays come from typically one of three processes: bremsstrahlung, synchrotron radiation, or inverse compton scattering. The first is by far the most common. The implications of this will lead right into the computing needs.

One first implication is that there are relatively few photons. You are high up in the spectrum. Each photon carries a lot more energy than an optical photon, and therefore as a general rule, you will find fewer photons. Secondly, they are hard to image. As you know, X-rays do not reflect very well. They tend to enter the surface. Therefore you end up with photon counting devices as opposed to large integration experiments. You tend to observe photons one at a time. You measure their position, their energy, their time of arrival, and you have lists of photons as opposed to true images. You construct the images later. The telescopes are constructed in a strange way. You saw the other day, when Ulrich Zimmermann gave his presentation on the ROSAT grazing incidence telescope, if the angle of incidence is more than about one degree to the surface, the X-rays get scattered greatly. You also have very strong energy responses. The reflection is a strong function of energy. The detection is a strong function of energy. It introduces problems in the analysis which I shall come back to. The calibrations as a result of several of these effects are very difficult to do.

A very important fact is that X-rays do not reach the ground. You need satellites. This introduces a whole other series of problems. Background is very variable around the earth orbit. The magnetic field varies. There are trapped particle belts above the earth. There is a day-night effect as far as the upper atmosphere and the near reaches of space go. You look at earth glow. A lot of mission planning is necessary in a satellite observatory. You must schedule efficiently. All the background effects, day-night, occultation of a source by the earth have to be taken into account. Finally, you are doing remote observing all the time. You cannot go and tune your instrument very easily. You have to have command links. You cannot recalibrate your instrument in orbit easily without taking a large amount of time, so you have special calibration requirements on the ground.

I tend to divide the data analysis or the computing requirements in about three categories. The first category I would call functional simulations to design instruments. Since these are very complicated instruments, you have to design mirrors and detectors with strong simulations. You are developing much more sophisticated detectors than have appeared in optical astronomy until very recently. As a result you are always using computers to control the experiments, to test them before they get integrated into a satellite and launched. There is a fair amount of computing involved in the mission planning. The experiments are very expensive and you want to use them efficiently. Finally, you are operating a remote facility and there is a fair amount of monitoring the instruments from the earth. Then there is the quasi-functional category, which is things like pipeline data processing. Again this is not really unique to X-ray astronomy, but it occurred there first, largely to an extent of some technical ideas like having to solve for the position of pointing aspect analysis, the jitter. And then you take these streams of photons coming in and you have to build up an image, it is not like having a plate right away. Finally, because these are large systems, there are not many facilities in the world, perhaps one at a time. You have to set up a user facility and in some way pipeline at least the first parts of the data processing.

ZIMMERMANN: Thank you, Ethan. We might come back to these kinds of problems later in the discussion. I would now propose going on to see what the specific problems with Gamma-rays are. Roland Diehl, could you please give your view on this field.

DIEHL: Typical Gamma-ray sources involve nuclear physics. Particle interactions and particle field interactions are the physical processes which generate gamma rays. Excited atomic nuclei which may be in the interstellar matter, or dust clouds interacting with the cosmic rays generate Gamma-rays. The accretion phenomenon around compact objects like neutron stars or black holes also makes up Gamma-ray sources. The typical source fluxes are in the order of ten to the minus four photons per square centimeter, a typical number in the MeV range. This means that large collection areas and long observation times are necessary; that is why all the Gamma-ray observatory instruments point to the same region of the sky for two weeks or so. Typical spectra are exponential, i.e. the intensity falls off quite rapidly with the energy. For the higher energies you really talk about only a few photons per hour or per day.

The observational status: about thirty sources have been detected so far, which is the Galaxy itself or nearby molecular clouds. The COS-B catalogue contains about twenty two galactic point sources, we have about four extragalactic sources and two others have been observed. The identification of Gamma-ray sources has to take into account additional features like the time structure or some correlation with optical objects, if possible. Essentially, we explore a new field of astronomy, which means we do not extract particular source characteristics, but we try to image the sky with a large field of view of the instruments. We need good resolution in order to identify the objects, and of course we have problems in background rejection.

The instrumentation of Gamma-ray astronomy is quite different from what you use in the optical and even up to the lower energy X-ray range. You detect the photons via photon particle interactions, and the observation is a set of detector events. You do photon event counting. Our detectors are a set of elementary particle physics detector modules, so they are not really telescopes, they are not focussing, they are detecting one event which is triggered by incoming photons. We measure, in general, many event parameters. Typical instruments are collimated scintillation detectors, where a detector unit is surrounded by some kind of anticoincidence detector unit. You read the signals from both these units and you make a logical combination between them. Another type of detector is based on spark chamber stacks, where the photon converts into an electron positron pair, and then you detect the tracks of the electron or the positron in a stack of spark chambers. Another quite popular instrument is the coded aperture position sensitive detector unit, where you also have two layers of detectors; the upper one is sometimes passive or active; the lower one detects the interaction of a photon and also the position of this interaction. So you measure again many parameters: energy, position, and, in addition, the flags of the anticoincidence. Another type is the Compton telescope which I mentioned before. There are two layers of detectors, and you detect in both layers, both energy and position of the interaction. In summary, you have non-focussing instruments, not a typical telescope, and the ground analysis of your data finally gives you incidence direction and incidence energy of the photon in a probabilistic way.

The problems in Gamma-ray astronomy are the following. We have in general overwhelming instrumental background, and this background again is a function of time when you have a satellite instrument. Usually the instrument gets activated and this leads to varying background which you have to understand primarily before you can form an image. I mentioned already the low source fluxes. Instrumental evolution, when you look at high energy physics or at the accelerator physics, always makes three different steps. The first step is you start with a simple instrument because you want to explore the field, you want to get started, you want to see what is out there in the scientific area. Then, your next step is you want to understand the source and the background parameters in detail. So you build a very complex detector. A third step then is: you have an intelligent front end in your detector system. In high energy physics for example, you have the very fast Camac based computers which make the decision at the front end, whether you have a useful event or not. I think Gamma-ray astronomy at the moment is at the second step.

The data analysis tasks are different, because you have a different type of instrumentation. Your input is a set of event parameters, and the events are measured with many parameters. Your tasks, in general, fall in three different categories. First, you have to connect the pre-launch calibration and your pre-launch simulation. Because your calibration is not complete, you have to connect those two to your actual instrument performance. Then you have to explore your features in the data, which means that you have to deal with a multi-dimensional data space, and you have to find features in that space. Your last step is then to apply an astrophysical model, plus your instrumental response knowledge, plus your background knowledge. You try to combine that and do essentially some kind of maximum likelihood fit to extract the scientifically relevant parameters.

ZIMMERMANN: Thank you, Roland. I think we got a quite detailed view on the specific instrument related problems in X- and Gamma-ray astronomy. In order to conclude the introductory remarks on this field, I would like to ask Livio Scarsi to give us his ideas on what specific roles X- and Gamma-ray astronomy play amongst the other branches of astronomy.

SCARSI: Yes, we have heard about two specific windows in astronomy: X- and Gamma-ray. But I must say that it is difficult, if not strange, to imagine that the activity of an object lying in the sky is limited to a particular window of conventional astronomy, let us say radio, optical, UV, X or Gamma.

It is obvious that broad band astronomy spanning from radio to Gamma is the reality, and information has to be transferred and correlated from one specific window to another.

The type of information that we are looking for to study a source is essentially: space information (where a source is located, its morphology, ...), time information (how the emission is depending on time) and energy distribution of the photons, i.e. spectroscopy.

In wanting to correlate windows as far apart as radio and Gamma, for example, we have to face a very different scenario about the forms in which the information is present or at least can be extracted.

First, let us look at the data collection rate. Typically, in the radio field you have data in a continuous string, while in the Gamma- and X-ray field, you do photon counting. This occurs also within the optical window itself, when you go from one extreme to the other following the visibility of the object. In this different way of treating data, for example in the photon counting case, statistics has to be called in as a very important tool in dealing with the "granularity" of the information.

A second point in which we find a lot of differences is the morphology of the object. We have to correlate for example images taken on the long wavelength side with angular resolution of a fraction of an arc second to Gamma-astronomy, where fraction of a degree point spread functions dominate the scene and where the image is a probability distribution. The correlation is vital if you want to make a source identification. Furthermore, the necessity arises of obtaining coincidental relationships connecting space to time dependence in the photon emission at different wavelengths for the same hypothetical emitting object.

And here to make time correlations at different wavelenghts introduces the problem of simultaneous observations with very different instrumentations in very different environments, with the difficulty of scheduling observing time slots not only contemporary but usually with widely different requests on exposure time.

ZIMMERMANN: Thank you, Livio. I think we are now well prepared to begin our major topic, the analysis methods. What kinds of methods are presently used in X-ray and Gamma-ray astronomy, and what can be learned from methods that are common in other branches of astronomy? We should now have an open discussion, and I would appreciate it very much if after each of the contributions the audience would make some comments, or put questions to our experts. Who would like to begin?

BUCCHERI: I will give a short overview of the problems encountered in X-ray and Gamma-ray astronomy and concerning data analysis. It is certainly not complete but it may serve as a starting point for a discussion.

Let us start with an item, already discussed somewhat during the last workshop, namely the discrimination of background in spark-chamber pictures in Gamma-ray astronomy. In the case of the COS-B instrument the problem was approached using a parametric pattern recognition method of analysis based on the knowledge of the physical process giving rise to the spark-chamber tracks. Non parametric methods, alternatively or in addition, could also be used. The problem here is that the shape of the pictures may change with time due to the degradation of the spark chamber gas, such that the parametric methods, fixed at the beginning of the experiment, could not represent the reality after some time. The second point is the derivation of maps for the study of the radiation from the sky. All of the methods used presently for mapping such as Cleaning, Maximum Entropy, Bayesian convolution have their problems in representing the real structures of the radiation emitted by the galaxy or in interpreting what is mapped. The most important problems are related to a good knowledge of the Point Spread Function of the experiment and to the estimate of the statistical significance of the structures observed. Up to now, most of the efforts have been directed to the display of the maps, in particular the use of false colours to better visualize the structures. Unfortunately measurements are always subject to the laws of statistics and therefore some of the visualized structures could just represent chance fluctuations of the emitted radiation, not necessarily localized radiation sources. In Gamma-ray astronomy, where the counting statistics is quite low, the estimate of the statistical significance of the detected structures is of vital importance because it is related to the detection of sources of radiation.

In general, when analysing Gamma-ray data, we are faced with two different topics: exploratory data analysis when the data are investigated for detection of previously unknown structures, and confirmatory data analysis for measurement of the physical parameters of previously known sources. Testing and estimation theory, as discussed by Jerry Friedman during this workshop, can provide the necessary tools for the analysis. The application of these theories to Gamma-ray astronomy need, however, a clear knowledge of the statistical distributions to which our data obey. In fact, due to the low counting rate typical at high energies, and also to the particular experimental conditions such as discontinuity in the data collection, periodicities intrinsic of the experiment, the data do not follow the theoretical probability distributions which must be found for each case in relation to the particular problem at hand.

Let me mention now some methods, that are related to different topics, all of which have problems as soon as they are applied to low counting statistics data.

First problem: the search for periodicities using (fast) Fourier analysis. Fast Fourier Techniques are widely used in astronomy but are certainly not advisable in Gamma-ray astronomy again due to low counting statistics and to the duration of the experiments which may be of the order of a few weeks. They would be unviable because exceedingly high computer time would be necessary. Also the normally good time resolution of the experiment would not be fully exploited.

Another problem: the study of the structures of the light curves of pulsars using histograms. Histograms do not lead to the best estimates of the parameters of the light curve, again the time resolution is degraded, so people are starting to suggest other ways of estimating parameters: kernel estimators, clustering techniques.

Clustering techniques in the presence of noise: the classical Minimum Spanning Tree, when applied in the presence of noise background, is not so simple as it is in absence.

Last item: simulations. Always extremely important, as they are needed, for example, for calibration purposes. The main problem here is to use good random number generators with exactly known properties to be able to apply them in a reliable way.

ZIMMERMANN: Thank you, Lino. That was quite a complete overview over the whole field. There should be a number of additions or comments from our panel members.

VOGES: Livio has already explained why combining data can be important for understanding the physics of the observed sources or interesting parts of the sky. A good reason for archiving the data in the future is that our instrumentations are very expensive and they are normally built only once per decade. I should like to say something about combining data from different wavelenghts and I will now try to sketch how we can accomplish this task.

Firstly, we have to know how and where we can find what other kind of data are available from a particular source or region of the sky we are interested in. Most of the data are already stored in national or international data centers. Information concerning these data can be obtained but should be made more easily accessible, in the near future, by the use of common networks. On the other hand, there are a lot of data stored only in local centers and institutes, which should be made available as well. At least general information about these kind of data should be stored at some central site and should be easily accessible.

Secondly, assuming the data we are interested in, is found and could be made available in unreduced, reduced or deconvoluted form. Now, depending on how sophisticated we want to do our analysis, and how we want to combine the data, is determined by the format of the data. Probably a lot of additional information and additional data sets are needed. Just to repeat what has been mentioned already by Ethan and by Roland, X-ray and Gamma-ray astronomy is not only dealing with images, but also with spectral and temporal data analysis. In particular, the temporal data analysis has just recently given us very nice results like the QPO's, the quasi-periodic objects. Therefore, information is needed about particular features of the instrumentation as well, and also observational constraints. Modes of operation, detector efficiencies, energy resolution, timing accuracy, pointing accuracy, dead-time correction, exposure integral, background contamination, calibration data, etc. These are just some of the things we need. Many of them are also highly time dependent. In addition, a lot of information is required to be able to use or to re-analyse data of other instrumentations. It is obvious that an extensive documentation is needed about the analysis software with which the data were reduced and analysed. Perhaps more important, would be a copy of that software package.

I want to close my contribution here with the following remarks: particularly in the X- and Gamma-ray field, we have to establish a way how to archive all the recorded data - in reduced and deconvoluted format. The secondary data, for example, calibration data, and the software are required to make it possible to re-analyse and to combine data. That transporting software from one computer to another is in itself a problem, has been stated during this conference a couple of times. We have to eliminate that problem.

It is also important to develop some new software programs, to be able to deconvolute, for example, spectra taken from different energy band widths, and to establish methods for testing different algorithms and for determining which algorithms should be used for the special analysis problems.

It becomes more and more necessary to simulate the behaviour of instrumentation and software systems and to check the functioning and calibration of both. Perhaps this is the right time for Jean-Marc Chassery to enter our discussion.

CHASSERY: For my contribution to this panel, I would like to orientate my discussion to prolems of image and data interpretation. Although of course I am not a specialist in astronomy, in this domain data analysis considered in terms of image interpretation has much in common with other disciplines such as bio-medical applications and fluid mechanics. In these domains, data interpretation is becoming more connected with image analysis and we are frequently confronted with the significance between acquired and analysed data.

But present technology permits access to new representations of data. In the particular case, where the data support is associated with a digitized image, we can circumscribe the problem of significance by the following questions:

- what is significant in the picture being analyzed;

- what can be predicted by using models;
- what can be expected of the elements which are not detected;
- what is the meaning of those elements which are detected; are they pertinent?

To attempt to answer these essential questions, we must examine the general scheme of Image Processing. Let us suppose that there are no problems concerning the acquisition phase and modeling of the observed phenomenon and image resolution.

The main purpose of image analysis is to detect, extract, describe and identify or recognize the information which is expected according to some predefined structural models. Sometimes image analysis may be used to explain and to formulate an interpretation of the oberved data or scene. But such an extension is essentially related to artificial intelligence with the definition of expert systems.

Let us return to the problem of significance. The manipulation of pictures can be approached in 2 ways: statistically and structurally. Each of these approaches is related to some test of the quality of the evaluation of data interpretation.

Using the statistical approach, an image is considered to be a multi-dimensional random variable. Classical statistical methods can be applied to the evaluation of estimators or manipulation of tests, for example. The problem of the significance of descriptors extracted from the image can be solved by using methods of computing the entropy or by the use of maximum likelihood estimators. Other techniques are also useful for image processing. These techniques are based on classification methods using multi-dimensional hierarchical analysis.

For the structural representation of an image, the problem is to characterize the relevant aspect of the different structures. In such an approach, the image is described in terms of features. The advantage of this approach lies in the ability to associate models with structures using similar terms and similar environment. If it is possible to formulate a model of the structure being sought, then it is possible to use a matching process to recognize the components of the image.

Often it is possible to access the decomposition of the image in terms of objects or features. In such a case, an interesting approach is the use of cannonical analysis to detect what is significant and what is not. This is a very powerful method to determine the pertinent features or objects which are present in the image support.

Before ending my discussion, I would like to mention an important remark concerning the observation problem. In astronomy as well as in other fields it is not always the common even that we want to detect and classify but it is rather the scarce event which we wish to observe, analyse and explain. The classical methods based on statistics are not really adapted to this problem. To consider such a situation, if it is possible to provide some data, also called knowledge, it is necessary to use the environment offered by new systems which are the knowledge based systems as, for example, the expert systems.

ZIMMERMANN: Thank you, Jean-Marc. I think it is now time for questions or comments from the audience.

HECK: I would like the panel members to be specific about which kind of services the existing data centers, like the CDS, could provide for best helping us.

SCHREIER: One of the problems that has been dealt with partially in the radio community, but not or very little otherwise, is the question of data storage and management, interchanging formats and so on for multi-parameter data sets. We have discussed that, for example in X-rays for a given event, you have positional coordinates, a time coordinate, energy information; in all a multi-dimensional array. FITS was designed to handle data of this sort, but there are no convenient ways since you have gigantic amounts of information. If you want to have a good archive which is remotely accessible, what are the standards that are going to allow you to access it and be able to pull out the information you want easily? Data base management systems are used extensively now in data centers for catalogue organization, not for data organization.

DIEHL: I would like to take up the comment that Ethan Schreier just made; we are dealing here with multi-dimensional data spaces, so all kinds of image processing which Jean-Marc Chassery was talking about is: you want to have a two-dimensional image in the end, but you start from a multi-parameter space, you do not start from a blurred, two-dimensional image, you start from a maybe six-dimensional data space, and you have to find features which then translate into two-dimensional objects in the end. That is the particular exception of X-ray and Gamma-ray astronomy.

Handling of multi-dimensional spaces is quite common in high energy physics so we expect that we can benefit from using software which exists at the moment in centers like CERN in Geneva. We are using these libraries, and you may be interested to know that CERN, at the moment, is developing front end interactive graphic displays to handle such multi-dimensional data spaces. This is based on workstations like Apple, Mackintosh or Apollo stations.

The other computational demand which Gamma-ray astronomy has, are statistical analysis tools as Lino Buccheri mentioned, where we have to do highly iterative methods like Bootstrap analysis for the significances and Maximum Entropy for image formation.

This is the demand which we have to the computer science side. If they can help us, then I think they most likely will help us with advances in faster transform techniques, but probably not so much in giving us parallel computer architectures which we have to program then in a more assembly type language. This is too inconvenient to be used.

ZIMMERMANN: Has everybody the same opinion that the solution lies more in the evaluation and optimization of specific techniques, for example the Maximum Entropy technique, than in general computer resources?

HECK: I would like to turn my question differently: What are you expecting most from other wavelength ranges?

SCHREIER: Can you clarify that a little bit ... what do you mean?

HECK: Well, I have the feeling that Gamma-ray and X-ray astronomy are more or less a self-standing community, while one sees that the UV astronomers are more and more interconnecting their data with optical data. So what can we do to relate new data with data already existing in the CDS for instance?

ZIMMERMANN: That leads us to the next point of discussion: how can we make more use of data of other branches of astronomy?

GROSBOL: I would like to remark that we in ESO developed the table file system during the last five years exactly to enable multi-dimensional analysis of heterogeneous data. I think this is exactly what you are asking for.

DIEHL: An answer to Andrew Heck's question: at the moment Gamma-ray data analysis probably would use something like source catalogues or CO maps as input for their astrophysical models. What we want are tables of already derived data. We would feed these into our analysis methods in order to solve the ambiguity problem. Because at the moment we always have many images which are consistent with our data, we have to find some way of choosing the most probable image. We do that either by introducing new mathematical algorithms like the Maximum Entropy criterion, or we use more astrophysical information which comes from neighbouring fields. So, we need the data centers to provide this knowledge.

SCHREIER: I have a comment which goes back a little. I would like to stress again that a lot of these problems that we think of as unique to X-ray or Gamma-ray astronomy are really not. Let us just take a look at optical astronomy. Has anyone there been investigating the techniques for proper storage of something like prism plates, of real two-dimensional spectroscopic data? The X-ray problem, that I was describing or the Gamma-ray problem of multi-dimensional parameter sets, starts becoming trivial compared with something like that because the density of information is so much greater. And yet, as far as I know, the essential storage mechanism is plates.

MOORE: Could I just go back one more point. There has been some suggestion that you need storage mechanisms for your data. Let me make an analogy between an X- or Gamma-ray event and a radio astronomy integration. If you substitute those two words, I think our storage requirements are remarkably similar. You store events which are multi-parameter sets of numbers, like time, position, energy. That is identical, I believe, to the radio astronomy technique of storing integrations which are multi-parameter measurements. They are tabulated against base line parameters as UBV or frequency. They are multi-parameter data sets basically stored in ungridded form on a random axis, and I think our storage requirements are remarkably similar.

SCHREIER: I had in fact mentioned that I thought there were more parallels with radio than with optical astronomy.

TANIMOTO: In geographical information systems, one is often trying to integrate data from Landsat, ground and aerial images and there are kinds of boundaries that have to be put into geographical information systems. Some of the people designing these systems have developed some interesting data structures that allow the integration of these different kinds of data which have inherently different resolutions. One problem you do not have in astronomy, maybe that is not true, but there is only one view point at any one point in time, when observing the universe. When you observe land, you may be in different places and you have to compute where you are looking from. Even if you are looking at it at the same time with different instruments they are seldom in the same place. Of course in medicine you also have a problem integrating data about physiology from different kinds of sources. The cat's skin gives you a certain kind of information, and the animal gives you other kinds of information. If you want to build a common model that incorporates all this information, it is a very difficult problem. So, the problems that are being faced in astronomy here are not unique, it is a difficult research problem in a lot of areas, but in some areas, I think, some progress has been made.

ZIMMERMANN: Don Wells, did you want to make a comment on this?

WELLS: I was just going to add a little to what Pat Moore said. One question I have is whether there is any precedent in the X-ray or the Gamma-ray field for a software package of one satellite project reading raw measurement data from another satellite project. That has actually occurred among radio observatories. In 1979 the radio observatories in the Netherlands and NRAO agreed on an external data format for the raw visibility files. As a result there is in the radio community an ability to transport the raw files, no images, but the raw, multi-parameter data sets. That agreement was subsequently passed in resolutions in the FITS standard committees. You should be aware that there was an original extension paper, published in the journal right after the original FITS paper, which was an extension design specifically for multi-parameter data sets. Can you exchange data between X-ray processing packages?

SCHREIER: I wish we had the problem. There have been instances with two X-ray satellites at once in orbit, but not that many.

ZIMMERMANN: We face perhaps not so much a parallel data problem; we face a time problem, in the sense that for every next X-ray satellite we start again to develop our software new, to define again our standards for archiving the data and so forth.

SCHREIER: But I think that is actually changing. Right now, in your own project, there is a lot of the corporate history of the Einstein data analysis system moving into the ROSAT data analysis system.

ZIMMERMANN: That is true and promising. The experience is transferred, but totally new structures are being built up, and its not sure at the moment whether this will become a standard somewhere, e. g. in Germany.

VOGES: May I say something to that point? I really think, that so far, we have not estimated how large the data base will finally be, which we would have to store in order to be able to re-analyse certain parts of the data. I thought I made it clear: we need a lot of additional information; information about the instrumentation itself, information about the particular observation, many of which are time depending. I think it is a huge amount of data we have to store. Is this really worthwhile? Should this information be stored in an international data center or rather at those institutes collaborating on that project? Certainly, you need a lot of experience, documentation, and software to do this job of re-analysing the original data again.

ZIMMERMANN: I regard this as a major problem. If you look at the American data center, the NSSDC, then much of the data is really of little use because essential background information is missing.

DIEHL: There are in principle two different ways in which you can use previous astronomical knowledge or knowledge from other instruments. One way of doing it I described before: you use de-convolved information, and feed it into your astrophysical model. But that means that you have to believe in their significances. There is another way of using knowledge from other instruments which could open a new dimension or insight into your data, and that is using data from other instruments combined with your own data. But that means that you have to have access to the raw data of another instrument and to their instrumental response. Then you would do a combined fit or combined likelihood analysis of the other instrument's data and your own data. This really requires storage of raw data plus response in a convenient form, and also asks for standard ways of doing storage and data handling.

MOORE: Could I re-emphasize Roland Diehl's point about the importance of raw data? I think radio astronomy data reduction has really gone through three phases: the first has been aperture synthesis instruments. The raw data were immediately Fourier transformed to the sky, and the raw data were thrown away. We then moved onto a second phase, where we did that initial transformation, and then we cleaned up those images in the image plane, and we did what I would call image processing on those numbers. We have now gone into a third phase, I believe, where the raw data is the important quantity. We no longer do image processing. We do simulations of the instrument. We have a trial picture of the sky, we put through all sorts of filters and distortions that our telescope applies, and we compare the results of all those distortions and all those corrections with the data. It is the raw data which is really the most important thing that we are measuring. We are trying to model our picture of the sky and compare it with the raw data. Most of our comparisons now are not done in the image plane, but are done in our raw observation plane where we really make the measurements.

SCHREIER: I would certainly back that. In fact, before your comment I was going to say that I did not know of any astronomy discipline which was not compulsive of keeping the raw data. I am surprised you went through that first stage. Certainly in X-ray astronomy in the past no photon was thrown out, they were too valuable, and once you deconvolve it in some way you are losing some information. I would like to make one other approach: this whole need of modelling in order to deconvolve instrument responses or modelling a source structure and comparing it with the data, is very important, because it leads to the necessity of a system approach to an entire discipline. You can use the same data analysis techniques that control instruments to analyse data later, the same simulations in modelling that go into understanding how the detector works go into the final data analysis. You cannot treat it in vacuum, the same pieces of the system have to be available all the way through from conception through final archiving of the data fifteen years later. One cannot get rid of that approach any more in any of the astronomy branches.

VOGES: I think, Ethan (Schreier), we should clarify one point; I do not believe that we are really meaning raw data, I really think we are talking here of somewhat reduced data already. We are not talking about the telemetry data coming down or of repeating the so-called level 0 analysis.

SCHREIER: Well, there is a lot of terminology that has been developed in the United States, in the NASA community, as far as data standards and levels of reduction are concerned. The basic point can be stated very simply: you want to store the data at the level at which anything you have done to it is reversible. As soon as you get to the point where you have applied some process to the data that cannot be uniquely reversed you have lost it. That is the final level at which you have to keep it. You also may want to go beyond that. If I put on my Space Telescope hat I would say that we are storing the data at the essentially raw level. Raw meaning we have got rid of telemetry artifacts, but it is still all the data, which can then be redone at different times in the future as we know how to do it better. And finally there might be really analysed forms of the data, but that is far in the future.

DIEHL: You want to get rid of the instrumental fluctuations first, but as soon as you have physical meaningful quantities, you should treat that as real raw data vector.

SCHREIER: No. You have to keep that before you get rid of the instrumental fluctuations. Because taking out the instrumental fluctuations may be wrong, you may have done it wrong.

WELLS: Your remark about the version you want to save is the version before you go too far and lose the original traceability to the instrument. In the very long base line array project, there has been a major effort in the last two years to design the data formats for that project. In fact you could argue that the data processing software design project really turned into a discussion about data formats, because the data formats would drive the algorithms ultimately that will be built. One of their goals was to present the data sets as calibrated data sets, however, with sufficient tables and information attached to the data set that the calibration could be removed. That means, the actual data processing algorithm, the pipeline processing algorithm in its early stages, is to be recorded accurately in the data structure, so it can be undone at a future date in case it proves to be wrong.

Another subject I wanted to mention: there was the argument that the data set has to know too much about the instrument because you have to fully understand the instrument. That is true of course. Basically for radio astronomy, the agreements were for a language for talking about the instruments and about the parameters that describe them. It is not just raw bits, auxiliary tables, and calibration tables are added to the files so that they amount to a full description of the instrument process sufficient to model it. That required an agreement between the key people in the different institutions on how to talk about their instruments. And that agreement, that language is independent of the software systems in either institution. In other words, it is a formalising of the knowledge about observational technique in the field, an agreeing on formal rules for describing it. I think that is, maybe, a useful model for thinking about these problems.

CRANE: It seems to me that the comments that Don Wells made and some of the comments that Ethan Schreier has made, make it clear that X-ray and Gamma-ray astronomers would be well served if they thought carefully about how data formats could be useful, not just for themselves but also for everybody else. X-ray astronomers are a growing group of people, and the ability to be able to have X-ray data in a format which is transparent to everybody and especially to the X-ray astronomers themselves, would be a great contribution to the field in the same way that FITS has been useful for optical data, and the extension to FITS has been useful for radio data. That could be really a major step forward in understanding even some of the details of the instruments and codifying it in a way that would be useful for you as well as for everybody else. I do not know whether that is something which you are willing to put the effort into, but I suspect it would be well worthwhile.

SCHREIER: I agree fully with that. The only thing you have to differentiate is that it may not be that easy to always archive the data in the same format in every discipline, as you need the transport mechanism that allows you to do that, but you also may have special requirements that make it infinitely less efficient to try to start the same way.

But that is not the point. If an archive is set up which uses a more specific form, there have to be the tools to produce a standard protocol and get you over a network. You need the interface protocols, not every piece of data must necessarily look identical as long as you know how to read it. That was what FITS has done.

CRANE: Codifying the information in some specific well understood way would clearly aid not just the analysis of the data the first time but at all times later. That should be seriously considered in a very constructive and useful way.

ZIMMERMANN: I think we all agree on the need of building up structures for archiving our data in a proper way. But I feel this problem is also connectd to the role that individual institutes play at that time within large X- and Gamma-ray projects. In the German ROSAT project for example three countries are involved, Germany, UK, and the US. The data access problem therefore looks much less difficult to them than to any other country.

We also face the fact that these large satellite projects are only conducted by one or a very few institutes, so that instrument developments in the X-ray regime will also be limited normally only to a few institutes.

A further problem area lies in the fact that up to now X-ray and Gamma-ray satellites live for typically a few years, say three to five years. That means that all the experience in management, in software production and maintenance, in the distribution of software, the guest observer service, the archiving and so on, that all this experience is still built up only for this special purpose. All this may be lost afterwards, if for the next project in this field, a different institute becomes the leading institute.

So, on the one hand, I see advantages that individual institutes are performing the satellite experiments because thereby scientific research and instrument developments are easily implemented in these projects. On the other hand, it may be problematic that long after the mission, we have to rely on the support which such an institute can give to other observers to analyse their data. Therefore I would like to put this question to the panel: are those structures, presently found in projects in these fields, all of the form we can use efficiently in the future?

SCHREIER: My point of view is very strong in that direction. I have a biased point of view, but I think that such institutes are essential. You need certain concentrations to maintain continuity. ESO exists, NRAO exists. You have to maintain some corporate identity of a field. It becomes even more important in the X-ray field, where the experiments are spread out so sparsely in time. And as the time extends beyond the lifetime of a graduate student and perhaps an astronomer, one can lose the field altogether and you can lose the technology you built up unless you have some corporate structure. There is a concept that was coined by the US committee on data management and computation, called the scientific data management unit, which to some extent said that scientists have to be involved with the data, and this all through the conception of instruments up to the archiving. If you separate that out, you lose dramatically and you end up with useless or not very useful data. The concept may differ from field to field, the kind of discipline, the kind of data that exists, but to some extent one has to come up with a viable institutional structure, that can maintain the history, that knows how the data is going to be used, and that protects the data.

ZIMMERMANN: Are there more comments?

BUCCHERI: I wanted to return to the analysis methods topic which did not receive, in my opinion, much attention. In particular I consider of some importance the evaluation of the significance of the detections in the Gamma-ray range in view of the poor statistics available at these energies. Perhaps there are not many statisticians interested in this problem but we all face it continuously in various aspects of the data analysis of Gamma-ray data.

The problem received some attention recently by scientists working at VHE (very high energy) and UHE (ultra high energy) Gamma-rays, where lots of detections have been claimed at low significance levels, say two to three sigma, and successively "confirmed" again at marginal significance level. When not confirmed, time variability has been claimed. I have tried several times to raise this problem, in view of the importance of it in the exploratory data analysis phase, where new discoveries are generally made. Unfortunately, formal statistics deals with theoretical probability distributions and optical astronomy deals, generally, with high counting rate such that the theoretical probability distributions fit well with the reality. For low counting rates one has to deal with probability distributions not usually known, to be derived by simulating the experimental conditions. These probability distributions become quite complex, and sometimes in a not well predictable way, when scanning the sky to search for sources or when searching for periodicities in a data sample. In this case, in fact, one has to face problems related with the number of trials used in the scanning process. The calculation of this number is not trivial because generally the trials are not independent and one cannot add them together in a straightforward way. I do not know whether there will be enough time to discuss this argument, but I feel that it cannot be ignored anymore, if one wants to investigate in fine details some aspects of Gamma-ray astronomy.

I hope that there is some comment, especially from statisticians, how to work out these problems in a systematic manner, such to establish agreed rules of behaviour.

ZIMMERMANN: Is it still a specific problem of Gamma-ray astronomy to get rid of the smell of being a three sigma astronomy?

VOGES: Well, from the historical point of view, in the past there were quite a lot of two or three sigma results; by using different instrumentations, different groups tried to confirm the data. These days are gone. We only have one instrument in orbit, and only this one instrument records data. We will have to see that we get the best information out of this data. I think what you asked is: Can we not develop commonly used methods to get out of the few events recorded the best information based statistically on firm grounds? Do you ask for a common approach to this subject?

BUCCHERI: Yes, I ask in particular that, as the statistical tools have improved in the cases of high counting rate, some effort should be put into the cases of low counting statistics, into the scanning processes where you have partially dependent trials for deriving significances with pre-defined and agreed procedures. I think the problem will arise soon also in optical astronomy, with the incoming Space Telescope satellite, when fainter and fainter sources will be looked for.

CHASSERY: I would like to return to the problem of data manipulation: storage, transmission and other communication problems. The data management problem is also encountered in the medical domain where the collection of information relative to the same person or structure comes from different environments. At present, two approaches have been introduced for data storage. The first approach is that of generalized data also called generalized documents. The second approch is the use of the concept of associative memory. Actually, we can see the introduction of new architectures to include the concept of associative memories. I do not know whether in astronomy such architectures are used, but I think that they are a very interesting approach to make data management convenient.

(here the recording of the panel ended)

PANEL DISCUSSION--PROBLEMS IN LARGE SCALE EXPERIMENTS

Ethan J. Schreier

Space Telescope Science Institute
3700 San Martin Drive
Baltimore, MD 21218   U.S.A.

Ethan Schreier:

The panel is about problems that arise in the design of very large experiments, and we will attempt to take an interdisciplinary view. We have Michael Duff, who will deal with problems from the large computer systems point of view, Preben Grosbol from ESO, dealing with large ground based optical facilities, Pat Moore from NRAO, representing a large ground-based radio facility, and Ulrich Zimmerman from the Max Planck Institute in Garching, representing ROSAT, an xray astronomy satellite project; I will chair the panel and can also represent both space-based xray astronomy and space-based optical astronomy.

There are a few general themes which I would like to introduce. Then I would prefer to let the speakers give their own interpretation of where they see the main problems.

I think one of the key ideas is the question of discipline dependence or independence; my thesis would be that perhaps there are more similarities than differences among the problems encountered in very large projects and large experiments. You have also already heard clearly several times during the school that large scale problems tend to require a systems approach. I think that's a fairly positive development in the treatment of astronomy problems.

There is also the mixed blessing that large scale projects involve large amounts of money. This in turn means that there are not many of any given category of large project. This may lead to a one-of-a-kind approach to problems and systems, but it also naturally suggests the need for user facilities; if there's only one of something, then you have some moral imperative to create a facility that other people can use. This can lead to coordination both within a discipline and between different disciplines. Finally, the large scale of the project can lead to a concentration of resources that allows for new developments, for advances in the science and in associated technology.

With these introductory remarks, I would now like to give the panel a chance to voice their own comments.

Zimmerman:

Let me begin, and I would rather like to speak a little bit about the problems which we faced when we began our project. Only one sentence to introduce our project -- the Rosat satelite is an X-ray experiment which is supposed to operate for three years, but we have also a preparation time of about four to five years regarding subjects (e.g. data analysis and operations) which we are discussing here, and we will have a certain time after the mission to analyse the results. The problems which are the most important with such a large project are really in the planning and the management areas, and not so much in the more technical areas such as software development and so on; it is really the management and coordination of all the tasks. There are different problems, and there are different schemes of how one can try to overcome these problems. You could say: "Okay, this kind of project has been done several times in the past, so why don't we take the same structures which have been worked out more or less well in the past and try to make them better." This would require that you have access to those structures. The best thing that you can think of would be that you have an institution. In Germany we have something like the DFVLR, which is the general research support facility, and one could imagine doing the whole project only within this scheme. In the States I think it would be Goddard, for example. What we have experienced in other projects, however, is that although these institutions in principle are very experienced in doing those kind of things, they face exactly the same problems as if you come totally new in the field. Their estimates are often not much better than the estimates from those people who try for their first time to get a good overview of these things. The structure which is used in these institutions often is so rigid, so inflexible, that you have enormous difficulties in order to build up good and workable interfaces.

I would like to go into a little bit more detail on the kind of problems that really exist. I think the problems begin already when you plan the whole project. You first have to collect the requirements -- the external requirements and the user requirements. What we found is that quite a large number of these user requirements are, in the early phases, more qualitative than quantitative. So, if you are acting as a manager or planning person, and looking for a general configuration of the whole system, you immediatly face a problem of how to derive from these more or less qualitative requirements your internal requirements: the requirements of cost, man power, time scales, and so on. Here we really saw a major problem. On one hand, there is not full knowledge of the consequences which may derive from the fact that you do not have full oversight into the requirements; on the other, you have to meet different boundary conditions, which also are not fully known at this stage to you.

One of the major problems is also directly related to the analysis methods you have to apply. In a relatively early period, before the general layout can be detailed, you already have to initiate prototyping in those areas where you are very unsure as to

what the requirements really are -- what demands there are on
hardware and software, and so on. This means that you immediately
build up two environments in such a project, a prototyping environment, and an environment which you think will be more professional
(structured), or at least is supposed to be more formal in order to
facilitate maintenance and updating of the final system.

The next point, which comes after you have defined the general
configuration, is to develop a scheme to subdivide the whole
configuration. You have to create groups, and we have already
mentioned several times here that perhaps groups of five to six
people are the maximum. So as you define smaller blocks which only
require this size group, you need to set up interfaces, and coordination. This is an additional and appreciable effort and it is not
very easy. When you begin to implement all that you have planned,
another type of problem arises -- your man power is increasing
enormously rapidly, and you have problems in recruiting these
people. There are totally new problems coming up because you have to
assure the information exchange between so many people, and deal with
the sociology of these groups -- this is often where you see that you
have totally underestimated your problems. I am speaking here mainly
about experiments to be performed in orbit; a principal difference
from ground based projects is the fact that we can't change instruments after launch. So, knowing how the instrument will behave is an
enormously important thing, and we have attempted to simulate how the
experiments will work as well as possible. So simulation, in our
view, is very important to verify that we can really do the mission
in the sense we want to do it. I think we will come later on to the
post-mission phase. I should close here the discussion of the
problems and look for solutions afterwards.

Ethan Schreier:

   Okay, good idea. Preben, would you like to continue.

Preben Grosbol:

   I should say that I don't regard optical astronomy, nor ESO, as
a large experiment. We have been here a long time and that is, in
principle, what is different and what changed the boundary conditions
for optical astronomy. We are a more continuous operation, and
therefore we don't have these peaks of development. We don't have
the peaks of trying to get man power. That means that we can plan on
a somewhat longer time scale, and we are not necessarily forced into
these excesses of manpower which are difficult to administrate.
Concerning the systems approach, I think the long time scale during
which optical observatories exist makes it vital that we have systems
on which we can base most of the data analysis and other development
work. One needs to ensure that the systems will live a long time, to
have good interfaces both to data structures and to devices and
software. The data interface we strongly support is FITS. Getting
data out of instruments in a specified format gives us a very good
interface to the data processing systems. We are independent in that
sense, more independent of the specific details of the instruments.
If we have a standard data format, we can feed data into a standard
system and that saves us a great deal of work. Likewise, of course,
software interfaces are of extreme importance. Because our group is
limited in size, we can build up the system and provide basic
services, but to a large extent we will depend on external users to

provide software to this system; use of standard interfaces is the only possible way in which you can integrate foreign software into the system and provide it to a larger community.

Pat Moore:

Well, I presume I've been asked to talk here to represent radio astronomy, but I don't wish to bore you with the technical problems that radio astronomers come up against. Instead, I'd like to concentrate on just one point, which is that perhaps what we should be aiming for is to maximise both the user and programmer productivity in this field. You've heard a lot about how systems can help us do that, but let me just take those two areas and give for each an example of how we may improve people's productivity in the future. Let's start with the user, the astronomer using one of these large systems which go with these large experiments. I was very interested to hear the talk by Chassery earlier today, showing what can be done to improve the interface between users (i.e. astronomers) and systems. I think there's an awful lot of research that can be done in this area to improve that interface, to make that interface more productive, so that an astronomer sitting down in front of a terminal with an idea about something he might like to do, can very easily convert those ideas to instructions to be executed on the machine -- to come up with some kind of answer. I'm sure this would involve some kind of expert system to pose the problem. I hesitate to suggest which language would be a suitable one to use, but I think the example that we saw earlier today was very thought provoking, with a combination of an expert system, an immediate display system and one of the new SUN work stations with multiple windows and all the menu- driven features that that allows. So, I think this is probably a productive area for designers of large systems to work in -- to seriously consider how they can improve the interface between astronomers and some of these large systems.

The other area where we'd really like to see an increase in productivity is in the programmers' building of these large systems. Very often you need many highly specialized people who have an intimate knowledge not only of computer science but also of the experiment that is actually being undertaken, the observatory details and how it works, and the algorithms you require to de-convolve or clean up the pictures you get. It is a fairly complicated field to work in, and I think anything we can do to simplify the work of programmers in that area would be very useful and would lead to significant gains.

One point I'd like to suggest we think about for the future is how we combine different elements to make one of these systems. Let me give you an example. The system often consists of many different parts. There are the number crunching parts which perhaps run on parallel processors or array processors or vector machines. There are organizational or structural elements which combine or utilize these operators in complicated algorithms, and then on top of it all there's some kind of user interface which interacts with people. An astronomer expresses an idea, it goes down through the different layers of the software, and ends up setting off a piece of number crunching code to do something. It is by no means clear that there is one single language which is suitable for expressing the different levels of that software. I suggested earlier that an AI language may well be the language to use for the user interface, Fortran is possibly a good language to choose for some of the number crunching code, and we've heard discussions about languages for parallel machines. Perhaps a point we should think about is how easily we can combine some of these languages together. Standards committees sit for hours and hours and discuss the technicalities of Fortran, C and Ada. I think it will be a fruitful area of research to concentrate on how you might mix these languages together into a large system -- the interfaces between those languages. It is technically quite a complicated area, not just how you call a subroutine in one language from another. There are many more points of contact between different modules in a language, there are global variables, there are asynchronous operations, there are a whole host of things going on. But I think it would greatly aid the coding of some of these larger systems if we could really use the languages which were most suitable for particular tasks, and having such interfaces between different languages I think would make that very much easier.

Ethan Schreier:

Is there any significant work going on in that now, as far as coordination between languages?

Pat Moore:

I don't know of any.

Ethan Schreier:

Well, maybe we should come back to that later, with the audience. But I have a question to ask before we go on to Mike, in regard to Preben Grosbol's statement about the systems approach. When Ulrich Zimmerman talked about the systems approach to space experiments, he stressed very long range planning. When ESO (and I guess the VLA as well) started the planning and design phases, were there plans for data analysis systems and for archives? In other words, was a systems approach applied only to building and operating it, or was there also significant planning for the data analysis aspects.

Preben Grosbol:

It is obvious that there was very little planning twenty five years ago, when ESO was founded, because it was simply not feasible to do these things at the time. (Of course, the plate archive was envisioned, which has in principle been achieved.) Therefore, the data analysis capability grew more gradually within ESO; it slowly developed from IHAP to MIDAS. It is an expanding thing, and that means that at a certain time you have to upgrade your requirements; but it is a more floating thing when you see that old requirements -- five, ten years old -- may not be adequate. You must go into and modify these requirements, and make them more flexible. I think we have done that in the step from IHAP to MIDAS, and perhaps we have to now go one step more, a MIDAS to a portable MIDAS. But we are prepared to do that and it is a gradual change.

Pat Moore:

I'm not sure I should really be answering this question since I'm a latecomer to the VLA. I'm sure there was indeed substantial planning for how things should be done. My impression was that its very very easy to underestimate the magnitude of some of the problems. You think you can perhaps get away with non-systems approaches or slightly cheaper approaches only to find that you really have got a very large problem there. Its something you can very easily underestimate, and I would like to stress that point.

Michael Duff:

I'm the sort of joker in this pack -- I think I'm the odd man out in that clearly I am not an astronomer, and I suddenly had the awful feeling I didn't know what I was meant to be talking about at all. So I thought what I'd do is just mention some parallel experience that may be of help. I started life as a high energy particle physicist of a sort, and by the time it got to be the late fifties, I was drafted onto trying to design image processing equipment which could be used in high energy particle physics experiments. In a way, there are a lot of similarities between the problems they had in those days and the problems I think you're getting into now in astronomy: masses of data, literally millions of pictures being generated which have to be analysed. I'm not sure you do things quite this way: teams of women, working in the semi-dark, presumably measuring the pictures. They didn't like this; it should have been computerized, but, of course, the trouble was that computers had barely been invented in those days. So people like myself were trying to build data analysis systems with post office relays and well, you know, eventually we got on to valves and even transistors, but the point is that then and now there was a considerable resistance to change. Everybody had got their methods working, they weren't particularly keen on anything which meant changing those methods. If you could treat it just as "plug in a machine, plug out a person," that was fine, but anything which interfered in the slightest way with the experiment was out. That's true now in medical physics; if you plug out a laboratory technician in a pathology lab and plug in a machine, its fine, but if you suggest that the material should be prepared in a slightly different way that's not fine. That may not be the situation in astronomy -- I don't know.

The second point is that we had fairly way-out ideas -- you might even say lunatic fringe ideas -- and the problem was to sell these ideas to the scientists. Unless they could still do the experiment in the old, traditional method, and not hold it up in any way, they didn't want to hear about anything new. We usually found that by the time we built equipment, and eventually software and so on, to meet the specifications for a particular experiment, the specifications had changed, and we were too late. You could say it was a bad balance between engineering effort and scientific requirements. So, that was the second problem, the drifting spec, which made me begin to feel that the universities were not the right place to do this sort of engineering development to support large experiments.

The third thing was that if you did manage to get a prototype system built, and get it into the hands of the scientists, the problem was to get the bugs out of the system. If something went wrong once, they lost confidence in the system and it was extremely hard to get online for debugging. And then the final point was that to get credibility, you not only had to build a prototype, you actually had to get a manufacturer interested, and no manufacturer is interested unless he can see a good consumer base for his product. We are still seeing this problem now. We've tried desperately for seven or eight years to find a manufacturer for our processor array. We got into this viscious circle where we could only get a manufacturer interested if we had customers, and we could only get potential customers interested after we had the manufaturer buttoned up so that they had a machine to buy. So, really, its a long extended and rambling plea for something that can be done for people who want to try something really new, and want to interact with big experiments and want to get a sort of sympathetic ear for their ideas, without being made to feel that they're wasting the time of the experimenters, and probably their own time as well.

Ethan Schreier:

I would certainly agree with much of that. I also came from high energy physics, and my early experience was that many astronomers were far behind the high energy physicists in dealing with computer-related issues. I think you have been exposed to perhaps the much more receptive astronomers in this school, people who have taken data analysis seriously. If you go out into the astronomy world in general, there is probably still a lot of resistance to change, and a lack of understanding of these issues. Another point of similarity is that astronomers also have a significant problem with manufacturer s' responsiveness. As an example we have discussed before in the meeting, we can't drive the market for image work stations as effectively as perhaps the medical field can.

At this point, I would like to summarize the three categories of problems we face; several examples have just been discussed. The technical problems, in my biased opinion, are not the major issue. We can take actions to get specific technical problems solved, whether they are actual scientific questions, or how to build a better detector, or how to develop a better algorithm. Conceiving new ideas may be hard, but it is why most of us are here, and what we are paid to do. In any case, the current discussions are not going to help the technical problems very much, except in as much as there are problems in common with other disciplines, where solutions already exist. Thus, for example, remote observing for ground-based astronomy should be able to take advantage of space astronomy experience, including data compression techniques. Similarly, rule-based systems are beginning to be used both in space-craft scheduling and in data analysis and interpretation; it would be silly not to compare notes.

The second category involves management problems, which I think are less straightforward; these are harder questions to deal with. First of all, as Ulrich Zimmerman mentioned, is the generally accepted fact that you can't put together a productive software group of more than half a dozen people. So how do you do a very large project that requires twenty people or thirty people or hundreds of people? You have to introduce interfaces; there is a significant overhead in that. Another very common problem, the problem that Preben indicated he didn't have, has to do with a long stable base. In a canonical academic research environment, you don't have extreme ups and downs, and you're not driven by any hard deadlines. In a space project or any really large experiment that involves a lot of money, large groups and complicated interfaces, there are schedules involved, and developing a software system or a new capability to a schedule is very different from the standard research methodology of moving along as you find the ideas and are able to implement them.

Another thing that we heard Ulrich Zimmerman discuss is user involvment. In any project, you have to determine user requirements at the beginning; if you don't, you eventually have problems. But because the early requirements are very qualitative, you also need user involvement during the development phases, to interpret the usually qualitative original requirements. The really quantitative requirements essential to design and build software or hardware systems entails continuing involvement of the people who are going to use them; they're the ones who know (or should know) what they really need.

And finally, there is the category of sociology issues. We've touched on this earlier in the meeting -- we've heard talk about the "macho programmer" who can't work in a team effort, and the macho-astronomer who wants to do everything himself or herself. Some scientists, and especially astronomers, tend to be against the concept of "big science", of projects that require alot of people working together, and I think astronomy is hitting that in spades with the Hubble Space Telescope. Astronomy tends to be among the more individualistic of the sciences: the old stereotype of an astronomer sitting on top of a mountain all by himself, and getting a plate, and owning that plate for the rest of his life. This is a culture that has to change somewhat, if we are to make use of large nationally or internationally developed facilities. In the software area, the team software approach has gained some credibility alongside the individual or macho-programmer approach.

One final point, also in the "sociology" category, is the not-invented-here syndrome. This is related to the issue of interdisciplinary communication, of people thinking that they have to develop everything themselves, as opposed to surveying the field for useful existing systems. I think the kinds of discussions many of us have been conducting during the past year or two in this area -- software standards and interfaces and networks and whatever -- are going far toward dealing with that problem, but it certanly is still a problem.

I would like to open the discussion up now. The panel may have comments on some of these points, or we can go to the audience and also get further questions or comments.

Zimmerman:

Ethan, I think the last point you mentioned was very important. You may find that there are structures available which essentially cover all the problems which may come up; individual things may have already been done in a more or less similar way. If you now are planning for the whole, you hardly have to consider the structures which have been invented before -- you can take them over, rather than invent them again. That means that you have to see what kinds of interfaces are necessary to do this, what the structures mean, and whether they are too rigid or flexible enough to accomodate the new problem. These kinds of decisions are not easy to make, and they can either make you underestimate the problem, or make you think that the problem can not be solved with the resources you have planned. Both things are very bad for the project.

I think the usual method which we have used until now is standardization. We try to standardize certain things, and the standardization can work well, let's say, in things like software development and data interfaces, and eventually in communications, although we are just touching this field. But, for example, standardization works very little and relatively weakly in the field of management, although you can think of standardized management structures. Standardizing also is a little bit of a problem in the area of planning. You have to do different kinds of planning: long term planning for the resources, but with these (space) observatories you also have to plan the daily observations which have to be performed. I see a very big opportunity for artificial inteligence especially in this regime. I think artificial intelligence could help enourmously to make these projects better, to make the projects more interesting and fun, and therefore also to influence the sociology of the whole thing. Artificial intelligence, in my feeling, is the ideal tool to handle planning problems, and to handle development problems in software. It could, for example, help overcome the problem which has been raised several times in the workshop -- the macho versus team problem.

I think that you can take better advantage of persons who don't fit in to the normal beaurocracy. You can give them an environment, an intelligent environment I would say, which allows them to follow a "macho approach" by automatically supplying the software environment, interfaces, etc. that this macho approach which in the present state is not usable in these You might really be able, with this approach, to take advantage of macho programming, of the ingenuity of the guy who has a special solution, and also improve the effectiveness of such a person.

Michael Duff:

I'm all for standardization in theory, but I have an awful feeling that in practise, it stifles innovation in a peculiar sort of way. As everything gets polished and made nice and and made to fit together neatly, at the same time it develops a sort of dead feeling. I've seen this happen in our own group, where we've got a Unix based system now. A lot of discipline set in almost without me doing anything about it. Everybody produced beautifully documented subroutines and all these heirarchical file structures; it was beautiful, and actually our productivity increased, but to me, the quality of the work has gone down a step, its all gone a little bit sort of dead pan.

Ethan Schreier:

Is this inevitable ?

Michael Duff:

Well, it may be that I'm just getting past directing a group or something, but there is a feeling that when people come up against difficulties, somehow the scientist seems to work better; the engineer works when the nut fits the bolt, but the scientist likes to hammer the nut through the bolt.

Ethan Schreier:

I thought you were going to say that the danger in standards is, as I have observed, that traditionally when managers think about standards they say "thou shalt write in Fortran," and "If everyone writes in Fortran everything will be perfect won't it?" They don't think about standards in terms of interface standards, and protocols that allow systems to interconnect, while allowing and even encouraging individuality in each piece. That's a little bit of a different concept.

Michael Duff:

Yes, but you see the point is that if you have a really standardized system, inevitably it seems less flexible than a system which is tailored to the particular field in which you wish to work. An example of this is the accounting system in my institution. Whenever you want any particular kind of information people just say, "You can't have that; the computer doesn't provide that." Its a beautifully standardized system, and it doesn't give you what you want. It has been very carefully engineered to provide standard information, and it isn't flexible enough for people to get into. I think standardization reduces flexibility.

Pat Moore:

That's why I said we should try to standardize in a way that allows lots of different languages. I think that's a way of not stifling new approaches to some extent. If you want to define a new language to tackle a particular problem, you have some freedom to do that. Not total freedom perhaps, but I would like to think that avoids the situation you were talking about, where things are totally stifled.

Zimmerman:

One point about standardization that I think is always seen is that as soon as you have made these standards perfect in your system, they are already obsolete, because the development goes so fast; you are bound by your own standards after a few years.

Phil Crane:

I have to disagree with Ethan Schreier's statement about the fact that we have to get rid of macho astronomers. You're going back to standardized astronomy.

Ethan Schreier:

I certainly didn't mean to imply that you have to get rid of macho astronomers.

Phil Crane:

You said "there's the question of big science versus macho-astronomers, and clearly that has to change." Now I don't know exactly what your implication was, but I certainly took it that way. I feel that certainly if standardization is going to reduce productivity and make things all sort of grey in the software area, I can imagine its going to do it in spades in the astronomy area. I really feel strongly that the system should support the individulistic astronomer. We must provide the tools for these guys to do whatever comes into their mind quickly and easily, and that's got to be the goal of the data analysis development schemes. You have to cater really to the macho astronomer, and not to the team man. Maybe somebody else would disagree with that.

Ethan Schreier:

I would not disagree.

Preben Grosbol:

I will basically agree with that. I think standards are good for two things. First of all, it provides you with some security in the sense of being able to maintain the software, and also move it from generation to generation of computer. Now, the other thing is that we really want to remove the trivial tasks from the astronomer -- we don't want them to have to think of how to put the data on discs, how to plot data. They should be caring about the algorithms; that should be individual, and there we should have a very large free approach, but we should provide the standards so they are relieved from the trivial tasks, that is the important thing.

Phil Crane:

I have the impression at least that that has been reasonably successfully done in the software systems that have been developed at ESO, but the thing that has not been done, and this is again from the systems point of view or the management point of view, is to figure out how to capture the output of the macho astronomer. He's not going to do anything about it, but you still don't even know how to capture it, and that's a big question which arises. I don't have any ideas, I think you provide him with the tools to do his thing, how do you capture it in the end, and that's a question which I suspect the panel members can help us with.

Michael Duff:

Well, really the point is, and I think we're all agreeing, is that standardization should be applied at the sort of lowest levels, the levels which are really levels where you don't want to know about what's going on. You just don't need to know. For instance, we find it extremely difficult to send an image to any other image analysis group, in any way whatsoever.

Ethan Schreier:

I think one answer about capturing the results is to provide not only the tools for easy software development, but (if Don Wells isn't going to say it) also the tools for easy communication. Any other comments?

Don Wells:

One of the things that limits people's imagination in research is turn around time. I think simply giving a fast, fairly consistent turn around time is one of the best things you can do on any computer system, regardless of what the man machine interface is.
Unfortunately most existing computers that people work with today are too slow. They don't react as fast as people, and particularly in large scale computing projects. I think one of the great dreams of the parallel, and the massively parallel research people is that by increasing the turn-around rate or speed of the hardware on these very difficult problems they will enhane people's creative powers. I personally think that's probably right.

Ethan Schreier:

Yet another tool to use...

Pat Moore:

I conjecture that we will always be in that state, because I think people will develop algorithms until they become intolerably slow and they stop. I think there's alot more algoritm research that's going to go on if we get an order of magnitude increase in computing power.

Michael Duff:

Yes, I'd like to comment on this. We moved from an environment in which we were doing image processing on an IBM 360, which probably dates me, and this was typically the situation in which you'd laboriously punch out an image on dirty stack cards, or whatever they were, and write your program and about a day later you'd get back an error message of some sort, which meant you'd left a full stop or a comma out or something of this sort. Anyway, you finally got the program working, and then you found that you wanted to try it on several images, and it took days to produce these images. Well, we built our first massively parallel processor in 1970, and it did a bench mark on one simple H5 ring operation, which was in those days quite a process on the IBM 360. Our machine ran ten to the seven times faster, and from that day we never looked back. Now we tend to think in terms of sitting at a work station with a sort of push button approach to algorithm development, and with the result that in a recent test some colleagues in the States sent a benchmarking task; they gave the timing on the machine that they built, and it was 70 seconds to process this image. I actually took the image in hard copy out of the envelope, sat down at our machine, developed and debugged the program, and ran it, all within that 70 seconds. The actual program took a few mili-seconds. So, the real point is that the program development -- the algorithm development -- becomes incredibly different when you're interacting at almost the speed at which you can think. You know I'd really like to see the effect of this on astronomy. People have been talking about recognizing shapes of galaxies. If we had a picture of a galaxy, we could just stick it into the machne and fiddle around with it; I guarantee you'd get promising looking results in about five minutes.

Zimmerman:

I wanted to make a comment to a different kind of turn around time which one should really aim to shortem in large systems. That is the turn around of the results which are produced by the large systems, influencing the system itself. I think its here we can really do something, if we try to plan our large system in such a way that research and the normal data processing are done in parallel, so that this kind of turn around influencing the ongoing program can be dealt with quickly.

Ethan Schreier:

I'd like to come back to Phil Crane's comment for a moment. Let's remember that the topic which the panel is dealing with is problems and solutions with very large experiments. The question is not whether there should be macho astronomers versus big science. There are both, and the real question is how best to let a macho astronomer use a big facility, if he or she wants to, and still allow the macho astronomer to do his macho astronomy. That's fine, I'm certainly for it. The whole purpose of anyone doing science is to do individual thinking, individual research. There are some projects where collaborations work (although big ones tend to get fuzzy,) but there are also some projects that require really massive amounts of support to get done; its the problems associated with those that we're trying to address, not the question of whether macho astronomers ought to be abolished.

Wolfgang Voges:

Don't you actually need more or less two different types of software people in an institute which wants to develop and run large experiments? If you want to develop hardware, you first need a prototype in order to see that what you want to do is feasible. I guess you also need these people in the software field, doing some kind of prototyping, and there you need these individuals. You can have the innovation you really need in order to produce the software which is suited to the analysis of your data.

Ethan Schreier:

I would agree fully. I think the only thing I can add is that I think if we provide the right tools for the prototyping efforts, the individual programmers will be developing things which, if they turn out to be right, will almost automatically fit into the final system as well. There's a great tendency to equate the 1962 graduate student writing his own program with a competent software person doing state-of-the-art prototyping. These are not the same thing. One can supply the environment and the education and the tools that allow that individual, playing totally on his own, to create a very good system that, if it works, is right there. You don't have to redo it.

Wolfgang Voges:

Yes, but I think that it has already been said a couple of times that you want to have motivated people in your group, and I think very often we don't motivate them, but we frustrate them, because they have to obey so many laws or they have to be aware of so many boundary conditions, that I really sometimes think that this is not serving the development.

Pat Moore:

I would chose that as a measure of how good the software system or the software environment is. I agree that can be problem, and people are often tempted not to work in a particular environment. But those environments frequently offer huge advantages: you can either run a program to do something, or look at the results, or plot the results, or you want to compare them with other things, or carry on processing with other algorithms. I suspect that people's software environments are going to improve over the next few years, and that problem is going to become less important.

Wolfgang Voges:

Well, I think if you are talking of tools which make it easier to program or to come to a result, then I fully agree with you.

Preben Grosbol:

Yes. One of the things we, in ESO, could consider at least, is to sort of help the astronomer by linking some kind of software support to him. That would give an advantage to the astronomer in the sense that he would not need, to the same extent, to learn the system. He's interested in the astronomy, but that software support would give him the possibility of directing, standardizing and securing the software somewhat. The main disadvantage is that it is expensive in software efforts, but the potential benefit is that you can integrate very vital packages into the system without so much complication.

Ethan Schreier:

Yes. These comments fall in the area that I would call enlightened self interest -- supplying the tools and making it easy enough so that the person realises that its to his own great advantage to work in this environment.

Michael Duff:

The thing that struck me very much in dealing with some computery type people, is that they tend to run a particular system and make out that its the best type of system in the world, and everybody has got to have this system. At the moment, its the Sun, you know, Sun, Unix, VME , and so on, and unless you have this you might as well pack up. In fact, somebody working here said that, "If we don't get these, there's no point in continuing, the project is finished." Now, I guarantee in a year's time he'll be onto some thing else, and something quite different. How do you cope with this, the need to standardize on the one hand, and this feeling that computers have got a life of about five minutes?

Ethan Schreier:

I don't know, it's education and communication -- whatever; its not just computery people. You can take non-computery people, a scientist who grew up on a 360, and expose him to an interactive system (in my experience, it was a Data General machine): all of a sudden, a system where you could really interact and see what was happening. That kind of experience leads to statements like: "I have the best system in the world now. You're not going to tell me anything different." They forget that evolution goes on. And there are people who regard data analysis systems in the same way. As an example, a system now that's popular in the States, IDL, which is a very, very user friendly system. It provides capabilities that are tailored to data analysis that are very nice, and therefore people who have tried this, who have never had anything else like it, can't conceive of the next step. It takes a while, and it takes education, and sometimes pushing.

Michael Duff:

But you see the trouble is that you build up these great blocks of software and all the rest of it, and really they just won't run on these new systems. I mean they will run, but they don't take any advantage of the new facilities.

Ethan Schreier:

Exactly. The whole idea of standards and portability is to allow people to develop systems that will be able to run on a new system. To think ahead, and to expose people to Don's crystal ball, and others, that say which directions things are going to go; that if you program this way instead of that way, you'll probably be able to take advantage of these developments more easily when they come along.

Peter Shames:

You know there are some partial models that do work both for device independance and hopefully for people independence. The whole concept of modular design is to isolate user interfaces from the gut level algorithms, from the specific details of the operating system. This kind of approach does give you the ability to change out the user interface when a new user paradigm comes along, a high resolution bit-mapped display, the mouse, the cluttered desk, or whatever else. The same kind of thing is the case at the bottom end. If you've defined your operations at the vector level, for instance, or at the matrix operation level, and some new whizz-bang piece of hardware comes along, you have a hope of being able to port over onto that new piece of hardware and take advantage of it too. This requires a certain level of discipline, and it really requires an open systems approach to systems design, as well as the open system approach to network design. I think that is the key. If you look at the network world, for example, that is one area where standards have been applied, but they have still succeeded in allowing for a certain degree of variability and flexibility and for a different kinds of things to be done without necessarily hampering what's done. The point is that standards are very important, but how you apply the standards and what kinds of standards you apply are even more important. A very careful choice needs to be made at the level at which you say okay, here we're going to lock it in but here we won't.

The choice of open system interfaces is one kind of standard, and the choice of Fortran 77 is a very different kind of standard, indeed.

Ethan Schreier:

That's certainly what I was trying to say before.

Don Wells:

Michael Duff, as a manager, if your guys come to you and say " Sun, sun sun, sun, sun," you jump them hard because that's bad thinking. They better be thinking about Masscomp, Apollo, and you can pick several other names. They better be looking for least common denominators. They better be trying to figure out where the main thrust of the evolution is going. If they jump on to one single vendor they're likely to get trapped in some peculiarity of that vendor. They ought to know better than that, and if they don't you ought to teach them.

Michael Duff:

I'll have to respond to that. In fact, to be fair to these people, we did need a Sun in order to host our particular machines. We had to deliver instructions at a particular rate, but I have a sneaky feeling that even if we didn't, they would still have made the same sort of jumping-up-and-down type noises. You know the truth of it is that people like Suns more because of the pretty pictures than anything else, I suspect (but also the fact that its Unix based, and VME, and the other buzz words.) It is a very real problem because

things are developing so fast now; you have to appear to have the
newest and best of everything. You not only have to satisfy
researchers, but you also must look smart when the funding
authorities come round. You know, you're still using black and white
and everyone else is usng color, and somehow they don't think you're
trying. So, it seems one has got to move a little bit with the
fashion.

Don Wells:

But how then do you hook in the next year's machines with the
current year's machines and keep your lab evolving with an
overlapping of hardware and software as it changes. How do you keep
the evolution going? And I think what you have to do then is rely on
the open system interfaces, and on least common denominator
technology so that the new things fit into a previous pattern, and
the old things are able to survive in the pattern.

Ethan Schreier:

Don't go too far to a least common denominator because that also
can be...

Don Wells:

Its an art.

Ian van Breda:

As a very small user rather than a very large user, one way
which we tried to cope with the problem of changing technology is to
stage things, so that we've taken one type of, say, eight bit
processor originally; we've then gone to sixteen bits only when we
couldn't do with eight bits, and taken a new generation of
processor. Now going to 32 bits, we've taken VME bus as our
standard, not necessarily 68020. Our software is such that we're not
terribly processor dependent, and we can add new devices gradually.
I think it is a mistake to try to keep up with absolutely everything
that's coming out new. You've got to say, "I want to do this work or
this astronomy. Can I still do it with that processor?" If you
don't do that, certainly for the small users, you just find yourself
trying to chase the latest processor every year.

Pat Moore:

I think there's a very big difference between doing computing in
astronomy, and doing computing in a computer science lab.
Astronomers only have to stay on the forefront of astronomy, they
don't have stay on the forefront of computing. We only really want
to use hardware which has got a definite advantage to us. We don't
necessarily have to go out and buy the latest new piece of
technology.

Michael Duff:

But are you not in a sort of competitive world , and don't you
find that if you don't have the latest toys somebody else will have
them and this is bad for you ?

Pat Moore:

They're not always the most effective way of doing things. They are sometimes. You have to judge them quite critically I think, and see whether its going to get you anything.

Michael Duff:

Yes, to be cynical about it, I'm not even sure that people look at the results too hard, they look at the environment in which people work, and this somehow builds up the status of an organization, in the short term. In the long term, it's the results which matter, but in the short term its how much of a show you can put on. Don't you think?

Pat Moore:

There's a certain amount of that, yes.

Ethan Schreier:

We're now moving into the PR or marketing field. But before we close, I'd like to take what Don Wells and Peter Shames were saying one step further. We have mentioned this before, but its important. If you look at the communications (sociology) aspects, you can alleviate the worry about lack of support for or recognition of the macho astronomer. In the computer science field, the effects of networking were incredible, and it did not stifle creativity. I think a lot of advances took place because of this very quick and easy communication, to the extent that people didn't even think twice about it. I think that was extremely important, and it is the direct opposite of the regimented team approach. It is just total, open involvement of everyone who wants to be involved. The correct kind of communications, standards, and open system interfaces, and reasonable kinds of coordination in "big science" can indeed facilitate creativity and provide new research opportunities.

If there are no more questions or comments, then we should thank the panel and conclude the session.

NOTE: The panel chair wishes to thank the speakers, and apologizes if any of the sense of their contributions was changed in editing for clarity.

TRENDS IN PARALLEL PROCESSING APPLICATIONS

Panel members: Philippe Crane, Stefano Levialdi,
Peter Shames, Steve Tanimoto and Michael Duff (Chairman)

The Chairman introduced the panel members, whose interests ranged from Astronomy to Computing, and explained that the discussion would attempt to find answers to the following five questions:

1. Is there really a computing bottleneck in Astronomy or is all this discussion an attempt by computer scientists to `sell` computing?

2. Why don't astronomers demand the newest and fastest computing techniques?

3. What new computer technology is actually available today?

4. What positive action should we take to advance the use of computers in Astronomy?

5. What `great new things` might happen in Astronomy if three orders of magnitude increase in computer power/speed were to become available?

The first question to be answered was: "Is there a computing bottleneck?"

Crane: I think that question is actually quite well posed. Let me say that I do not agree with the fact that it is an attempt by computer scientists to sell computing; that is probably a misstatement. It is certainly true that there is a wide range of problems that astronomers do not even think about because they do not have enough computing power to tackle them. The people that are perhaps least guilty of that are the radio-astronomers; they need the computer power just to see what they are doing. The optical astronomers and, I think, to a somewhat lesser degree, the X-ray and gamma-ray astronomers, really have not explored the capabilities that one can find in computers to extract new and interesting ways of looking at data. I know, in my own case, I do not often do a deconvolution of an image, even when I have the capability of doing it, just because it takes so long.

There is another aspect to this computing bottleneck which is really a corollary: almost all instruments these days are producing data in digital format. We also have a data bottleneck, so just getting through looking and seeing what your data looks like also represents a bottleneck. Not only do we have a computing bottleneck but also an I/O bottleneck. It depends on your particular job which one of those things is limiting you, but both are limitations.

Duff: Do you think it is a limitation that the astronomers are feeling themselves. I mean, are astronomers jumping up and down and saying: "we can't compute fast enough"?

Crane: Well, I do not think they are jumping up and down saying that, but they have not explored what they can do with new capabilities. You know, you do not know what a pleasure it is to drive in a Ferrari till you have done it, and it opens up a whole new range of capabilities. I can see that in many of the things I am doing, had I the capability to massage the data in a reasonable way, in a reasonable time, then I would explore new areas. There are many other people in the audience who probably feel the same way. We are getting along, as it were, with what we have got, but we are not taking full advantage of the data that we have.

Duff: Peter, you are a systems builder, how do you react to that?

Shames: Well, in thinking about this problem I came up with much the same conclusion that Phil Crane did, that there really is both a computational and an I/O bottleneck, and many of the ideas that people are talking about, things like distributive processing, distributive workstations, are a way to try and tackle some of the computational limitations that have been handled basically by applying the canonical VAX system to the canonical image processing problem. That is all well and good and it is a way of getting more computing into the hands of people but it is not going to fully solve the problem until we get the I/O capabilities to handle five or ten megabytes of data at a time. In the area of modelling and the area of radio-astronomy there is an enormous need for far more computing power than we have right now. As Phil Crane commented, there are probably a lot of imaging techniques and analysis techniques that are not applied because the computing resources are limited.

I would make another comment that addresses your last question: what great new things might happen? We can point to all the things that we understand right now that would be helped by computing resources, but I would argue that there are a number of things that people have not thought of yet because they do not have the computational resources available to them. There is a great big question mark there as to what kinds of new things would happen if there was a three orders of magnitude increase in computer power. We certainly experienced that before in the radio field and the aeronomy field when array processors became available and, all of a sudden, people started applying new techniques to the data that just were not possible previously. I think we have got some answers. We know that there are, in fact, bottlenecks in a couple of specific areas right now, but we do not really know all of what might happen if we had additional resources available.

Tanimoto: One of the areas in which the potential has not been achieved is the general area of applying Machine Vision techniques, higher-level things than image processing, in Astronomy. A few people that I have talked to here are experimenting with some of these things, and Adorf mentioned morphology by Monique Thonnat at INRIA. This is perhaps an example of something that is leading the way in the higher-level kinds of vision in Astronomy. Generally there is a bottleneck in Machine Vision.

Certain kinds of applications can be handled by very simple kinds of things involving binary image processing and so forth, but the kinds of higher-level, more complicated Machine Vision, dealing with grey scale images and applications requiring some sort of real-time performance, are not doing so well, because of the limitation in computing resources.

Also, another area is the Artificial Intelligence area. If we had all of these parallel PROLOG machines and so forth that some people are trying to develop, then you would find a new kind of experiment in lots of sciences, including Astronomy, being done. Again this is something that not many astronomers, in my limited experience here, are 'climbing the walls' to get going on. Nonetheless, it is the Ferrari syndrome again, that if people started doing it they would see the potential of it. They cannot start doing it until the hardware, software and general methodologies are widely available.

<u>Levialdi</u>: Astronomy can be used as a testbed for new approaches in computer science. The relationship between the researcher and the computer is quite a complex one. I remember many years ago when we were comparing computer technology in the eastern countries - and, of course, they are generally behind - they had very good research in many areas. There is a duality in this fact, that if you have a lot of computers and a lot of computer power, then you can try new ideas. On the other hand, it can also stop you from inventing or enlarging your fantasy.

Turning to Artificial Intelligence, I think that perhaps the tools are not ready yet. For example, we have tried to design an expert system in a bio-medical application; the main result was that in formalising the rules we understood more of the problem. In other words the effort to transform knowledge into a set of rules, and therefore formalising this knowledge, was very helpful for the doctor because he could understand what he was deciding upon. He could also use it as a pedagogical tool for teaching others. Perhaps these ideas could be translated into Astronomy as well. The fact that you must formalise your knowledge is a good influence. Yet it could be that PROLOG (and the first order predicate logic) is not as flexible a tool as one would like. Again, I also feel that what we need is a more comfortable system - systems in which you can relax, as it were, and which do not require a lot of handbook pages to be studied before you can use them. I think there is always this difficult choice: if you have a new tool, how much time should you spend in learning how to use the tool? If that time is above a certain threshold, then you will not use that tool. So, perhaps new systems that do not really require an enormous amount of investment in terms of time and money to learn how to use them, could be useful for astronomers or other communities. But tools which are complex, in which if you make one little mistake then nothing works, and where there is a lot of documentation that you must browse through before you can really get to use them, are a hindrance more than a help.

<u>Duff</u>: I have had no experience in dealing with Astronomy imagery, but I have worked with medical imagery, and the impression I get is that all the push really comes from people on our side of the fence. The medics, or more particularly their technicians, say they would like more automatic techniques and so on, but when they are actually produced, they do not necessarily want to use them as they are used to interpreting their data in a particular way. When they see their data displayed in another way, they do not like it. For example, when Halley's comet came over, we saw a nice dichotomy between the television presentation to the general public and the response from the people who were in the control centre, because all the information came through in false colour. There was an awful feeling of anti-climax; everybody watching at home saw all this false

colour and could not interpret it in any way. It looked like terribly modern art! Everybody wanted to know where was the centre of Halley's comet. Where is the block of ice? What does it really look like? The data had been changed. The feeling was it did not look anything like what we were seeing on the screen. Various people tried to comment on the pictures. The interesting thing was that some of the astronomers there, people like Alex Boksenberg, looked at the false colour and said words to the effect: "I cannot say anything about that at all - I don't really know what it means." I am just wondering if it is the sort of thing that might be happening generally in Astronomy. All these techniques are available and, apart from rapid data transmission techniques, anything which modifies the data, so it does not look exactly how it was presented before, might not be so popular.

Crane: I do not think I necessarily agree with that. I think that false colour has its usefulness, and it is perhaps not in visualising shapes and sizes, because our human interpretation of false colour is a very difficult thing to quantify. We interpret colours in different ways than we do intensities. False colour is good for certain things and not good for others. It is only useful in relatively few areas. It looks spectacular and you can do all kinds of fancy things with it. I do not think people are reluctant to use these techniques. I think that they are not often the correct techniques to use, and that is the bottom line. People are still finding false colour pictures spectacular to look at, but not very easy to interpret.

Pat Moore: Could I extend the first question you asked? I quite agree with Phil Crane in identifying two areas of bottlenecks. I see users sitting in front of terminals reading books. I see users coming to blows over tape drives and disc drives. But there is a third area of bottleneck, which has not been mentioned, and that is the new software bottleneck. There are a lot of programs that need to be written, and any work that can be done in easing writing those programs is also very important. I get a lot of people hammering on my door saying: "Why is there not a program to do this?", or: "Why can't I do that?", and very often I just have to turn around and say: "No, you can't do that; we have not written that yet." That is possibly an even bigger bottleneck than the other two.

Duff: Is this a lack of people to write the programs, or a lack of know-how?

Moore: It is the difficulty in writing the programs. It is a fairly substantial effort to write some of these programs and, if we could ease that difficulty, it would be one way of addressing that bottleneck. I am not sure how we could do that, and I am wondering if any of the people here would like to comment on ways we may do it - whether the new AI languages are going to be easier to program with, in addition to being easier for people to use.

Levialdi: Help could be gained by having a specification language and then automatically producing programs. Could this be one way? In other words, having a fixed protocol for giving the specifications, then giving them to a program which returns a well-written, formally correct program. There are some people working in this area, so this could be one possibility. Another possibility is to have completely different languages, which are more natural for astronomers - that is, for astronomical objects, in a sense. I do not know if there have been attempts to produce special languages for Astronomy?

Shames: I would comment that it may not be so much that there need to be languages specifically for Astronomy, but that there certainly are languages in some of the extensions that we have heard talked about that have built-in array operators and things like that that would make specification of programs on large blocks of data much clearer. Stepping back from specificational languages, there are some things that are closer to current reality in terms of production facilities that would help an enormous amount: good interactive debuggers and syntactic editors are fairly obvious things that are starting to become available and should bring substantial benefits in terms of generating programs.

Levialdi: And perhaps also visual languages...

Shames: Yes, exactly.

Duff: Is it true that all the functional and applicative languages run so slowly that they are almost unusable if you have real data?

Shames: It tends to be interactive debuggers and interpreters that have that problem. But for development of algorithms and testing of algorithms they can still be awfully useful.

Don Wells: Regarding the question about languages for Astronomy, there is, in fact, an outstanding example of this. It is the language Forth, invented by Charles Moore when he was employed by National Radio Astronomy Observatory (more than ten years ago) and extensively used in Astronomy worldwide. Most people think of Forth as a language for data acquisition. Actually, over the last ten years, there have been a number of important projects that have used it for data analysis. The proponents of the language, and it is known that I have been one in the past, have argued that the key point about the language was its ability to define new syntaxes. I would claim that the difficulty in programming is basically that in a large scale problem - the difficulty of programming effort required is related to how many symbols you have to enter to define the program, whereas ideally what you would like is to write one statement: "do this". But, of course, we cannot; you have to break down the problem into sub-steps to be done and, unfortunately, in most of the languages you have today - like Fortran - you are only a few steps above assembly language and you have to break down into an interminable number of steps. I believe that there lies most of the difficulty in programming, and so the idea is to abstract your problem into a minimum number of steps to make it easier to comprehend as well as to generate. Now, in Forth it was always possible to define your own syntax which was particularly suited to the problem you were solving and, in fact, it is common among really skilled Fortran programmers to invent syntax on the fly while they are coding. Thus they manage to express their problem in a minimal number of symbols, and claim this really shortens development time. This is rather analogous to the kind of claims that LISP people make for why their language is superior.

Now, putting on another hat, a management hat, I will say that this approach historically has led to severe management problems. For which reason, for the last five years, I have advised senior management to discourage Forth use. The management problem has to do with the fact that if every programmer invents his own syntax, you are then unable to maintain those codes.

Tanimoto: In general, there is the problem that in order to make programming easier for some, it becomes more difficult to manage things and to have some sort of compatibility. I believe that standardisation is more important than making programming slightly easier for a few people, because it is standardisation that provides the incentive, or a large part of the incentive, to write a program. If you can write a program or if you can write prose or write a paper in a language that many people can read, over the long run it makes a larger contribution to the growth of the field. Now, whether Fortran should be the standard, I do not know, but, de facto, I guess that is what it is. If someone is going to spend a lot of time developing a new kind of program, it seems that it is actually good for the community as a whole if it is written in the standard language. Hardware, of course, has the same problem, that programs that run on one particular machine often do not run on another machine, especially the state-of-the-art kind of machines. If there is an effort towards providing standard interfaces, ways to plug in the same cards in new machines, more compatible cards in the machines and so forth, that will help this problem of standardisation.

<u>Shames</u>: There is a slightly different view you can take of the same problem space that touches on these two topics, which is to take a tool-orientated or a package-orientated approach. I would suggest that one of the reasons why Ada is as interesting as it is, is that it does have this package concept, and that the tool-orientated concept is also something that appears very obviously at the surface in Unix where there is the notion of pieces or tools that you can string together in interesting ways. This is in itself an interesting paradigm for approaching data analysis problems. There is another activity out in the computing field now that takes much the same kind of approach, which is the Gibbs language. This essentially takes a lot of specification of problems and solutions for problems in a more or less natural language, and identifies packages of functions which can then be tacked together in interesting ways. This kind of approach is one way of reducing programmer activity, but it is also a method, from the top down, for giving the user a way of coming up with a canonical "do it" function, where the user says "do it" and the function is performed by a hidden collection of operators which have been tacked together. It does not do away with the need to invent or define new algorithms, but if those algorithms are done in a more or less generic way, as much as possible, or generalised, as much as possible, with clean interfaces, they can then be applied to other problems. This is a very powerful way of extending the power that people have available to them.

<u>Tony Reeves</u>: My experience of managing a software package to provide convenient tools for Vision, is that scientists have minimal interest in spending the care and effort needed to develop a package that can be used by others. They want to get their science done. If they are really forced into writing a piece of code, then they will want to write it in the most expedient way possible. Certainly, this is not the goal of the system manager who has the almost impossible task of trying to assimilate that piece of code into a more general tool. If you bring in a language such as Ada which imposes a tremendous amount of structure on the way you do things, this would be rejected by the scientists because it is more difficult to use. They want something that is as easy to use as possible to solve their problem and then to forget about it.

<u>Shames</u>: You are absolutely right, but I was approaching it from a slightly different point of view in that there is a fairly large class of programmers who are out there writing programs in support of scientists. I happen to be one of them. There are a number of other people who work in that regime, and one of the things that we try to do is to build tools in a generic way so that the astronomer can then tackle his problem very quickly by stacking these pieces together in this tool-orientated mode.

Crane: But that is not what Pat Moore had in mind at all. I think he had in mind a man sitting at the terminal saying: "Hey, I need to turn every left-handed photon into a right-handed photon and, although I do not have that option, I want to do it now." I think that is exactly the point that Tony Reeves points out, that when you want to do something, you want to do it fast. You are going to do it the quickest way and the easiest way, and if that happens to be a dirty way, that is not important because you can do your job and get your science done. You can later walk over to the local system manager and say: "You need a program to turn left-handed photons into right-handed photons, and here is mine, but it might not be of any use generally." I am sure that that is the kind of environment which most scientists would like to have, and it is not this great structured environment.

Shames: Isn't that situation exactly the one that you would like to have: an interpretive language that is more or less orientated towards the problem space, so that when you do not have the tool at hand you can build one very quickly that solves that problem and, because this is quick and easy to do, it can be a 'throw-away'? Now, it may be terribly computationally intensive, in which case perhaps you lose as much as you gain in the long run, but you have got the quick facility for putting the piece together.

Ian van Breda: As a Forth user, I would like to at least speak in its defence. I use Forth in the laboratory and, by using that language, I have been able to go through three generations of microprocessors. I do not have any professional programmers working for me; I have detector experts. I am absolutely convinced that I would have had to have had a professional programmer to get the same software base that I now have for detectors available to me in the lab, if I had gone the Fortran way or any other way. The other thing about Forth, in particular, is that it allows me to write compilers, that look like Forth themselves, so as to be able to drive bit-sliced processors the same way as I drive image processing. You have got to be very careful with standardisation in that the language can lock you down. You have very limited structures, very limited ability to put in new ideas, completely new ideas, into things like Fortran. For me Forth does that. While I fully acknowledge that that is not the way to run a major image processing system (and I am very excited about the possibilities of dovetailing Ada with the sort of work we do) and that astronomers would want to write in Fortran and the like, nevertheless for the sort of work in our laboratory it is easily the best and most efficient way for us to work, and I think you do have to take the most efficient way available to you.

Shames: I just wanted to second this statement. My criticism was of the use of Forth in things like data analysis systems where the software project gets very large and where they are used for diverse purposes, and lead to management troubles because of the longevity of the code and maintenance problems. In a real-time environment, Forth has a long history and a lot of success to its credit.

van Breda: I fully agree with that; it is just that we must not lock out Forth because you say you have only got these other standard languages; there are other things to think about.

Duff: I would suggest the answer to the first question is: "Yes, there is a computing bottleneck but that not all astronomers are completely aware of it; some are very intensely aware of it and, at the moment, the pressure is not just from the computer scientists to sell computers to astronomers, in any sense." What is also coming through very strongly is that the astronomers feel the problem is in the software. So, it may be that we have already answered the second question: "Why don't astronomers demand the newest and fastest computing techniques?" The answer is: "Because they feel they are not going to get the software to drive them." They really want faster techniques and so on, but they know that somebody is going to have to program these machines and they are damned sure it is not going to be them. Is that a fair statement?

Crane: I think it may be a bit unfair, because although I would say that your interpretation of software being the critical thing is correct, the solutions to a lot of these problems are often unclear. Exactly what piece of software you need, or what technique you need to approach a problem, is not often clear. Many of the problems which you are attacking, you attack better on a very small workstation system than you do on a super-computer. This is true at least for the area of research data analysis as opposed to production data analysis. Once you have decided what it is you want to do, in very much the same way that the radio astronomers have decided that they have to do Fourier transforms or cleaning algorithms and so on, then I think computing needs are very strong. It is only in the algorithm research area where it is not useful to have a super-computer. You do not really know what it is you want to do to begin with.

Duff: What is the balance between the amount of effort needed for production computing and for development?

Crane: It depends very strongly on the particular problem. For example, at ESO we have a package that does reduction of echelle spectra. After the initial effort of setting up the parameters, which is an interactive task, it becomes quite compute-intensive, whereas in developing the approach to this business and in setting up the procedures which are appropriate, it is extremely interactive. In research algorithm development, it is only when you finally have figured out what the algorithms are that you can turn the program loose on ten or fifteen spectra and get out your results. So it depends to a certain extent on the maturity of the particular problem, and on the level of infrastructure that you already have.

Duff: But, you know, we find that the major advantage in having a very fast computer in the CLIP system is for algorithm development, not because you have got to do a hell of a lot of computing, but because you want to do short bursts of computing very quickly. When you want to try an algorithm, it is very nice to be able to flip through different algorithms, almost as with a push button, and see what gives you the effect you want. When it actually comes to production 'churning-out' of results, then you can leave the thing running and wait a bit, but if you are sitting there, and you press a button and then have to wait twenty seconds for anything to happen, it is maddening! But if it all happens in less than a second (and if it is wrong you can immediately try something else), then you can develop your algorithms very quickly.

Crane: I understand what you are saying. Maybe we have not reached the sophistication in the algorithms that we are talking about to need that kind of speed. It is easy to write down an algorithm mathematically but when you actually go through and do the calculations, it is often not trivial, and you have to worry about all kinds of edge effects. You do not need a lot of computing to do that. You just have to try a little bit of it. Maybe I am wrong; maybe the algorithms are sophisticated; but at least in many of the algorithms that we are worrying about they are unsophisticated and/or we have not approached some of the more sophisticated techniques which are available, possibly because we have not been shown the way by people such as Steve Tanimoto!

Duff: Are you going to respond to that, Steve?

Tanimoto: Well, listen, that is my answer to question number four!

Duff: Okay, we will leave that till question four then. You see, the reason I posed the question in this form is that I am always a little bit suspicious of a situation in which nothing is happening very fast. Usually, if anybody is keen to do something, things begin to happen. Now, the point is, are astronomers jumping up and down and saying: "We must have more tools", or is this an illusion? The message that is coming to me, as a complete outsider, is that a few people are saying they would like more tools, but that the body of Astronomy as a whole is not, at the moment, that worried about it.

Levialdi: For instance, could it be that, as in many other areas, there is a lot of investment in existing programs, and also investment in the fact that a lot of technicians already know and are skilled for specific computers, and specific terminals and specific procedures, and to change all this takes time, effort, and money. So there is always, in the change of technology, a sort of a barrier or inertia, like in other completely different fields where you may have ten million lines of code of Cobol programs which would be very difficult to change. Yet, a lot of commercial firms or software houses speak of new structured programs, new techniques for management of software projects and so on, but they are not using them because change is very difficult. In Astronomy, although I think a very important borderline is the one that Phil Crane was mentioning (between research and production), and although this borderline might be fuzzy in some places because some consequences of a research project might directly impinge on production, yet I think the requirements, the needs, and the barriers are different. I am referring to those in a research environment and the others in a production environment. Perhaps it is difficult to move because one has: a) an investment which must be protected, in some sense; all that has been done must be used to the limit; and b) one must be careful to choose the right computer, the right system to adopt, particularly since computer technology is moving so quickly.

Shames: I think I would tend to agree with that and I would make a further comment. There are a number of areas or examples we can look at in the past in Astronomy. Array processors are a key example; they are something that is fairly well accepted right now and are used in a number of different places. There was a great deal of difficulty in getting them to be used; those few people, who saw that they had a need, were willing to invest in learning how to micro-code these 'beasts', had some great successes and other people followed on after that, who are using the same codes or developing new codes, both in post-processing and also in real-time data acquisition. But they have now come up against a bottleneck and that is that if you want to apply these same techniques to newer machinery you have to suffer the same costs of reinvestment all over again, because the codes are not portable. That is a very important point in thinking about how to use cellular arrays, parallel architectures, systolic arrays and so on. Tony Reeves made the point very clearly. If the user or the programmer, every time he wants to use a new architecture or program a new architecture, has to think entirely differently about how to tackle the problem, there is going to be a great deal of difficulty in gaining acceptance for these machines. It may fundamentally be the case that there is no choice, but I think a very fruitful area for research and a problem that is going to have to be resolved before these new architectures are really easily or readily accepted by the community, is in coming up with some sort of high-level language.

Crane: It is very important to get high-level languages available to make use of these new techniques. I could not agree with you more; we will not really find them useful as general tools until we have the languages.

Ethan Schreier: I would like to go back to your question of a minute ago. You cannot make a uniform statement which is pertaining to whether or not astronomers are demanding better techniques or whatever. In fact, they are. However, the level of sophistication in the Astronomy community differs widely. You have people who have been exposed to sophisticated computing for twenty years, if you can define sophisticated computing as having existed in the mid-60s, as you certainly can in some ways, and there are people who have come on to it much more recently. Therefore the trend I have been seeing is that as more people have become exposed to better computing, they tend to want to go in that direction. As the problems get more sophisticated, the programs also have to get more sophisticated; it turns out that the software people that are employed become more sophisticated themselves. The astronomers do realise the need to get more sophisticated programming done so it is getting to be more and more the case that the 'Macho' programmers ['Macho' meaning individualistic, non-conforming - ed.] are moving much more in the direction of not only turning out Macho code, but turning out good Macho code, which is portable and modular, code that does tend to live. The result is that in more and more places, systems are evolving.

I have divided things up into two levels in the Astronomy community. The first major breakthrough came with systems being developed which were used by many people rather than by a single person. These were the first-level systems. The next-level system seems to come when the sub-discipline makes a new jump and it becomes a major user facility, where many people have to use it; furthermore, it is often used in several places. At this point, the level of resources invested in making such a system is large enough for people to realise that they cannot all duplicate it; you cannot have new systems developed in every observatory and every Astronomy department around the world so you are going to be forced to agree some standards; it is an inevitable course we are on. What we can do is to try and help things go in the right direction, and the more sophisticated systems will be pulling in inputs from computer science. You see that happening right now.

Duff: If I can quote our experience at University College London, we have a processor array, a cellular logic array, which we developed in our group. It runs under the control of a PDP-11/34 (which we are just exchanging for a SUN workstation) with a UNIX operating system. The point is that the different users all use the system with different levels of sophistication. People like myself, who do not get much time to program these days, need a menu, push-button system; if the problem gets more difficult than this approach can handle, then one calls for the help of a graduate student who understands how to write the code for the next level of program sophistication, and so on. The essential point is that we can all do something with the system. For much of the time, I do not know, when I use the system, whether a program is running in the array processor or in the host. Now, is the future for Astronomy, with a big, mixed user community, an environment in which everybody who logs-in finds a system which he regards as user-friendly, at his own level of sophistication?

Reeves: I would like to comment that I do see in the systems coming in that UNIX in itself has become a de facto standard for workstations. So now you do not have to worry about what software environment you will find when you go on to a workstation; you know to a large extent what it is going to look like. With the very first generations of the new Hypercube

machines that are coming out, there is a strong move to enforce that the
languages will be C and Fortran. The message-passing primitives have
been established and standardised, so that any software developed should
port between all these different types of machines, although maybe not
with optimum efficiency. Anything that is developed should be able to be
ported now. That is a very different situation from that when Floating
Point Systems came out with their array processors. I think this trend
is going to apply more in the general community and that areas such as
Astronomy are going to attach on to the tail-end of that movement and find
that a large part of the environment is defined for them. This would be
very beneficial; you will have programs that work on your own system and
when new hardware technology comes along, you can expect your programs to
be invisibly ported to the new computer architecture, or whatever, and not
have to know the details of exactly how that is being done.

Moore: Could I just get back to a point that Phil Crane made earlier when
he said that you can do a lot of algorithm development on smaller
machines? I would like to argue fundamentally with that point. One of
the things that we have discovered time and time again is that a lot of
the algorithms that we have are very poorly founded mathematically; they
have peculiar instabilities, unexpected behaviour, and almost every time
you 'push back the frontiers' and go on to a slightly larger image or a
slightly deeper image, you come up against different problems. Almost
every time somebody does some image processing, they are in some sense
developing algorithms because they are discovering new problems that were
not known before. That is one of the main reasons we have been pushing
for a single super-computer, so that it is possible to interact with the
process, to see how it is going, and to do algorithm development as the
astronomers are doing their regular processing. The problems that we are
coming up against are sufficiently complicated that they are not solvable
by an astronomer on his own, or by a computer programmer on his own, but
probably require both of them working together and conceivably even a
mathematician working alongside them. They involve very, very subtle
interactions of complicated algorithms, so I do not think you can do any
of that work on very small machines.

Crane: In the particular case of the kind of things that you were talking
about I would totally agree with you. I was talking more about my own
experience in optical data processing, where the situation is somewhat
different.

Tanimoto: I am a little puzzled in one respect. I have not heard astronomers demanding integrated workstations for Astronomy. In a number of
other fields there are clear efforts to develop workstations, like a music
workstation for all kinds of musical research, composition and so forth.
Obviously there have been lots of efforts toward integrated office automation workstations that present word processing, mail, and all sorts of
things. In Mathematics, for example, we have systems like MathLab and its
successor Maximum which are intended as work-environments almost entirely
for the people that do that kind of thing. So far I do not think I have
heard anyone say that there should be an astronomer's workbench system
that perhaps integrates electronic mail, word processing, image processing
and data analysis libraries support for image database analysis...

Crane: Tell us how to do it, tell us how to do it!

Tanimoto: Well, you have to want it before someone is going to develop it
for you...

Duff: Is the Starlink system approaching that in any way?

van Breda: Can I just make a point, not on behalf of Starlink, but there was a problem when we asked if we could connect our 68000 or 68020 machines into Starlink. It is always a problem with VMS that you run into a brick wall with this sort of requirement.

In my opinion, astronomers do want the sort of facilities we are discussing, but I think there is an educational problem; they do not always realise they want them; they know they want to do the Astronomy but they do not know how it can be achieved. With a lot of these imaging activities there is an educational problem as much as anything. Some of the concepts are very sophisticated and astronomers are not used to that. I mean, you have still got astronomers around who worked with brass knobs in the past...

Shames: Well, let me just comment on that for a minute, too. The point that was being made much earlier, relating to the computational loads and the I/O loads, was directly addressed to that. You have to understand that the canonical image processing workstation to date has been a VAX-780 with a DeAnza or Ramtek display on it and that that is the way people think. They have been thinking: "I need to do image processing, therefore I need the 780 and I need a high resolution display and I need a tape drive on it and I need a disc drive on it, and that is what is required". Well, it turns out that you can essentially do all that with a Micro-Vax II or a SUN-3 or whatever. If you have got five or ten megabytes of data per image scene, you have got some I/O problems using the Ethernets that exist right now.

Computationally, if you are trying to do things like convolution or spatial filtering, then there is a little bit of a lack in computational resources, but at the 90% level the workstation really does provide adequate computational resources. What has happened is that it is only just now that we are starting to have the software packages and the hardware together where we can start providing workstations for astronomers to use. We think we can get workstations at the 50,000 dollar limit, which is very much more interesting than the 350,000 dollar VAX, but we are still going to have I/O bandwidth problems. It is going to take the new fibre optic technology or the much higher bandwidth networks that are becoming available to make that really viable. People are starting to move that way.

The demand is not there because the understood demand has been a 780-class machine.

Tanimoto: But isn't there also the problem of providing an integrated front to the user? Clearly, you can go out and buy a lot of these components and put them on a workstation, but I am not sure that makes an astronomer's workbench; I think there is again some kind of overall consistency that is required in building one of these systems, consistency from the point of view of user interface and all that sort of thing. I do not know if anyone is sufficiently interested in this for it to have been done, and in a way it is a little surprising.

Shames: I guess I would say that we are doing it. At least, I think we are, and I believe this is the same direction that is, to some extent, being followed by at least a couple of groups (MIDAS, IRAF and SDAS). These activities are not complete by any means, but there certainly is the concept of an integrated user-environment in which you can do optical processing, image processing, spectral processing and to the extent that the algorithms are made available and plugged into the environment, radio processing. Also, you can access underlying operating system capabilities and, using command language shells that are separated from the underlying computational capabilities, people wanting to apply some of these new icon-based systems on the front end can do so. Certainly there are directions being pursued along those lines.

Schreier: I wanted to say also, partly as Peter Shames has just said, you are wrong in thinking that some of these things are not being done, there are several groups working in this direction. There is a problem which is that, although astronomers know what they want, the capabilities do not exist yet, because the astronomer's workbench, the single thing that most astronomers would like to have now as a local environment, including fairly sophisticated image display and analysis, is not available in cheap workstations. We are starting to get there but the workstation market mainly has been driven by manufacturing, by Cadcam and, as you were saying, astronomers have very little leverage in this market.

Therefore it is not that technology does not exist because, in all the fancy Cadcam workstations, the same technology could be used for deeper images. Existing systems developed from simpler graphics systems currently can handle 8- or 10-bit images but are not yet able to cope with holding several 12- or 16-bit images. I think Astronomy is not a big enough market to drive the technology. You will find that the few companies that have had a closer relationship with scientists, have now developed systems that are near to what is wanted. The SUN workstations are getting to be very popular in Physics and Astronomy departments and are coming very close to supplying the facilities that are needed. So, we can work on it from the system standpoint, saying what kind of user interfaces and lower level standards we want, but until the hardware reaches the required sophistication, we cannot really expect the Astronomy community to be able to drive industry to create the product we need.

Don Wells: Of course, I generally agree with you, except for one little detail. You said that SUN workstations were becoming very popular, 'SUN' being generic technology here. I do not quite agree with that. If you mean there is one in every major institution and even in a lot of minor institutions, yes. But if you are implying we have reached the point where a senior astronomer demands one as part of his rights in the working environment, no, we are not there.

I would claim that the 'personal computer' mentality has stood as a major barrier to the advance of the technology that Steve Tanimoto suggests we need. I agree we need it, and obviously Peter Shames does as well. In fact, a number of people in this room who are involved in R and D software, computer hardware, hardware R and D in Astronomy, are very well aware of what is possible with the multi-window workstations and sophisticated software environments. We know what the implications of this are; we have a sales job when we go to our senior management and ask for the dollars. Unless they are very imaginative management who can see through the history of the past and see into the future, or unless they happen to trust us unusually well, they will not fund the projects. So normally these workstations are insinuating themselves into institutions on other justifications than their real justification. They are coming in for other reasons. Now, what I hope is that they will be successful, that the pilot projects will demonstrate a style, a spirit and a capability which will be attractive to the senior astronomers and make the revolution happen. One point that I would express here is I think the computing budgets of many astronomical institutions are set too low and that is going to be a barrier also.

Duff: We have drifted into the third question, which asks what new computer technology is actually commercially available today. The SUN work-stations have been mentioned repeatedly, particularly SUN-3s, I presume, with full colour and all the rest. I understand one could buy an MPP, although I do not think one has yet been sold. ICL have progressed from selling their rather unsuccessfully large and expensive Distributed Array Processor to a smaller and cheaper machine, MINIDAP, which can be used as a co-processor to a conventional small computer. Our own system, CLIP, is now manufactured and marketed by a company in the UK. Floating Point Systems have been selling array processors for many years and Sternberg's CYTOCOMPUTER has recently appeared on the market through several companies in the United States. There is also quite a lot of activity in Sweden stemming from the research at the University of Linköping (GOP and PICAP) and commercial versions of their research machines are now available.

It was requested at this point that the report of the panel should include some further information about the systems mentioned. The following addresses may be of interest to readers:

Dr Caldwell McCox (MPP)
Code E1, NASA Headquarters
WASHINGTON, DC 20546
USA

ICL (UK) Ltd (DAP or MINIDAP)
Bridge House
Putney Bridge
LONDON   SW6
England

Stonefield Systems PLC (CLIP)
Lawson-Hunt Industrial Park
Guildford Road
Broadbridge Heath
HORSHAM
West Sussex   RH12 3JR
England

Floating Point Systems Inc (Various array processors)
Box 23489
PORTLAND
Oregon 97223
USA

Machine Vision International (CYTOCOMPUTER)
325 East Eisenhower
ANN ARBOR
Michigan 48104
USA

Contextvision (GOP)
Teknikringen 1
S-58330 LINKOPING
Sweden

Teragon Systems AB (PICAP)
Teknikringen 3
S-58330 LINKOPING
Sweden

Duff: To the best of my knowledge, most of these systems have a limited amount of software available, but the impression I have is that it will not be angled particularly well toward the astronomical user. So, are there any systems, which anybody here knows, which are commercially available systems which you think could be recommended immediately to astronomers?

Levialdi: Some of these firms trade-in software for hardware. If you write some programs of the sort you find in image processing packages, then you get a discount. You can then get access to the whole user-community group and can try the machine and so on. I think this holds for GOP and also for certain image processing packages, GIPSY, for example.

Shames: There is one more machine that you did not mention and that is the Sky Warrior. There are some other Sky floating-point processors that are available to plug into workstations. That class of machine, I suspect, is a rather fruitful area for investigation. You start thinking of several megaflop machines that you can buy for 10,000 dollars and plug into a workstation, turning what is a very powerful user interface with the appropriate software, into a number cruncher. This can be very exciting, but I want to reiterate one more time that the I/O bottleneck is still something that needs to be solved before all of these things can become really viable. I suspect the solutions will come from manufacturers two years from now in the form of FDDI and things like that, but they are just not here yet, so we still do have bottlenecks in the data path.

Levialdi: Another class of machines which, however, may not be particularly suitable for Astronomy, are those built as special-purpose boards for personal computers. I am thinking particularly of work done in association with Serra of the Mathematical Morphology group in the School of Mines, Fontainebleau. They have come up with a new version of a machine in which they have coded the language MORPHAL, most of their algorithms being in a structured form. Another machine is based on Fabbri's software called GIAPP. It is interesting because you can try out a great number of algorithms at very little cost. Of course, for Astronomy, this may not be sufficient, but, at least for trying things, it might be adequate in a restricted, research environment. This, therefore, represents another class of machines: low cost, PC-based, well documented, standardised with a lot of research behind them, so you can also contact people for more information.

Tanimoto: Yes, I think that is a really significant development. Things are happening not only top-down, big machines getting smaller or cheaper, but also it is becoming feasible to do some certain kinds of work on PC-based systems as we get fancier and fancier systems. Sometimes you can buy a graphics board for an IBM PC or a compatible, which costs maybe 10,000 dollars, whereas the computer costs 2,000 dollars, but which really makes the power of the whole system comparable to something that used to cost 50,000 or 80,000 dollars. Someone recently brought to my attention a new board that has been advertised by Matrox. It plugs into the IBM PC AT, and apparently for 2,500 dollars you can get a 1024 x 1024 frame buffer with digitising capability, display capability, and a convolution engine and that is a lot for that kind of money! It seems that this trend is likely to continue and that people should be looking at systems like this. There is another thing that people in the States are excited about: there is a family of image processing boards for, I believe, both MULTIBUS and IBM PC-BUS from the Datacube Company. They have a whole long line of different boards for different operations in image processing, and I think they are all priced in the under 5,000 dollar range.

Crane: Steve, I agree with you; these are all very exciting developments, but I would question whether or not they are really appropriate today for the kind of problems which many astronomers are trying to solve. In spite of the fact that you have a 1024 x 1024 display, living colour and black and white, digitising capabilites and what have you, many of the problems that we are looking at are not just compute-orientated but are also I/O-bound. These small beasties are not yet at the level of being able to handle even four or five of Peter's five megabyte frames.

Levialdi: Can I just add that the video disc is coming very soon, and that these will be connected to PCs in the very near future. This technology is nearly mature in the sense that it will be possible to record and to read data very quickly. I just wanted to mention this as a way of having cheap memory and for inputting or outputting images rapidly.

Crane: Yes, sure enough, I understand that the technology will be there and I suspect that next time we have a meeting, the whole range of capabilities will be changed. Today and now, I think we are still forced for our minimum system to go the MICRO-VAX, SUN, or MASSCOMP route, in order to get: a) the generalised computational abilities and b) bigger data storage capabilities that I think are needed.

Duff: We are approaching the end of this session, and there are two questions left, the first one being: "What positive action should we take to advance the use of computers in Astronomy?" Nobody has said very much about budget availability. If these new systems do prove powerful, and useful to the Astronomy community, is there a feeling that money would be available to fund them, or is Astronomy running very tight on its budget at the present and therefore it is all 'blue sky' stuff?

Crane: I think you have to differentiate if you are talking about money. Money does not necessarily differentiate between institutions, but in effect it does, and if you look at the type of institution that Peter and I represent, we are quite handsomely funded by comparison with some of the University environments such as Ivan King and Harlan Smith are from. If I am slighting anybody, I apologise! But we are big public institutions and we are in a much better position to take advantage of relatively more expensive technology than some of the University departments. Therefore the availability of big parallel processors, DAPs and CLIPs, would be more appropriate for the kind of institutions which we represent than for University departments, at least within the next generation of hardware development. Whereas the kind of thing that Steve Tanimoto and others were mentioning are more appropriate for University departments. However, there is always a money crunch in that you would like to buy more than you can afford. Nevertheless, the spectrum of what various institutions can afford is pretty broad. Money is important on the two levels that I mentioned, at least two levels, probably a lot more.

I would guess that in our institution, ESO, were a 40 or 100 megaflop type of machine available in the price range of 500,000 dollars, we might be in a position to take advantage of it. Maybe those numbers are off by factors of two but, in any case, that is more or less the range in which we are able to operate.

Duff: What about positive action? Is there anything, following a discussion like this, that one ought to go away saying, such as: "Well, now we have heard the story, what are we going to do about it? I know, we will..."? For example, should one try to carry out a survey of all the available equipment, or draw up some specifications, or whatever?

Crane: You are a hardware man so I am sure you are thinking in the hardware area, but let me come back to the point that Peter Shames raised, which I think is crucially important and will bring these new developments much closer to the everyday man in the street, that is the software. We need to have developed the techniques in the software, in the high-level languages, to take advantage of these advances in hardware capability. Unless we do that, the techniques are going to be available only in very special cases and very special places.

Duff: But who is going to develop the software? You see, you will not get hardware manufacturers to develop the software unless they have got a customer.

Crane: I would put it to the hardware manufacturers that they are not going to get customers unless they develop the software. There has to be a joint effort.

Don Wells: I was looking at your positive action suggestion and I do not know what to say about putting massively parallel computers into Astronomy. Phil Crane is just right that until it is apparent that we can apply them easily, we will not try. Now, I will give you a strategy, an overall strategy which I think will optimise our move toward new technology. I call it the 'Open Systems Interface Strategy'. What it says is that, in your procurement actions, you try to choose components and systems which conform to national or international industry standards. The implications of this are, both in hardware and in software, that when you want to use new technology, a vendor who wants to, can build a sub-system that conforms to these standards which will be plug-compatible with the existing systems. Thus we can evolve, with not too many discontinuities, from one generation to another. Now, the kind of standards I am talking about are like the operating system UNIX's agreement on network protocols, agreement on language interfaces, agreement on subroutine libraries. A very important point is agreement on I/O interface standards, so things like VME or MULTIBUS promote innovation in the market. In fact you see this in the PC. The PC is a lousy design! It is poor, it is an obsolete design, poor chip, poor architecture, the operating system is mediocre at best, and yet it is an astonishing success. The reason is, it created a de facto standard, which caused a great deal of innovation to occur around those interfaces within the PC. Effectively, the PC interfaces have become open system interfaces. Actually, Digital Equipment Company did the same thing with the UNIBUS. They probably did not intend to do it, but they created an open interface which has stimulated all the third party vendors of hardware, and in fact software as well, surrounding the PDP-11 and the VAX. I claim that standards, actual frozen designs, far from inhibiting innovation stimulate it and permit it to enter the market place profitably. It is a good strategy for astronomers to follow.

Shames: I would make one point, again related to the software issue. Probably the toughest single problem that is going to have to be dealt with is to somehow provide a language that gives access to array processing and parallel processing capabilities, that is sufficiently attractive for people to be willing to stop trying to do things in Fortran. I think Ada is not the one because of its complexity. Whether Parallel Pascal or Parallel C or something similar is the potential answer I am not sure; but certainly languages like that are far more accessible than Ada would be. We clearly do not want to have to force people to try and write programs at the micro-code level and to the greatest extent possible we would like to hide the details. It is the whole concept of information-hiding that you want to hide the details of these underlying architectures to the greatest extent you can. In trying to get the scientific community, which is fundamentally a very conservative group of people, out of the Fortran rut, they have got to be shown that there is some better tool that they can use. It has got to be the responsibility of the machine vendors, I won't say the hardware designers or the software designers, but these two components, the hardware and the software, have to work together to provide the kinds of solutions that will really make these machines accessible and useful.

Hans-Martin Adorf: I would like to throw into the discussion a point which I have not seen discussed up until now. It is related to an old astronomical instrument where the data reduction rate does not cope with the data production rate. This is the Schmidt telescope with which you can acquire a deep survey plate, say, in between one or two hours, having a gigabyte of data which cannot be processed in real time.

Shames: The data processing problem is still a substantial one and all I can do is reiterate the comment that I made before, we need both computational power and I/O bandwidth in order to tackle that problem head-on. The computational power we see coming; the I/O bandwidths are coming but they are a couple of years away.

Crane: I would also add that the photographic plate is an extremely efficient storer of archival data. Despite the fact that data might not be processed, it is at least archived efficiently.

Shames: That is data storage but not data retrieval; this is the fundamental problem.

Duff: We have actually run over time so I would just like to ask the last question and I am asking it hoping for one-minute answers, if anybody has a one-minute answer! Sometimes when there is a big advance in technology, things do not just increase, they change qualitatively. The purpose of this last question is to ask you to make a wild guess and say if there is likely to be a qualitative change in Astronomy if computing takes a quantum leap, which may be three orders of magnitude either in speed or in ability to handle more complex computations. Now, this will be a wild guess which nobody will hold you to in ten or twenty years time.

Crane: Well, I think I alluded to that originally when we said: "Why aren't we demanding?" We will start applying far more sophisticated concepts of how to do things - the kinds of things that Steve Tanimoto was talking about yesterday as well as low-level operations such as Fourier transforms and deconvolution. Many of the things which we are doing today are limited by the computing power which we have at hand and we get by without doing some of them or by doing them in a poor man's fashion. There is no question about the fact that the way we approach image data will change when we have more and bigger facilities available to us. It is already obvious in the way we have approached things in the last five years. Qualitatively, we are doing things in a very different way now. If you handed us a thousand times more computing power tomorrow, we might not know what to do with it, but five years from now we certainly would.

Duff: But you cannot guess what might happen? You are not going to suddenly discover new universes or something?

Crane: We might be able to codify our pictures in a very different way and have completely different ways of looking at them. We are going to have different wavelengths or we can have different ways of representing the data which would aid us in interpreting what is there. We have not yet really been able to do that. We extract parameters from what we think is there.

Reeves: I have a comment which relates both to the previous question and to this question. We are talking about higher-level languages. There is a lot of software effort that goes into developing them and it has to be an efficient higher-level language to effectively use a highly parallel processor. Now, the vendors that are producing the super-computers today do not have the resources to also provide a speciality software package for every application that comes along. A classic example is the system coming out from FPS, where I think it is being recognised, at least at Cornell, that we have to develop our software from scratch. If you want three orders of magnitude without a similar increase in cost, then the way you have to do it is to put the effort into it. It is the responsibility of the astronomers to come forward and say: "Yes, I will develop the software environment that I need because I need that performance". That is the way it is evolving at Cornell, where a group of researchers are getting together because they acknowledge that current resources are not adequate and that a ten- or twenty-fold increase will not solve their problem. They are going to have to go forward and integrate with people who are in the computer science area and say: "This is how we want to specify our software; this is how we want the machine to work; we are going to be prepared to help develop that environment because we need it." If you do not know that you need something, then you cannot expect someone to give it to you on a plate!

Don Wells: I will just try speculating. Your parallel machines are going to happen and I think they are going to be important, but I do not think they are going to change science radically; they will just 'up' the compute rate. The same is probably true of the useful vector machines that we already have. However, there is a little hidden thing that is going on: Artificial Intelligence. Parallelism applies in AI as well although a different kind of computer architecture is needed. I believe that when that comes in, with not just a thousand but probably a million times greater speed than we see today, then the changes are likely to be very great indeed.

Unfortunately, because the discussions ran for longer than had been anticipated, the recording discontinued at this point. In fact, neither the panel nor the audience were willing to commit themselves to predictions of major new advances resulting from vastly increased computer power. The general feeling expressed was that the same sort of work would continue but at greater speed. Perhaps this is the essential nature of scientific research: if you can say where you are going then it is development rather than research that you are doing.

Clearly, all the speakers felt that immediately available and soon to be available advanced computer technology will have to be adopted by Astronomy if proper progress is to be made. It was generally agreed that this will call for some substantial degree of effort on the part of the Astronomers themselves - an effort which would be well rewarded.

Spoken English and written English are two very different languages. I hope that my unskilled attempts to translate the one into the other, involving a certain amount of compression and rewording, has neither impaired the quality of the original discussion nor misrepresented the views expressed by the speakers. Should this have occurred, then I hope those concerned will accept my apologies.

Michael Duff

INDEX

Active optics, 45, 53, 145
ADAM (Astronomical Data Aquisition Monitor), 109
Adaptive optics, 145, 150
Artificial Intelligence, 138, 274, 285
Automatic classification, 31

CAD/CAM workstation, 138
Catalogs, 182, 195, 306
Catalogs systems,
   DIRA, 309
   IRAS, 306
   SCAR, 309
   SIMBAD, 181, 183, 309
   STARCAT, 195, 311
CCD, 18, 21, 45, 47, 53
Cellular machines, 203
Cine-CCD, 53
Classification, 31, 62, 97, 110, 112, 277, 280
Cluster analysis, 12, 87, 97
Clustering methods, 31, 36, 87
   ISODATA, 33
Coded aperture systems, 78, 82
COMPTEL telescope, 171
Computer vision, 285
Confirmatory data analysis, 5
COS-B satellite, 87, 96
Crowded fields, 45

DAP (Distributed Array Processor), 248
Data analysis, 3, 127, 141, 155, 273, 280, 315
Data archives, 163, 193, 306
Data bases, 128, 132, 143, 152, 181
Data processing, 157
Data reduction, 3, 49, 166
Dataretrieval, 305

DBMS (Data Base Management System), 161, 175, 308, 310
   INGRES, 161
Discriminant analysis, 31

EINSTEIN satellite, 157
Estimation theory, 6
EXOSAT satellite, 157, 181
Expert systems, 109, 273, 276
   MYCIN, 279
   PROSPECTOR, 279
Exploratory data analysis, 5

Faint objects, 45, 53
Fast vectors computers, 226
FITS, 133, 139, 167, 310
Flat fielding, 47, 51
Folding, 87

GAPP (Geometrical Arithmetic Parallel Processor), 240, 248
Graphics systems,
   AGL, 168
   GKS, 168, 176
   NCAR, 168
Gridding, 187
Grondbased telescopes, 141, 144

Image analysis, 157, 286, 291, 294
Image decoding, 77
Image degradation, 46, 51, 54
Image processing, 175, 294
Imagery, 49
Indexing algorithm, 187
Inference engine, 275, 281
Interpherometry, 151
ISMAP device, 259

Knowledge base, 275, 278

Languages for parallel processors, 225, 263, 270
Level vision, 217, 253, 290
LISP machines, 254
Low resolution stellar spectra, 61

Mapping technique, 12, 187
Maximum likelihood method, 8
MDBST (Multidimensional Binary Spanning Tree), 37, 38
Method of moments, 6
Microdensitometer, 109
Minimal Spanning Tree, 32, 89, 98
Multicomputer architectures,
    Hypercube, 208
    MIMD, 225, 235, 238
    Multi-SIMD, 214
    Pyramid, 208, 263
    SIMD? 214, 225, 239, 244, 264
Multiprocessors, 214, 254
Multivariate statistics, 4

Nonparametric statistics, 5, 10
Numerical mapping technique, 47

Operating systems,
    COS, 321
    UNIX, 135, 139, 227, 265
    VMS, 135, 191, 265
Optical image compression, 71

Parallelism, 315, 322
Parallel computing systems, 212
Parallel computers, 225, 253
Parallel machines,
    Alliant, 317, 319, 321
    CDC-Cyber, 212, 317
    CLIP4, 205, 254
    Convex-C1, 124, 317, 321
    CRAY, 125, 212, 317, 321
    ILLIAC, 203
    MPP, 228, 239, 242, 254
    PAPIA, 211, 263
Parallel PASCAL, 228, 236, 243
Parallel processors, 225
Parametric statistics, 5, 7
Pattern recognition, 30, 292
Phase histograms, 87, 89
Photoelectric photometry, 20
Photographic plates, 47, 49
Photometry, 146
Photon-counting detectors, 45, 54

Pipeline, 316, 322
PIPE system, 259
Preprocessing, 112
Principal components analysis, 13
Processing elements, 238
Production rules, 278
Programming languages,
    ADA, 225
    APL, 228
    C, 227, 236, 264
    FORTRAN, 190, 225, 236
    LISP, 138, 228
    PASCAL, 270
Pulsars, 87, 97
Pyramidal computation, 212

ROSAT satellite, 155
Rule-based systems, 109, 297
    ACRONYM, 297

SAPA (System for Analysis of Pyramid Algorithms), 270
Scalar computers, 315
Scene complexity, 293
Segmentation, 110
Scanning mode, 45, 51
Schmidt cameras, 18, 109
Space Telescope, 27, 58, 127, 134, 143, 165, 193
Speckle interpherometry, 150
Spectroscopy, 148
Stellar photometry, 24
Surface photometry, 26, 54, 53
Symbolic computation, 254
Systems for data analysis,
    ADAS, 130, 133
    AIPS, 124, 321
    COMPASS, 171
    DAOPHOT, 24, 25
    IDL, 130
    IRAF, 139, 156, 165, 195
    MIDAS, 40, 54, 64, 130, 134, 156, 165, 195
    RIPS, 156
    SDAS, 130, 134, 165
    STARLINK, 169

Template matching, 62
Trainable classifiers, 112

Univariate statistics, 4

Vector computers, 315, 317
VLA radiotelescope, 119
VLSI (Very Large Scale Integration), 213, 227, 239, 250, 258